IET TRANSPORTATION SERIES 09

Autonomous Decentralized Systems and their Applications in Transport and Infrastructure

Other related titles:

Volume 1	**Clean Mobility and Intelligent Transport Systems** M. Fiorini and J.-C. Lin (Editors)
Volume 2	**Energy Systems for Electric and Hybrid Vehicles** K.T. Chau (Editor)
Volume 5	**Sliding Mode Control of Vehicle Dynamics** A. Ferrara (Editor)
Volume 6	**Low Carbon Mobility for Future Cities: Principles and applications** H. Dia (Editor)
Volume 7	**Evaluation of Intelligent Road Transportation Systems: Methods and results** M. Lu (Editor)
Volume 8	**Road Pricing: Technologies, economics and acceptability** J. Walker (Editor)
Volume 12	**EMC and Functional Safety of Automotive Electronics** K. Borgeest
Volume 38	**The Electric Car** M.H. Westbrook
Volume 45	**Propulsion Systems for Hybrid Vehicles** J. Miller
Volume 79	**Vehicle-to-Grid: Linking electric vehicles to the smart grid** J. Lu and J. Hossain (Editors)

Autonomous Decentralized Systems and their Applications in Transport and Infrastructure

Edited by
Kinji Mori and Takashi Kunifuji

The Institution of Engineering and Technology

Published by The Institution of Engineering and Technology, London, United Kingdom

The Institution of Engineering and Technology is registered as a Charity in England & Wales (no. 211014) and Scotland (no. SC038698).

First published 2018

The Institution of Engineering and Technology
Michael Faraday House
Six Hills Way, Stevenage
Herts, SG1 2AY, United Kingdom

www.theiet.org

British Library Cataloguing in Publication Data
A catalogue record for this product is available from the British Library

ISBN 978-1-78561-281-7 (hardback)
ISBN 978-1-78561-282-4 (PDF)

Typeset in India by MPS Limited
Printed in the UK by CPI Group (UK) Ltd, Croydon

Contents

Preface

ADS (Autonomous Decentralized System) concept proposed in 1977 has been progressed and advanced in technologies of control, communication and computer in these 40 years. Their technologies have been continuously applied to the various 'Engineering' fields such as manufacturing, transportation, utility and information. These applications in the real world have been operated well, and they show the effectiveness of the ADS systems.

The book on the ADS published by John Wily & Sons, Inc. in 2014 with the tittle 'Concept-Oriented Research and Development in Information Technology' is focused on the concept-oriented design, technology and application. It is stressed in this book that the consistent and persistent innovation is achieved on the basic unchanged concept such as the ADS, and then the chain of technologies and markets is attained. Otherwise, only one innovation of technology often remains as a spur-of-the-moment casual idea.

Recently around 2010, we have been confronted in the world with the structural changes of socio-economic fields in society of declining birth rate and aging, economy of globalization and value of diversified personal life-style. These structural changes have been accelerated through the mutually and globally connected networks. Here, these structural changes are progressed more in system of systems with multiple heterogeneous systems in objectives and functions than by single system with multiple homogeneous subsystems. The new technology is driven to be evolved under the socio-economic changing situations, and its technology offers to support the structural change. The mutual and close relation between socio-economic structural change and technology are now driving to achieve the innovation in life. Conventionally, the technology has the stronger leverage in the innovation, but now the structural change in society and life is pushing more and more to encourage the innovation. But in the unpredictable situation in technology, society and business, the research and development of technology may be the tentative approach. The sustainability in society and technology cannot be attained without their concept. The ADS approach includes viewpoints both of technologies for society and of society through technology.

In the technology fields of the ADS, they cover control, computer, communication and internet to cloud, IoT (Internet of Things) and AI. Moreover the fusion of these technologies can be created on the basis of the ADS system architecture, in which the ADS system is well designed to combine the autonomous technologies of communication, processing, filing and utilization of knowledge.

In the business fields of the ADS, the engineering products have been rapidly commoditized under the global competition. The competitiveness in the business is gradually shifted from the product to service. Especially the infrastructure system continues its operation during a long period of 20–40 years. In this period, the system is requested and responsible for continuing its operation and maintenance and customers' life service, and then the service and its technology have been gradually and highly competitive in business. The service system area of the ADS research and development covers the infrastructures of train, air traffic, robotics, ticketing service, smart city, smart society and banking. The ADS has been recently accelerated to be recognized and evaluated as a primary concept and technologies not only in engineering but also in service, especially for infrastructure.

I proposed the ADS concept 40 years ago in 1977 through biological mimic. Molecular biology's viewpoint is that living thing is composed of uniform cells, each of which includes intelligence of DNA (deoxyribonucleic acid). In the living thing, the body is defined as the integration of the cells and it includes the 'Abnormality', that is, it almost anytime includes the incomplete cells, but it continues to live. On the analogy of the living thing, the ADS is defined as the integration of the autonomous subsystems. The basic standpoint of the ADS is that, as well as the living thing, system integrated by physical and logical subsystems includes the 'Abnormality' in the system and almost anytime it includes the inoperative subsystems with failure, under construction and in repair. Each of the subsystems has its own objective and function and their integration makes a system. The totality of the system cannot be previously defined, and the system exists in the process of integration. The autonomy is defined by two properties of autonomous controllability and autonomous coordinability. Two properties mean that even if any subsystem fails down or is inactive, any other subsystem continues to control itself and coordinate with the remaining subsystems.

That is, the ADS is a decentralized system composed of components designed to operate independently but also to interact with each other to perform the functions of the system. This design paradigm enables the system to continue working even if any component fails. It also allows for maintenance and repair of components while the system as a whole remains operational. In a large and complex system, such as an infrastructure system, it is not an option to stop operation at any time. Even if a part of the system fails, is being repaired or in maintenance, the system has to keep functioning. This results in requirements of on-line properties for fault-tolerance, on-line expansion and on-line maintenance. Then the ADS design of system is an approach of bottom-up where at first subsystem is clarified and system is successively constructed by integrating subsystems.

It is difficult to satisfy these on-line properties requirements by conventional technologies based on the centralized, hierarchical and functionally distributed systems concept. Dynamic and unpredictable changes in social and economic situations demand next-generation systems to be based on adaptive and reusable technologies and applications. Such systems are expected to have the characteristics of living systems composed of largely autonomous and decentralized components. Thus, ADS technology is an answer.

The ADS has been persistently advanced in cooperation with the international group including industries, universities and public sectors, most of which have been actively involved in the research and development through IEEE-sponsored international conference: ISADS (International Symposium on Autonomous Decentralized Systems) founded in 1993. This book, 'Autonomous Decentralized Systems and their Applications in Transport and Infrastructure', is contributed by these ISADS members. It is organized in four parts.

Part 1: Introduction – Paradigm shift by ADS:

- In ADS concept, at first, subsystem exists and its integration makes system. It has the proposition that almost anytime system includes abnormality, but it remains operational against abnormal subsystem.
- Architecture of ADS consists of subsystems and communication field with which subsystems are connected, and then they autonomously select to receive necessary data, process it and send out processed data. During these processes, subsystem judges by itself to control itself and coordinate with others.

Part 2: Infrastructure – Infrastructure of engineering systems of railway, air-traffic, manufacturing:

- Railway system assures its operation without stopping during partial failure, expansion and maintenance, and flexible schedule of transportation is arranged in delayed and emergency situations.
- Air-traffic system is guaranteed to avoid collision of flights in the sky and at airport under congested situations.
- Manufacturing system for various kinds and quantities on basis of customer and market-demand is realized, and flexible schedule of production is realized.

Part 3: Life – Ticketing service, robotic service, information service, social service:

- Ticketing service of railway controls gates for guaranteeing fluidity of passengers in walking pace of 0.2 sec passing through entrance/exit gates of station at rush-hours congestion to check passenger-card and calculate complex transport fare.
- Robotic behaviour is supported by its autonomy under cloud computing.
- Information service for passenger of railway is to provide in real-time congested traffic situation and detour information.
- Social service for residents is discussed to achieve resilience under emergence situation.

Part 4: Business – Blockchain, bank, railway service, smart system:

- Blockchain is utilized as distributed autonomous ledger technology for cryptocurrency and other application fields.

- Distributed clouds system guarantees fault tolerance, expandability and maintainability as levels of customer preferences and access of data.
- Banking systems are globally connected for achieving real-time and mission-critical services.
- Train operation services include vertical integration of maintenance and operation and passenger and society services for improving their quality of life.
- Smart system such as Smart city, Industrial 4.0/Industrial Internet and cyber-physical systems for the future include problems not only of technical matters but also of governance, design of security and of business models.

4/17/2018

Kinji Mori
Waseda University

About the Editors

Kinji Mori is a Distinguished Researcher of Green Computing Systems Research Organization, Waseda University since 2017. From 1997, he has served for 15 years as a Professor in the Department of Computer Science, Waseda University, as well as in the School of Management of Technology, Tokyo Institute of Technology, Japan. He also worked at Hitachi as Researcher and Director for 23 years since 1974.

Dr. Mori has been a Life Fellow of IEEE and Fellow of IEICE (Institute of Electronics, Information and Communication Engineers, Japan). His contributions were also recognized by the many highly prestigious awards conferred upon him. One of the most prestigious awards he received include the Ichimura Industrial Prize for contribution to the Science, Engineering, and Industry with the cash award of 10,000,000 Japanese Yen (=US$100,000), conferred by the Honorable Prince Mikasanomiya. He donated this cash award to the IEEE Computer Society to found the IEEE annual Kanai Award for outstanding contributions in the area of distributed computing systems. He also received the Japan Patent Awards, the Research Achievement Award in Japan, and others.

Dr. Mori's outstanding achievements and extraordinary research impacts also include more than 350 patents registered internationally and over 500 papers in archived journals, books, and international conferences.

Dr. Mori made fundamental contributions to the field of reliable and maintainable distributed computer systems. In 1977, he proposed a completely new system concept named Autonomous Decentralized Systems (ADS), and on the basis of ADS concept, consistently contributes to ADS architecture construction, technologies invention, product designing, business planning, implementation, de-facto standardization, and foundation of international conferences. The ADS concept-oriented technologies and innovative applications have been expanded in the field of not only control systems but also information systems. For example, Japanese Shinkansen Bullet Train (High Speed Railway) control and information system, Autonomous Decentralized Transport Operation Control System(ATOS) for Tokyo metropolitan railway system, IC-Card system (Suica) not only for fare collection by railways, busses, and the other transportation in all over Japan but also for e-commerce, Steel Production Process Control System for Kawasaki Steel Co., which are among the largest industrial systems in the world, and many other control and information and service systems were developed by applying the ADS architecture and technologies. They have been operated well and shows the effectively of the ADS technologies. The sales volume of the ADS is now over

US$5 billion. Even now since the proposition of the ADS in 1977, the ADS technologies and applications have been expanding in the world.

He founded the International Symposium on Autonomous Decentralized Systems (ISADS) sponsored by IEEE and three Japanese academic societies of IEICE (Information and Computing), IPSJ (Information Processing), and SICE (Control) in 1993, which has been hold every 2 years in the world. He has been the steering committee chair of ISADS. Especially for the Asian countries, he also founded two international conferences, IWDCCA (International Workshop on Distributed Computing, Communication and Applications) in 2000 and IWADS (International Workshop on Autonomous Decentralized Systems) in 2000, both of which are sponsored by IEEE. He has contributed to found the international conference HASE (High Assurance System Engineering Symposium) sponsored by IEEE in 1996.

He has been professors of five Chinese Universities and one Pakistan University. He was appointed as Distinguished Specialist among 1,000 people in the world in 2012 for the "Thousand Talents Program", China.

He founded the School of Management of Technology, Tokyo Institute of Technology in 2005. In 2012, he founded the new course in innovation of global research and development of technology and business at Waseda University for educating not only students but also professionals from Government, Industries, and Institutes.

Takashi Kunifuji is Principal Chief Researcher of Research and Development Centre of JR East group at East Japan Railway Company (JR East), Japan. He has engaged in research and development of railway signal control systems since 1998. He engaged in maintenance, design, and construction of signaling systems from 1992 to 1998. He has joined JR East since 1992. He has received Ph.D. degree in Engineering from the Tokyo Institute of Technology, Japan in 2011. He graduated from the Tsukuba University, Japan, in 1992 with a Master's degree in Electrical Engineering.

He is a Fellow of IRSE and a member of IEEE, IEICE, IEEJ, and IPSJ.

Part 1
Introduction

Overview

With the advancement of science and technology, the rapid and continuous changes in the social and economic environment have been affecting the life and business on a global scale. On the other hand, changing and unpredictable society and economy accelerate to make new needs of the human lives. As information can be easily shared through networks, such as the Internet, the needs of people and societies are quickly and globally influenced and changed by this information. Even if they share information, their needs are not unified. As shared information increases, their needs are diversified and change quickly. Therefore, with the rapid progress in the construction of the information society, it becomes more difficult to predict the changes, spreading speed and impacts of innovations in the market.

A changing society stimulates change in users' needs, and society changes due to these needs; therefore, sustainability is necessary. Sustainable technological innovation is inevitable for adapting to changing society. As an example, in this part of the book, we will introduce the concept of autonomous decentralized systems (ADS) and its paradigm shift of research and development (R&D) for information technology under changing and unpredictable market.

Since the concept of ADS was proposed in 1977, ADS R&D has been consistent for creating innovative technologies and developing new applications. As a result, a chain of technologies and applications is consistently generated with this concept [1]. The ADS concept-oriented technologies and innovative applications have been expanded in the fields of not only control systems but also information systems [2]. For example, the Japanese "Shinkansen bullet train's" (high-speed railway) control system, "autonomous decentralized transport operation control system" for the Tokyo metropolitan railway system, steel production process control system for Kawasaki Steel Corporation, "Suica" IC-Card system for not only fare collection by Japan Railway East and private railways but also for e-commerce, and many other industrial automation systems, which are among the largest industrial facilities in the world, were constructed using the ADS architecture and technologies. In these applications, ADS improved life-cycle cost, efficiency, software productivity, resilience, and sustainability. The sales volume due to ADS has surged to 5 billion US dollars, and over 350 patents have been registered internationally.

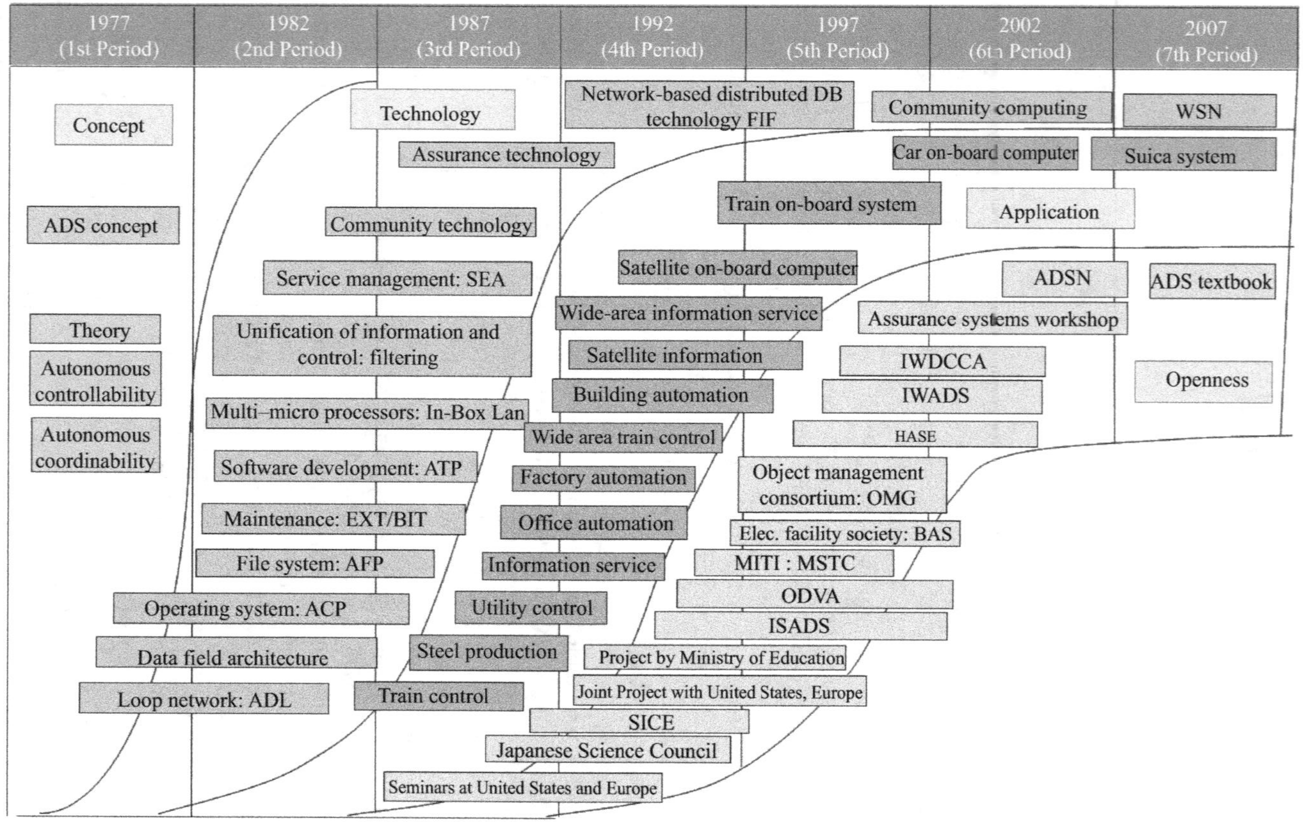

Figure P1.1 History of R&D on ADS

ADS technologies have been accepted as the de facto standard in consortiums such as the "Open DeviceNet Vendors Association," "Object Management Group," and Japanese "Building Automation Systems Association" [3–5]. Moreover, this concept and the resulting technologies and applications have been discussed at the "International Symposium on Autonomous Decentralized Systems" founded in 1993, sponsored by the Institute of Electrical and Electronics Engineers and the three Japanese societies, Institute of Electronics, Information and Communication Engineers Society, Japan; Information Processing Society, Japan; and Society of Instrument and Control Engineers, Japan. Several international workshops on ADS have been founded in Asia.

"The technologies and real applications" based on this concept and "the openness and standardization" are shown for every 5 years for 35 years since 1977 in Figure P1.1. From this figure, we can see that the paradigm shift of ADS concept-oriented approach for sustainable technological innovation was clearly achieved. In this book, not only ADS-based technological innovation in the railway system but also ADS concept-oriented new applications for value creation in different fields will be introduced.

References

[1] K. Mori, "Concept-Oriented Research and Development in Information Technology," John Wiley & Sons, Inc., Hoboken, NJ, 2014.

[2] K. Mori, "Introduction of Autonomous Decentralized System," Morikita Publishing, Tokyo, 2006 (in Japanese).

[3] http://www.odva.org/default.aspx

[4] http://www.ieiej.or.jp/standard/index.html

[5] http://www.omg.org/docs/orbos/99-11-05.ppt

Chapter 1
Autonomous decentralized systems and its paradigm shift

Kinji Mori[1] *and Xiaodong Lu*[2]

Abstract

The advancement of technology is ensured by step-by-step innovation and its implementation into society. Autonomous decentralized systems (ADSs) have been growing since first proposed in 1977 by Kinji Mori, and then the ADS technologies and their implementations have interacted with the evolving markets, sciences, and technologies. The ADS concept is proposed on biological analogy, and its technologies have been advanced according to changing and expanding requirements. These technologies are now categorized into three generations for different missions according to the heterogeneous requirements of customers and systems. These technologies have been widely applied in manufacturing, telecommunications, information provision/utilization, transportation, and so on. They have been operating successfully throughout the world. The paradigm shift of ADS technologies is shown in this chapter.

1.1 Background and requirements

As computing and communication resources have been gradually decreasing in cost, they have been widely applied and their roles in society and business have become more important. Moreover, according to the continuous growth of practical applications and large-scale networking of systems, systems have expanded and become increasingly complicated. In large and complex systems, such as railway, factory automation, steel plant, financial/stock, and wide-area information communication, it is not permitted to stop operation at any time. Therefore, replacing an entire system at once is impossible; and thus, step-by-step construction without stopping operation is required and its system property is called "on-line expandability." Even if part of the system fails or needs to be repaired, the application has to keep functioning; these properties are called "fault tolerance" and "on-line

[1]Green Computing Systems Research Organization, Waseda University, Japan
[2]Department of Surveillance and Communications, Electronic Navigation Research Institute, Japan

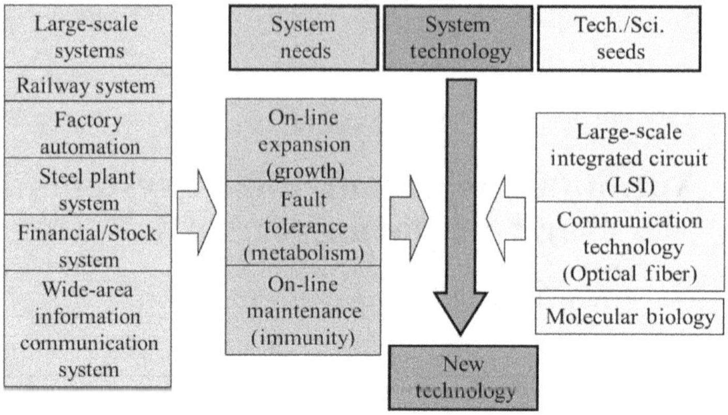

Figure 1.1 Background and needs

maintenance." Less restriction on computing and communication resources results in more requirements for "on-line properties" of on-line expansion, fault-tolerance, and on-line maintenance (Figure 1.1) [1].

Conventional computing technologies have been developed under a centralized system concept. Even hierarchical and functionally distributed systems are based on the centralized system concept with the viewpoint that the total system structure and functions have to be determined in advance [2–9]. This viewpoint itself is inconsistent with the system; the structure and the functions change continuously in the system, although the hardware and software structures are fixed and have little flexibility.

Many application systems have confronted with the changing situations. In the manufacturing systems, the multi-kinds and few quantity productions driven by the customers' demands have been gradually required rather than the few kinds and large quantity production on the basis of the schedule. It is so much difficult to predict the customers' requirements and to make the long and middle-term production schedule. The production processes have to be quickly modified and expanded according to the frequent change of products. Then, the organization for the production has been restructured to make the boundary of the departments of the management in the office and of the operation in the production process vague and partially vanished. The integration of the control system for the facilities in the production process and the information system for the management at the office make it possible for the production line engineers to have the flexible, adaptive and quick decision, control, and management by grasping the real situation in the production process and utilizing the management information at the production line instead of at the office. Then, the integrated control and information systems can improve the utilization of the human resources especially in the office because the short-term flexible management jobs can be placed under the production line engineers. While the office workers can also easily grasp the current process state and use its information for sales. Therefore, the assurance property is crucial for the manufacturing system under the changing markets.

In the train systems, recently, the service-oriented train regulation has been required as well as the safety and punctuality. The highly reliable, fault-tolerant, and real-time computing systems have been applied in these 30 years. As the competition among transportation becomes severe, the more services have been provided for the passengers to supply the train tracking information and to guide them the appropriate routes. When the information on the current sales, accidents, and congestion is received, the train schedule is rearranged, and the trains are regulated. The operation time of trains has been extended for almost 24 h, and then the test and repair of the facilities have to be performed during the train operation. Conventionally, the control system for the facilities with real-time and reliability properties and the information system with high throughout property have been constructed independently. But now, these two types of application systems need to be integrated to meet the service-oriented train regulation. The operation for the control system is permitted only to the licensee, but the information system can be easily handled by anybody. Therefore, the train application system is required to satisfy the high assurance property, while preserving the respective properties of control and information systems.

In the telecommunication systems, the highly reliable and fault tolerant networks and their management systems have been developed. Recently, the new services for multimedia, connection, quality, database, and so on have been added and modified day by day. Moreover, some services are customized according to their individual utilization and contents. The networks themselves have been continuously extended. The network management and the service management cannot be designed independently. Under these changing situations of the networks and the services, the assurance of the telecommunication system has been pursued.

To achieve the abovementioned on-line properties and high assurance in a changing system, the autonomous decentralized system (ADS) concept and its architecture was proposed in 1977 on the basis of a biological analogy [10–12]. Since then, the ADS concept has been applied to various technology fields such as networks, communications, multi-computers, software, control, and robotics. Its architecture and technologies have also been developed and applied in many fields such as factory automation, transportation, information systems, telecommunications, and e-commerce [13–27]. Over the past 35 years, the market and user needs have changed and diversified. At the same time, ADS technologies have also advanced according to these evolving situations [1,32,33,39].

1.2 The concept of ADS

1.2.1 Biological analogy

Organisms operate effectively by virtue of their biological functions. The biological system can be understood from two standpoints, cell and organ (Figure 1.2) [1,29].

In an organism, each cell has intrinsic information called DNA, which is uniform in structure. The fundamental functions of a cell, such as metabolism, growth, and immunity, are attained using the local information surrounding it. No dominant cell exists and there is no master–slave relationship. These characteristics of

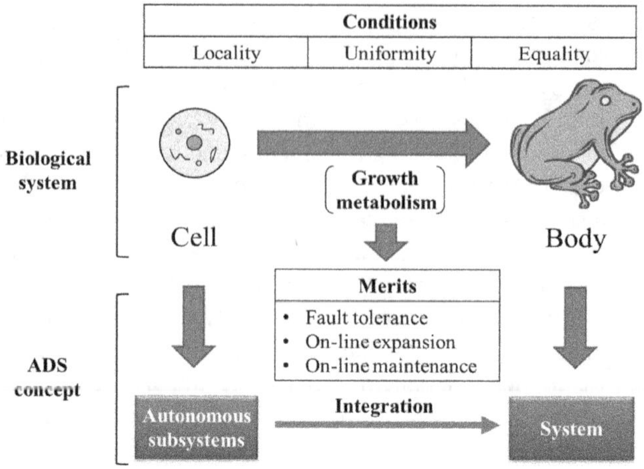

Figure 1.2 Biological analogy

uniformity, locality, and equality make it possible for organisms to survive. The functions of metabolism, growth, and immunity refer to the on-line properties of fault tolerance, on-line expansion, and on-line maintenance of a system.

The system of organism is based on a cell, and the cells are homogeneous in structure with same DNA, function without dominant cell and information on each surroundings. The organism consists of homogeneous cells. When an organ cannot adapt to the rapidly and largely changing environment internally and externally, homeostasis for stabilizing the biological system consisting of organs is achieved through cooperation among cells without mutually violating their functions. Therefore, the functionality of the system is ensured.

The totality of organ's structure and objectives cannot be defined. In the growth process, an organ is gradually formed from the integration of homogeneous cells performing their heterogeneous functions by linking with other organs for adapting to the environment through cooperation, ensuring the agility of the system. When a cell malfunctions or cannot use its function effectively, another cell takes on its functions, that is, the functions in a living system are mobile. Cells are autonomous in communication among appropriate cells and in functioning to achieve the abovementioned on-line properties.

1.2.2 Concept

Opportunities and challenges for creating highly complex, efficient, and dependable business-and-control systems are steadily increasing. They are driven by continuous growth in the power, intelligence, adaptability, and openness of technologies applied in computing, communication, and control systems. Dynamic changes in social and economic situations demand that next-generation systems be based on adaptive and reusable technologies and applications. Such systems are expected to exhibit the characteristics of living systems composed of largely autonomous and decentralized

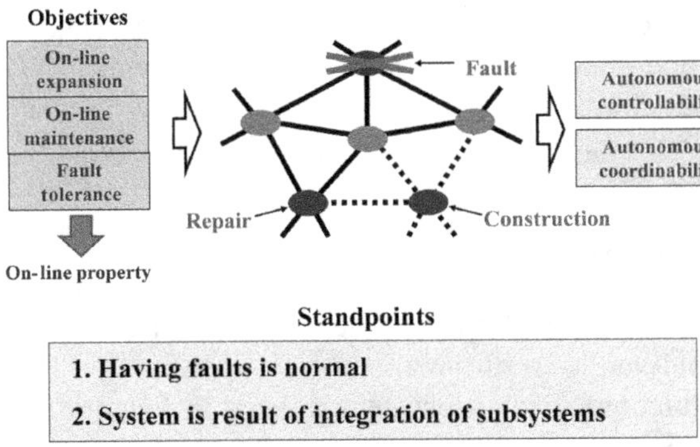

Figure 1.3 Concept of ADS

components. Such systems are called ADSs. As shown in Figure 1.3, such a system is characterized by the following two standpoints:

1. In the system, being faulty is normal (abnormal condition is normal)
   In the conventional system, if some components are failure or during construction or maintenance, the system changes from the normal to abnormal mode. However, for the living body, it is always changing and growing without stopping. And the metabolism of the cell and the growth of the organ are normal conditions. As a result, to achieve on-line property, in the ADS concept, the abnormal conditions in the conventional system are normal.
2. The system is the result of the integration of subsystems
   For conventional system construction, the structure and functions of whole system should be previously defined. However, with the scale of the system becoming larger and the functions of the system becoming more complex, it is difficult to clarify and define the structure of the entire system at beginning. Moreover, for the living body, with unstopping growth, we cannot define the final state and structure of the body from the start. Therefore, in the ADS concept, the objectives, structure, and functions of each subsystem can be clearly defined, and a system is the integration of subsystems.

The objective of the ADS concept is to achieve on-line property. From the abovementioned two standpoints, a system is defined as an ADS if the following two properties are satisfied:

1. Autonomous controllability: Even if any subsystem fails, is repaired, and/or is newly added, the other subsystems can continue to manage themselves and function.
2. Autonomous coordinability: Even if any subsystem fails, is repaired, and/or is newly added, the other subsystems can coordinate their individual objectives among themselves and can function in a coordinated fashion.

These two properties ensure the on-line properties of fault tolerance, on-line maintenance, and on-line expansion. They suggest that every "autonomous" subsystem requires intelligence to manage itself without directing to or being directed from the other subsystems, and to coordinate with the other subsystems [10,12].

An ADS is possible with autonomous controllability and autonomous coordinability; therefore, each subsystem is required to satisfy the following three conditions (Figure 1.2):

1. Uniformity (in structure): Each subsystem is uniform in structure and self-contained; therefore, it manages itself and coordinates with others.
2. Locality (in information): Each subsystem manages itself and coordinates with others based only on local information.
3. Equality (in function): Each subsystem is equal in function. No master–slave relationship exists.

These three conditions are completely different from centralized system concept. In the centralized system concept, such as mainframe-based system, the totality of system needs to be predetermined. If there is any fault in the mainframe computer, the operation of the total system will be stopped. To prevent the whole system from stopping, many fault avoidance technologies have been applied.

In the hierarchical system, some functions of the central or upper-layer computer are located in the lower-layer computers. If the computer in the upper layer is down, the computers in the lower layer cannot continue to provide functions without instructions. In the functional distributed system, the functions of the system are divided and provided by the different subsystems. In the system, specific master or coordinator subsystem has to get the global information on the entire system to manage or coordinate all other subsystems. The hierarchical system and functional distributed system concepts are formulated from the same conditions at the centralized system concept:

1. Multiformity (in structure)
2. Globality (in information)
3. Inequality (in function)

1.3 System architecture

1.3.1 Data field architecture

An ADS is possible based on "data-field (DF)" architecture without any central operator or coordinator. The DF architecture is composed of two technologies: content-code communication for autonomous coordinability and data-driven mechanism for autonomous controllability [11,12]. As shown in Figure 1.4, each subsystem has its own management system, namely, an "autonomous control processor (ACP)" to manage itself and coordinate with others. Each subsystem, called an "atom," consists of application software modules and an ACP. The DF in the atom is called the "atom DF."

Physically, the DF corresponds to the network or the memory, and the atom is the computer unit. In the network, the broadcasted message is physically deleted by its originating subsystem or the terminator in the network after the message is

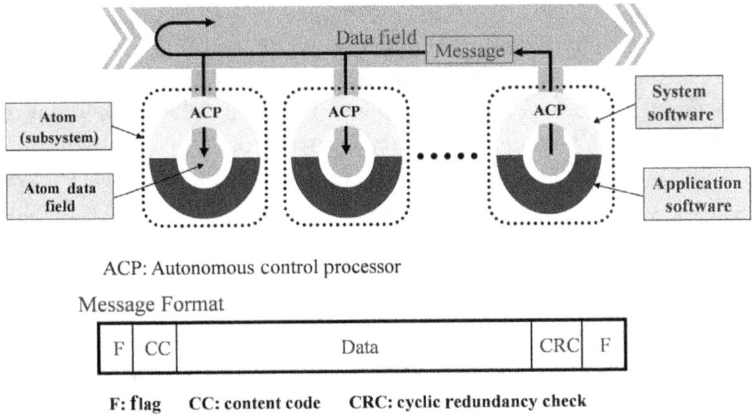

Figure 1.4 Data field architecture

transmitted over the entire system. When the DF corresponds to the memory, the first-in–first-out type memory is used, and the message in the memory can be deleted after all subsystems check to accept it or not.

1.3.2 Content code communication

All subsystems are uniformly connected only through the DF with a uniform interface; all data are broadcast into the DF as messages (Figure 1.4). An individual datum includes a content code defined uniquely by content. A subsystem autonomously selects to receive a message based on its content code (content-code communication). Each subsystem is not directed to receive the data by the sender specifying the receiver's address. This content-code communication enables each subsystem to be autonomous in sending and receiving data, that is, no master–slave relationship exists and equality among subsystem is attained. That is, each subsystem need not to detect the relationship among the sources and destinations but must specify the contents codes necessary for the application modules in the subsystem to process their attached data. This feature of content-code communication ensures the locality of information, which is necessary for each subsystem to coordinate with the others; thus, autonomous coordinability is achieved.

In the conventional systems applied in peer-to-peer (P2P) communication, the number of messages grows with increasing receivers. However, under DF architecture, all nodes in one communication can receive the broadcast message. Compared to the P2P communication, the number of messages can be reduced in the condition of multiple receivers. With the increase of subsystems and communication traffic, the system can be divided into small DF structures.

1.3.3 Data-driven mechanism

Each application software functional module in a subsystem starts performing after either all necessary data (AND execution) or its part (OR execution) are received (data-driven mechanism) [11,12]. This mechanism loosely couples modules. Each subsystem autonomously determines and controls its own action. Required content codes (CC1, CC2, etc.) for application software functional modules (M1, M2, etc.) are preregistered in the executive management system ACP, which can dynamically and autonomously assign necessary content codes in accordance with the situation in the system. The subsystem does not need to inform other subsystems if the content codes assigned to the ACP are changed. Every ACP has functions for autonomously managing the data, checking the data, and supporting the test and diagnosis. The function of the application software module is characterized by the relationship between the content codes of the input and output data (Figure 1.5). Therefore, autonomous controllability is achieved.

In a conventional system, data-flow architecture is based on a centralized controller, such as a distribution network. The sequence and time of software execution are determined ahead of time. In ADS, each subsystem autonomously implements the software according to the content-code data in the DF.

Even in the client-server model, the client and server subsystems are autonomous, and they need not know the relation between them. The relationship between them can be defined by the data. In an application software module, all required content codes cannot be previously specified but during the operation. When the required data for the application software module becomes clear during its operation, the application software module in the subsystem (client) needs to request the data with the content code to the other subsystems (servers), and it has to wait to restart its execution until the required data, which is sent out from the servers into the DF, can be obtained. The application software module in the client registers the required content codes into its ACP as soon as they are specified on the way of the processing.

The module in the server is driven by the data which is broadcasted from the application software module in the client. The application software module in the

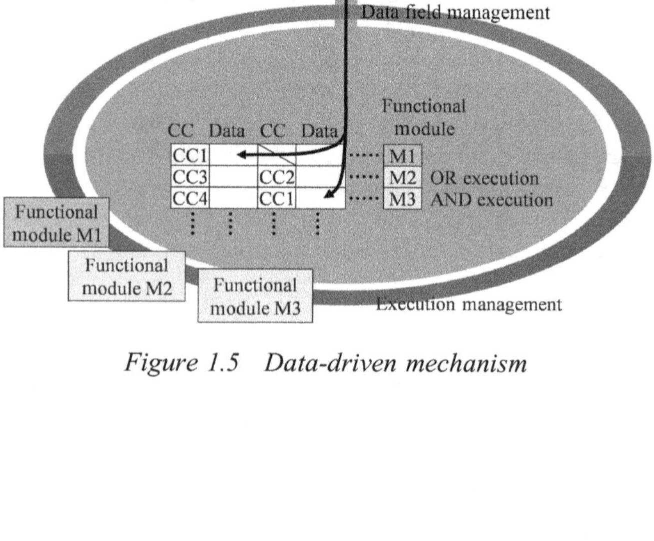

Figure 1.5 Data-driven mechanism

client can restart its execution when the required data broadcasted from the servers is received from the DF (Figure 1.5).

1.3.4 Software productivity

The data-driven mechanism need not have the linkage with the other modules. The application software module is completely encapsulated by the input and output data only via DF and it has not any direct relation with the other modules [11].

These characteristics of the software make it possible to produce the application software module independently of the other modules. This distributed software development can contribute to improve the productivity especially for the software design and test. The software development support tool is called "ATP (autonomous terminal processor)." The ATP has the following five functions: the data formatting, data flow generation and flow correctness check, module integration/division assistance, remote loading, and fault-propagation check to interactively support the software designer.

The first function of the ATP is to support to make the data format for each content code. In one data, there may be several data items. If the data of the content code includes many data items, the application software module has to delete most of the data items except the necessary data items in the received data. While the data contains small data items, the application software module needs to receive several messages for its execution. But the data format for the content code is basically determined by the application user as whether the content code corresponds to a meaningful group of data items for application.

The second function of the ATP is to generate the data flow among the application software modules. Each application module is defined by the relation of the input and output data through the DF and it can be developed independently of the others. The ATP can make the relation among the application software modules based on the input and output data. In this data flow generation process, the ATP can detect the incorrectness of the flow. It means that the input data necessary for some module will not be generated by any other module. Moreover, the ATP can check an infinite loop that may exist among the application software modules.

As the third function, the ATP assists whether some application software modules would be preferably integrated into one or an application software module would be divided into several. When the output data from one application software module is used by some specific application software module but not by the others, the application software module may be integrated with this specific application software module. If the application software module produces many kinds of the output content codes, it may be divided into several modules, each of which will be some kinds of the output content codes.

The rest of the ATP functions are the remote loading of the application software module and the fault propagation check. When the application software module sending out from the ATP attaches its corresponding content code, it can be selectively received to store by the necessary atoms. The data-flow model generated by the ATP can be easily utilized to detect the propagation flow of the fault caused by an application software module. Based on this analysis of the data flow, the modules on the critical path of the data flow may be better to be replicated to attain the fault tolerance.

1.3.5 Agility

In the conventional applications, the relationship among the systems is assumed to be fixed. But the agility is achieved by varying the relation among the systems. For example, in the information service application systems, the number of the service providers increases day by day. The preferences of the users vary person to person and they change day by day. Under these situations, the relationship among the users and the service providers should be flexible without restricting the coverage of communication and cooperation among the systems. But as the coverage expands, the risk of getting the inappropriate data increases and then the response becomes low. In the industry field, the systems for manufacturing, sales, logistics, and so on have been gradually connected to promptly respond to the user's requirements. But it is meaningless to combine them beyond the appropriate coverage. The technology for agility is how to manage and to change the linkage among the systems according to the situations.

Here, each subsystem utilizes content-code communications for establishing the flexible linkage under the evolutionary situations. The mechanism for the agility of the relationship among the systems includes the functions of the mutual content among the systems and of adjusting the other relationship. This mechanism is different from the conventional name server for connecting the links by the addresses, which is independent to the user's preferences, between the senders (users) and the receivers (service providers).

1.3.6 Mobility

In the conventional applications, the functionality of the system is assumed to be fixed. But the functionality is preferred to be performed at the suitable system for the prompt provision of the services and for the reduction of the communication among the several systems. Under the changing situation, it is requested for the mobility of the functionality that the system utilizing the functionality need not know which system originally generates it and whether the other systems are utilizing it. As well as the data in the DF, the functionality itself with it corresponding content code can move around the systems as the mobile agent and each system can accept the agent according to its situation based on the content code. Then the system can emerge the function transferred by the agent. This is a technology of mobility.

The agent has to have the autonomous navigation mechanism to select the route in the network to reach at the appropriate systems. The cooperation among the agents and among the agent and the system is necessary for this navigation. Each system should communicate with the agent for its navigation to inform the situation around it.

The systems have the heterogeneous information. The agent can utilize it being stored in the different systems, which may not be most appropriate for the agent. But under this heterogeneously distributed information environment, the agent can achieve to reduce the navigation time and to improve the fault tolerance against the failure of the systems.

1.4 Paradigm shift of ADS

There are three generations in the paradigm shift of ADS. The first generation is for continuous operation by constructing nonstop control and information systems. The second generation is for different lifestyles by providing customized services. The third generation is to achieve sustainability for the development of society, economy, and life.

1.4.1 Paradigm shift from operation to service

From the view of the total system, conventional systems were required to achieve high performance, be highly reliable, efficient, and so on. However, with the scale of the system increasing so rapidly, the total system cannot be determined ahead of time, and the system structure itself is also changing constantly. As a breakthrough in systems, ADS was proposed to achieve on-line property.

Along with advanced information technology and changing demands, the technologies based on ADS concept and architecture are also evolving gradually. In Table 1.1, the paradigm shift of ADS from first to second generation is seen from the view of system requirement and system structure. In homogeneous systems, the ADS architecture and technologies are applied to deal with continuous operation at system level. Moreover, with the advancement of information technology, the ADS has focused on the integration of heterogeneous systems to assure the system operation. However, along with changing requirement, the ADS was extended to the information service level to provide customized service in different applications. Currently, the ADS architecture and technologies have been evolving in the service infrastructure to improve the end-user's quality of life.

1. Requirement: from operation to service
   Requirement has changed from systems' operation to user services. As the users' requirements have become more diversified, it is difficult to achieve user satisfaction even if the system operation is guaranteed. Therefore, the system's service itself has to be evaluated.

Table 1.1 Paradigm shift from operation to service

	First generation	**Second generation**
Mission	Resource for (**control** + **information**)	Resource as (+ **service**)
Customer	Specified majority at **city and intercity**	Heterogeneous personal of **community**
System requirement	Online property assurance	Fair service unconscious service
Time frame	1980s–	2000s–
Applications	• Steel production process control system • ATOS (autonomous train operation)	• Data center • Suica
DF structure	Homogeneous and heterogeneous DFs	Local and timed DFs

2. Structure: from homogeneity to heterogeneity
   The system structure composed of homogeneous components shifted to a system consisting of heterogeneous components. To meet users' increasing demands, business in various fields has to provide new services. In this business innovation, the system has to treat the diverse kinds, quality, and quantity of the components and functions. As a result, the heterogeneous requirements can coexist, and the system can adapt to changing situations.

1.4.1.1 On-line properties

Background

With globalization in the 1980s, intensity in competition increased. Therefore, it was not sufficient for companies to win a place in a highly competitive market simply by reducing costs and improving quality. For example, in the steel production process control system, it is necessary to meet the various types and quantities of production demands in global scale. Moreover, in the systems which dealt in seamless manufacturing processes, from raw material to the final product, nonstop system operation was required. The ADS technologies were originally proposed to achieve on-line property [12]. The consumer requirements, technologies, and application shown in Table 1.1 are described in the following subsections.

Requirements

A system may need to change according to user requirements. However, its operation cannot be stopped anytime since the system becomes economically and socially important. As a result, the role of on-line property of on-line expansion, on-line maintenance, and fault tolerance in a system, meaning that a system can continue operation during partial expansion, maintenance, and failure, became more and more important.

Architecture and technologies

In the 1980s, as shown in Table 1.1, the ADS was targeted to be applied in the control field. The structure of the control system was composed of homogeneous components. Therefore, the structure of DF(s) was also homogeneous. On-line property is attained by this DF architecture, in which all data are broadcast, and each subsystem selects to receive the data necessary for its application modules based on the content codes (Figure 1.6). This feature makes the modules loosely coupled. Even if some subsystems are under construction or fail, the system can continue its operation. On-line expansion, on-line maintenance, and fault tolerance technologies were thus proposed.

1. On-line expansion
   There are three levels of on-line expansion: module, subsystem, and system. In module level expansion, the application software module and database in one atom are newly installed into or moved to another atom. Then they need only to register their necessary content codes into their own ACPs and do not need to inform the others. In subsystem level expansion, a subsystem can be constructed, modified, added, and deleted during operation of the other subsystems.

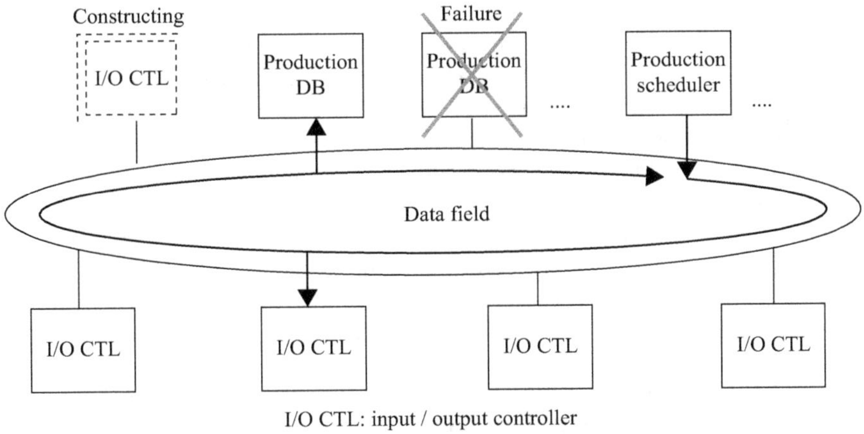

Figure 1.6 Homogeneous data field. Copyright © 2009 IEICE [34]

The subsystems do not need to know the direct relation with others and need not inform others upon their addition to or deletion from the system.

In the system level expansion, different ADSs are integrated into one. Two types of the systems integration are designed. In the first design, the DFs of different systems are combined into one DF. In the second design, the different systems are connected by a gateway, and two different DFs are combined through the gateway. The gateway connecting DFs (DF-A and DF-B) is regarded as one atom in DF-A by the atoms in the DF-A. While, by the atoms in DF-B the gateway is regarded as one atom in DF-B. That is, the gateway selects to receive from the DF-B the data attaching the content codes necessary for the atoms in DF-A and it passes the data into DF-A.

The data that is necessary for the atoms in the DF-B is passed through gateway from the DF-A to DF-B. The gateway has to register the content codes necessary for the atoms in DF-A to pass through from DF-B, while for the atoms in DF-B to pass through from DF-A. During the registration of the content codes in the gateway, the atoms need not stop their operation.

2. On-line maintenance
DF architecture makes it easier for application software modules to be tested while the system is operating. There are two kinds of modes for each module: on-line and test. On-line data and test data coexist in the DF. There are two kinds of approaches for an on-line test. In the first approach, the module with on-line mode uses the test data to do the test. In the second approach, the module with the test mode uses the on-line data to make the test. Online test is supported by a BIT (built-in tester module) in each ACP and by an EXT (external tester module) which is an application software module. The BIT module in the subsystem sets its application software module in the test mode, and then it generates test data and checks the test result. The application software module in test mode receives data from the DF and processes it. Then it

broadcasts test result data with a test flag to the DF. The BIT of the system in test mode prevents the signal from being sent to output devices such as controllers. The EXT monitors test data and test result data in the DF. By correlating test data with test result data, the EXT checks fault occurrence in the application software module in the test mode and broadcasts fault detection. The BIT independently decides whether to change the test mode to on-line mode based on test results. This test mechanism makes it possible for both on-line and test modes to coexist in the same system, at the same time.

For the associative module test, the modules, while they are in the normal mode, are driven by the test data in the DF. The module broadcasts the test result data into the DF, but it prohibits sending the signal to the output devices. The test result data is successively used to test the other application modules. With this mechanism, the application modules can be successively tested while they are operating. The EXT monitors the test data and the test result data successively output from the application modules and it detects how the fault propagates among the modules.

3. Fault tolerance

   The ADS architecture and its data-driven mechanism makes it possible for the subsystems and application software modules to run freely and asynchronously. The subsystems and application software modules are replicated according to the requirements and their level of importance. Replicated application software modules run independently and send out processed data with the same content code to the DF. Faulty data are also sent out to the DF. The ACP in each subsystem receives all data with the same content code from replicated modules and selects the correct data from them. Here, the data consistency management module in the ACP identifies the same data both by content code and event number induced with the data. Correct data are selected from the same data through majority voting logic, which is flexibly adapted to the predetermined time interval or the total number of received data.

In the conventional k-out-of-N redundant system, the redundancy is only in the subsystem level, and the voter detects the fault as a centralized controller. In the ADS, the redundancy is not only in the subsystem level but also in the module level. Each module autonomously detects the fault based on majority voting logic according to relative redundancy. Under this logic, fault occurrence is detected, and each application software module avoids being affected by fault propagation. A subsystem with a replicated module can intercept any data broadcast from other replicated modules. If the subsystem includes a faulty application software module, it detects the internal faults via this interception. As a result, even if an application software module is faulty, the subsystem continues operation by using correct data received from other replicated application software modules and recovers this fault by itself.

In the case of "AND execution condition," an application software module needs more than two data to execute that can be resolved by using the event number. The data consistency management module selects all data with the content codes satisfying the "AND execution condition" of the application software module, and it arranges the data in the order of the event number set by the original module.

With this arrangement mechanism, the data consistency management module can detect whether all necessary data have been completely received and whether some data are missing, and it logically synchronizes the data that originated from the other atoms.

This data consistency mechanism to assure fault tolerance is easily adopted to system reconfiguration, without stopping the operation. When there are the replicated databases as the servers in the system, each replicated database is independently accessed. It asynchronously stores the necessary data into the DF and sends out the data with the content codes requested from the client. In this case of the database access, the multiple data are broadcasted into the DF from the replicated databases. The application software modules independently select the correct one among the multiple "same data" by the data consistency management module.

After the fault detection, the faulty application modules and databases are required to be recovered. In the ADS architecture, any data broadcasted from the application software module or the database can be intercepted by the replicated modules or the replicated databases through the DF. Even if the atom includes the faulty application software module or the faulty database, it can detect the fault within itself by intercepting the data sent from the replicated application software modules or databases through the DF. When the application software module is faulty, the atom continues to its operation not by using its generating data but by using the correct data received from the replicated application module. If there exist faults in the database, the ACP has to copy the data from the replicated database and simultaneously to receive the current necessary data for recovering the faulty database. After all data are completed to copy from the replicated database, the ACP merges the current data in the copied database.

Application

As one application of ADS, a steel production process control system was proposed. To improve steel quality and to reduce cost, the software needed constant modification, revision, and testing. In this system, on-line expansion, on-line maintenance, and fault tolerance technologies were effectively utilized not only for the hardware system but also for the software system.

Production schedule data, broadcast into the DF from the steel production scheduling module, are received by real-time in/out control (I/O CTL) subsystems. Each I/O CTL has its own responsible control region, and it autonomously adjusts its own schedule according to the situation by communicating with other I/O CTLs through the DF, as shown in Figure 1.6. In this system, the application software modules are replicated according to their level of importance. Each module is driven only by the correct and necessary data received by the ACP. This autonomous data-driven mechanism makes it possible to expand, test, and repair the component during operation [30].

1.4.1.2 Assurance

Background

By the 1990s, nonstop control system operation had already reached high levels of efficiency. However, maintenance was manual and not yet automated. Moreover, labor costs had risen drastically. Therefore, not only the cost of operation but also the life-cycle cost, mainly consisting of maintenance cost, became a major consideration. Information systems to assist maintenance personnel required more and more details. It then became necessary to construct a heterogeneous system in which the facility's control system and maintenance information system were integrated. This structural shift brought about the trend from homogeneous to heterogeneous operation system. For example, the Autonomous Decentralized Transport Operation Control System (ATOS) for Tokyo metropolitan area railway system is such a heterogeneous system in which train control system and information system are integrated.

Requirements

An integrated system consisting of heterogeneous systems is required to keep operations safe and stable under heterogeneous properties and evolving conditions. This requirement is called assurance, which includes heterogeneity and adaptability [31].

Architecture and technologies

The main feature of this architecture is not only that the structures of the DFs are heterogeneous but also that various kinds of data of different quality levels are flowing through the DF simultaneously (Figure 4.2). The heterogeneous data in the DF need to coordinate with other systems to achieve nonstop operation. The on-line property among the heterogeneous systems can be achieved by the following technologies:

1. Heterogeneous integration

   Heterogeneous systems and content-code data coexist in the same DF. In this situation, to meet assurance requirements, especially for a mission-critical application in the integrated control and information systems, atomicity of the transaction process must be realized. Heterogeneous integration technology was proposed to guarantee atomicity by making coordination between heterogeneous subsystems both in control and information system. When performing a transaction process, each control/information subsystem autonomously checks the atomicity of the processed data by cooperating with other correlated control/information subsystems. Based on transaction data flowing in the DF, each subsystem judges the completion of the transaction and commits the process autonomously [13].
2. Assurance

   The assurance of the systems is to adapt to the evolutionary situation, while preserving their objectives by utilizing the heterogeneous information and functionality [31]. The assurance of the systems is realized by filtering the data among them and adjusting to perform their functions. The typical topic of the assurance problem is the atomicity among the related systems. The atomicity among the heterogeneous systems is dependent on the relation of the systems qualities. The high qualified system may ignore the low qualified system, but the low qualified system is largely dependent on the high qualified system.

   In the recursive model of the integrated systems with their respective missions, they can cooperate by using the data passed through the gateway.

When the data for the transaction in the system S1 passes through the gateway to the system S2, it may cause the transaction in the system S2. Then, the atomic action for the transaction in the integrated system (S1, S2) is defined as follows:

(i) Local atomic action: The transaction in S1 is committed only when the transaction including the application software modules are completely performed in S1.

(ii) Global atomic action: The transactions in (S1, S2) are committed only when both transactions in S1 and S2 are locally committed.

(iii) Semi-global atomic action: The commitment of the transaction in S2 is followed under the commitment of S1, but the commitment of the transaction in S1 is independent to S2.

These atomic actions can be achieved by the functions of the gateway. Now consider the case of the integration consisting of two systems, S1 and S2.

(i) The gateway registers the condition of the atomic action for each transaction.

(ii) The data in the DF of S1 is passed through the gateway to S2 for its transaction.

(iii) The application software module in each system sends out the executed result data into the DF. The executed result data for the transaction in each system is received by the gateway.

(iv) The gateway can judge the local atomic action to be attained when all executed result data for the transaction is received.

(v) Under the condition of the global atomic action of (S1, S2), the gateway sends out the commitment of the transactions to both of the DFs, if the gateway detects both of the local atomic actions of S1 and S2.

(vi) Under the condition of the semi-global atomic action of S2 for S1, the gateway sends out the commitment of the transaction in S2 to the DF of S2 when the GW detects the local atomic actions of S1 and S2. But the GW cancels the local commitment of S2 if the transaction in S1 is not completed.

Autonomous data filtering technology is for assuring the different response-time requirements in the integrated control and information systems, and not interfere by each other in situations of change. In message suppression technology, the gateway monitors both DFs and judges whether to pass on the message based on the system's workload [14]. When the gateway receives transaction data, it calculates the estimated response time based on both systems' workload and compares it with the requirement. If the gateway decides not to pass it on, it sends suppressed data to the DF. In function filtering technology, to avoid the data function of control system violated by the data of information system, each gateway runs autonomously to avoid passing through unnecessary information messages to the control system.

Application

As shown in Figure 1.7, ATOS is an application of ADS for integrating two heterogeneous systems: control systems such as route and traffic control, and information systems such as schedule and passenger information [15–18]. This system has

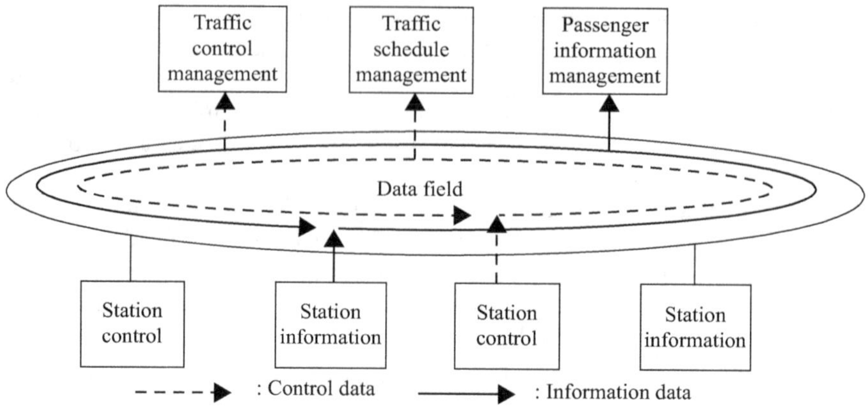

Figure 1.7 Heterogeneous data field. Copyright © 2009 IEICE [34]

been in development over the past 10 years, and some of the current parts will gradually be replaced even before the entire system construction is completed. Therefore, the inherited system, test system, and new system coexist at the same time. This system must be constructed step-by-step without stopping train service and disrupting operation of the currently installed parts of the system.

The system is composed of different regional DF structures. Each train line is composed of station subsystems, train-line traffic schedule management subsystem, and a train-line information service management subsystem. The network connecting both the station subsystems in the train line and the train lines are utilized for both control information missions. The control system is for real-time application, while the information system is required for high performance. In the station subsystem, the computers for control and those for information are divided and connected according to their own mission-oriented networks, namely, the control Ethernet and the information Ethernet through the gateway [19,32].

In this system, the control system should use the train schedule and train delay data, which is generated by the information system. However, the control system is running in real time and has to assure the safe train operation. By using the autonomous data filtering technology, the control system can autonomously judge when and how to utilize data from the information system. Meanwhile, the information system can utilize control data for monitoring the train condition and rescheduling at any time.

1.4.1.3 Fair service

Background

By the late 1990s, on-line property of system operation was achieved. In addition, the Internet became an attractive alternate source of information. The advent of the Internet has generated new requirements for services from users [34]. Therefore, the concept of requirement has shifted from operation of systems to the user services. Moreover, the access from the users cannot be predicted. Each of service providers has a different service level, and the transactions change rapidly and are unpredictable. Such a dynamic and heterogeneous environment has made it difficult for each service provider to manage computer systems independently. As a result, the data center comes forth to provide outsourcing of computers and management service to service providers. The task of a data center is to manage many computing resources for various service providers.

Requirements

A service provider must provide fair service, which means it must keep the same service level for all users with the same service level agreement without stopping its operation under evolving situations.

Architecture and technologies

The homogeneous structure of DFs is constructed for the data center system so that the information system may provide fair service to users. In this system, the same service level, especially response time, is required for each computer. Therefore, each computer needs data to be able to grasp the situation of other computers. The characteristic of this system is that not only data but also information on the situation of each subsystem is broadcast into the DF. The subsystem exchanges its load information with the other subsystems and decides whether or not the subsystem should join the group to process requests for service (Figure 1.8).

1. Autonomous resource allocation
   Autonomous resource allocation technology was proposed to provide and utilize fair service. To assure response satisfaction and avoid measurement delay, the load difference of each subsystem, which is the difference of necessary computing resource and actual deployed resource, is shared in the DF. Each subsystem works asynchronously and makes a decision autonomously according to the different sets of load differences. Autonomous load

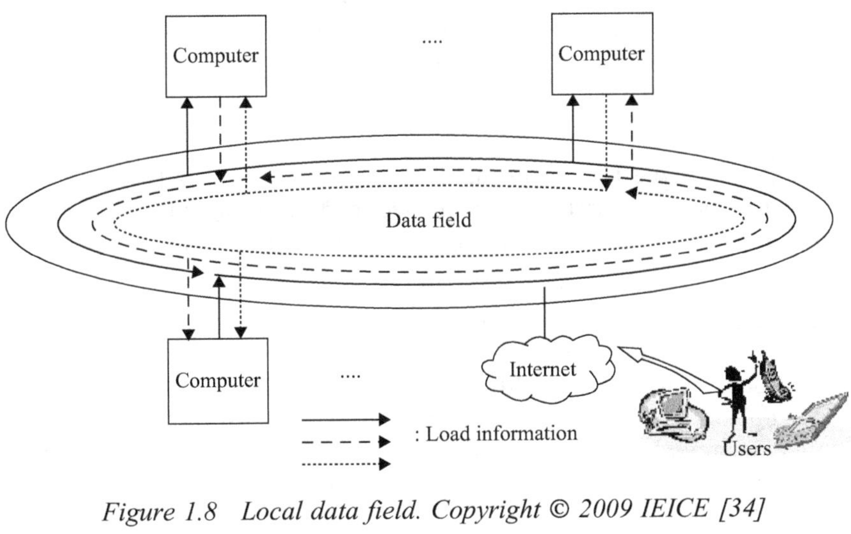

Figure 1.8 Local data field. Copyright © 2009 IEICE [34]

tracking measurement and control achieve quick response time by communicating load difference among subsystems and estimating total load using limited information gathered within a limited period [19,20].

2. Autonomous stabilization
   As situations change frequently, measurement and decision would not be accurate. To raise the level of response satisfaction, it is more effective to track load change. However, frequent change of subsystems makes the system unstable. A trade-off relationship exists between response satisfaction and stability. The system can be stable despite such errors since autonomous stabilization technology converges them according to feedback information [22,23].

Application

In a conventional data center, it is possible to schedule computing resource allocation because load is predictable. Nowadays, however, many online applications on the internet, such as electronic ticket selling systems, have difficulty in predicting the users' requests in advance.

The size of data centers has been increasing due to growing number of customers. Thus, a more efficient utilization of computing resources is required. Moreover, the customers demand many different service levels, with many users at each service level. However, transactions cannot be predicted in the internet environment. Unpredictable peaks can arise within a short time. Autonomous load tracking technology is effective to distribute the load of the system among subsystems autonomously, and fair service can be achieved.

1.4.1.4 Unconscious service

Background

By the beginning of the twenty-first century, services that are to be provided should not only take user convenience in account but also their quality of life as well. Service providers should offer appropriate service to users according to their situations. To realize this purpose, a heterogeneous system with heterogeneous requirement levels became necessary. This has made a change from homogeneous mass service to heterogeneous unconscious service.

Requirements

In this system, a large number of the users utilize the system. The unconscious service, which means the users take for granted services unconsciously according to their own situations, is required for improving the quality of life.

Architecture and technologies

Because there are different service contents, the process levels are also different. To meet different service process requirements, it is required to divide DF into heterogeneous levels and adapt different service processes (Figure 1.9). In addition, it is difficult to implement the DF with high response and reliability at the same time. Therefore, different levels of heterogeneous timed DFs are constructed for different process levels such as high-response low-reliability and low-response high-reliability.

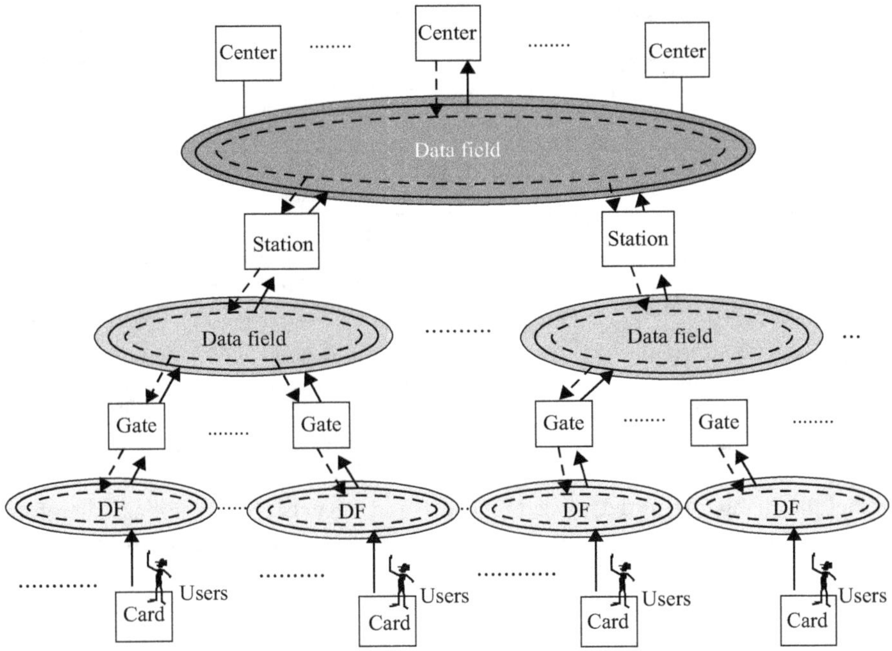

Figure 1.9 Timed data field. Copyright © 2009 IEICE [34]

To achieve high-performance and highly reliable processes, different functions are distributed into heterogeneous DFs. Each DF autonomously executes the functions and coordinates with other DFs to continue its own operation even if data inconsistency occurs.

Application

This architecture and technologies have been applied in Suica, the ICCTS, introduced by JR East in November 2001, and can be regarded as the second infrastructure to combine transportation with e-commerce. The contactless IC card has made it possible to integrate smooth passenger flow (through real-time control of gate devices) with reliable information processing of fare calculation [35–37].

1.4.2 Paradigm shift to society and economy

With the advancement of science and technology, the rapid and continuous changes in the social and economic environment have been affecting the life and business on a global scale. On the other hand, changing and unpredictable society and economy accelerate to make new needs of the human lives. As shown in Table 1.2, the third generation of ADS is to achieve sustainable development of society, economy, and life.

Through the globalization, the trend of social and economic evolution transforms from the high-speed growth with the competitive relationship to the sustainable development under the global cooperation. Moreover, with the rapid progress in the construction of the information society, it becomes more and more difficult to

Table 1.2 Paradigm shift to society and economy

Mission	Third generation			
	Nature	**Management**	**Economy**	**Aging**
	Disaster control	Strategic unit	Finance transformation	Health care/Labor force
Customer	Community (climate change)	Organization (restructuring)	Internet market (value and money)	Society (social security)
System requirement	Resilience (disaster prevention and recovery)	Sustenance (unpredictable market)	Security (exchange)	Fairness (generations gap)
Examples and applications	East Japan Earthquake and Tsunami, IoT, Big Data	Strategic (business unit)	Blockchain	Automatic vehicle, health-care network

predict the changes required, spreading speed and impacts of innovations. Therefore, it is a necessity that society and economy can appropriately transform themselves from an old stage to a new one in their life cycle. The ADS concept-oriented approach is a way to achieve the requirements of following missions [1].

1. Nature
   The frequency, social, and economic impacts of natural disasters show an exponential increase in recent decades. Cities and countries around the world have begun to realize that these events are no longer "hundred-year" storms but repeat within a few years. Not all of these effects are related to climate change, but many, including coastal and river flooding as well as hurricanes and typhoons, are becoming more violent as ocean temperatures rise. The new methods are necessary to enable urban and regional societies to improve their resilience to such events. We are powerless to prevent such extreme weather events and the exact time of their occurrence may not be predictable, but we can improve the systems for the impacts of these risks and their knock-on consequences, and instantaneous and long-term response to an event.
   With the advancement of IoT and Big Data, the approach of technology has changed from a single system to system of systems. The integrated and autonomous systems have major roles to play throughout such lifecycles, and urban, regional, and national infrastructure needs to be adapted to incorporate these new capabilities.
2. Management
   With the transformation of social structure, the industry structure transforms from the high-volume manufacturing to ecosystem, long-term service provision, and continuous operation. The business approach has also changed from localization to globalization. As a result, the management has changed from individual operation to alliance, the market has changed from closed to open environment, and the manufacture has changed from distributed to networked model.
   Toward the future, it is extremely important to adapt to the rapidly changing management environment such as shrinking market caused by low birth-rate

and aging society, strongly interrelated global economy, development of ICT, advance of globalization, environmental issues like greenhouse effect, and frequent occurrence of the natural disasters. The ADS concept-oriented approach has been approved that it is a way to achieve the chain of technologies and markets for sustainable development.

3. Economy
   The development of society and economy reduces the material living gap among countries. As a result, the individual lifestyle transforms from the uniform substantial individual to the diversified individual lives. In the future, the multiple individual styles will coexist in the same region. The cyberspace will transcend time and space to connect the persons with the same individual lifestyle in different regions.
   As an example, blockchain is an autonomous distributed software to provide secure and uninterruptible service. The applications of blockchain technology have been beyond currencies, as a disintermediating force in various industries, such as finance, healthcare, publishing, and software, including the potential for widespread distributed, autonomous organizations, based on smart contracts built upon blockchain technology.
4. Aging
   With the development of society entered a mature period, the problem of the declining birth rate, the increasing aging population, and the urbanization should be faced not only by the developed countries but also by the developing countries. The cost of pension, medical services, and elderly care services will become the main part for sustainable development of society. Therefore, for continuously and rapidly changing situations, the society transforms to the flexible structure in coop with the situations, and the infrastructure in society is of ultimate importance.
   The rapid transformation according to changing situations is required to achieve sustainability in society, economy, and life. The ADS concept-oriented approach will become more and more important not only on the innovation of technology but also on the value creation in society, economy, and life.

1.5 Conclusions

In the near future, users will require more mutual coordination in services. This requirement arises from unexpected situations and emergent events in the complex environment of modern society. This marks a shift from "service utilization" to "service creation" implemented by the input of users who share preferences and/or similar situations.

This requirement differs from the Internet requirement, which provides services to anyone, anytime, and anywhere. Under rapidly evolving situation, users with similar preference cooperate with each other not only share the services but also create the services. Such services are characterized by "right me," "right here," and "right now" and are provided/utilized in accordance with the cooperation of users.

To meet these requirements, users with similar preferences organize a community. Each user can autonomously and actively form a local community with other users based on physical locations, time, and the kind of service [38]. This communication field constructed by community users is called active DF, in which each user broadcasts information into it and shares information with other users. Moreover, users create services through the cooperating with each other based on the shared information. The DF changes by time and place, and the members of the community also interchange.

The concept, system architecture, technologies, and applications of ADS have shown the trend of the ADS concept based research and development through actual operations. The background of the trend from homogeneous service to heterogeneous services is common worldwide. Furthermore, the new trend in community has accelerated in the world for achieving sustainability of human life, company activities, and environment. For example, smart city and grid are considered parts of communities. In the future, different types of communities will be required for sustainability throughout the international cooperation and coordination. Therefore, the ADS concept-oriented approach has been a driving force for technological innovation and social transformation.

References

[1] K. Mori, "Concept-Oriented Research and Development in Information Technology", John Wiley & Sons, Inc., Hoboken, NJ, 2013.

[2] M. Athans (Ed.), "Special Issue on Large-scale Systems and Decentralized Control", IEEE Transactions on Automatic Control, vol. AC-23, no. 2, pp. 105–371, 1978.

[3] W. D. Barritt, "Centralized Control System for Appliances", IEEE Transactions on Industry Applications, vol. 24, no. 2, pp. 328–331, 1988.

[4] B. Ciciani, D. M. Dias, B. R. Iyer, and P. S. Yu, "A Hybrid Distributed Centralized System Structure for Transaction Processing", IEEE Transactions on Software Engineering, vol. 16, no. 8, pp. 791–806, 1990.

[5] M. Jeng and H. J. Siegel, "A Distributed Management Scheme for Partitionable Parallel Computers", IEEE Transactions on Parallel and Distributed Systems, vol. 2, no. 1, pp. 120–126, 1990.

[6] M. Athans (Ed.), "Special Issue on Distributed Computing Systems", Computer, vol. 24, no. 8, pp. 12–78, 1991.

[7] P. Jalote, "Fault Tolerance in Distributed Systems", Englewood Cliffs, NJ: Prentice Hall, 1994.

[8] R. Adler, "Distributed Coordination Models for Client/Server Computing", IEEE Computer, vol. 28, no. 4, pp. 14–22, 1995.

[9] G. Coulouris, J. Dollimore, and T. Kindberg, "Distributed Systems, Concepts and Design", Wokingham: Addison-Wesley, 2001.

[10] K. Mori, S. Miyamoto, and H. Ihara, "Proposition of Autonomous Decentralized System Concept", IEEJ Transactions on Electronics, Information and Systems, vol. 104, no. 12, pp. 303–340, 1984.

[11] K. Mori, "Autonomous Decentralized Software Structure and Its Application", IEEE Proc. of FJCC86, pp. 1056–1063, 1986.

[12] K. Mori, "Autonomous Decentralized Systems: Concepts, Data Field Architecture and Future Trends", IEEE Proc. of ISADS93, pp. 28–34, 1993.

[13] I. Kaji, Y. Tan, and K. Mori, "Autonomous Data Synchronization in Heterogeneous Systems to Assure the Transaction", IEEE Proc. of HASE1999, pp. 169–178, 1999.

[14] I. Kaji and K. Mori, "A Gateway Filtering Technique to Maximize the Transactions in Heterogeneous Systems", IEICE Transactions on Communications, vol. E84-B, no. 10, pp. 2759–2767, 2001.

[15] T. Kobayashi, O. Iba, H. Ebine, and S. Aoyagi, "Advanced Train Administration and Communication System Based on ADS Technologies", IEEE Proc. of ISADS1999, pp. 388–391, 1999.

[16] K. Kera, K. Bekki, K. Fujiwara, F. Kitahara, and K. Kamijo, "Assurance System Technologies Based on Autonomous Decentralized System for large Scale Transport Operation Control Systems", IEICE Transactions on Communications, vol. E83-B, no. 5, pp. 1085–1093, 2000.

[17] T. Aizono, K. Kawano, H. Wataya, and K. Mori, "Autonomous Decentralized Software Structure for Integration of Information and Control Systems", IEEE Proc. of COMPSAC1997, pp. 324–331, 1997.

[18] F. Kitahara, K. Kamijiou, Y. Kakurai, K. Bekki, K. Kare, and K. Kawano, "Phased-in Construction Method of ATOS", IEEE Proc. of ISADS1999, pp. 415–424, 1999.

[19] M. Matsumoto, A. Hosokawa, S. Kitamura, D. Watanabe, and A. Kawabata, "Development of the Autonomous Decentralized Train Control System", IEICE Transactions on Communications, vol. E84-D, no. 10, pp. 1333–1340, 2001.

[20] T. Masuishi, H. Kuriyama, Y. Ooki, and K. Mori, "Autonomous Decentralized Resource Allocation for Tracking Dynamic Load Change", IEEE Proc. of ISADS2005, pp. 277–283, 2005.

[21] T. Masuishi, K. Shibata, Y. Ooki, and K. Mori, "Autonomous Decentralized Load Tracking Techniques and Evaluation", IEEE Proc. of 2nd International Symposium on Dependable, Autonomic and Secure Computing (DASC 2006), pp. 69–76, 2006.

[22] T. Masuishi, K. Shibata, Y. Ooki, and K. Mori, "Techniques to Improve Tracking Ability in Autonomous Decentralized Load Tracking System", The 4th International Conference on Computing, Communications and Control Technologies, 2006.

[23] T. Masuishi, K. Shibata, Y. Oki, and K. Mori, "Autonomous Decentralized Load Tracking Systems and Evaluation Criteria for Response and Stability", IEEE Proc. of ISADS2007, pp. 255–262, 2007.

[24] K. Mori, "Autonomous Fading and Navigation for Information Allocation and Search under Evolving Service System", Proc. IEEE Conf. on APSITT, pp. 326–330, 1999.

[25] H. F. Ahmad and K. Mori, "Autonomous Information Service System: Basic Concepts for Evaluation", IEICE Transactions, vol. E83-A, no. 11, pp. 2228–2235, 2000.

[26] H. Arfaoui and K. Mori, "Autonomous Navigation Architecture for Load Balancing User Demands in Distributed Information Systems", IEICE Transactions on Communications, vol. E84-B, no. 10, pp. 1085–1093, 2001.

[27] X. D. Lu, H. Arfaoui, and K. Mori, "Autonomous Information Fading and Provision to Achieve High Response Time in Distributed Information Systems", IEEJ Electronics, Information and Systems, vol. 125, no. 4, pp. 645–652, 2005.

[28] K. Mori and H. Ihara, "Autonomous Decentralized Loop Network", COMPCON Spring, pp. 192–195, 1982.

[29] H. Ihara and K. Mori, "Autonomous Decentralized Computer Control System", IEEE Computer, vol. 17, no. 8, pp. 57–66, 1984.

[30] K. Mori, H. Ihara, Y. Suzuki, *et al.*, "Autonomous Decentralized Software Structure and its Application", IEEE Proc. of FJCC86, pp. 1056–1063, 1986.

[31] I-Ling Yen, R. Paul, and K. Mori, "Toward Integrated Methods for High-Assurance Systems", IEEE Computer, vol. 31, no. 4, pp. 32–34, 1998.

[32] K. Mori and A. Shiibashi, "Trend of Autonomous Decentralized System Technologies and Their Application in IC Card Ticket System", IEICE Transactions, vol. E92-B, no. 2, pp. 445–460, 2009.

[33] K. Mori, "Trend of Autonomous Decentralized Systems", IEEE Proc. of FTDCS04, pp. 213–216, 2004.

[34] D. Oppenheimer and D. A. Patterson, "Architecture and Dependability of Large-scale Internet Services", IEEE Internet Computing, vol. 6, no. 5, pp. 41–49, 2002.

[35] Y. Shirakawa and A. Shiibashi, "JR East Contact-less IC Card Automatic Fare Collection System 'Suica'", IEICE Transactions, vol. E86-D, no. 10, pp. 2070–2076, 2003.

[36] A. Shiibashi, "Autonomous Decentralized High-speed Processing Technology and the Application in an Integrated IC Card Fixed-line and Wireless System", IEICE Transactions, vol. E88-D, no. 12, pp. 2699–2707, 2005.

[37] A. Shiibashi, Y. Maruyama, M. Yamana, and K. Mori, "Multi-layered Data Consistency Technology in IC Card Ticket System", ADSN, Toronto, Canada, pp. 58, 2007.

[38] K. Mahmood, X.D. Lu, and K. Mori, "Autonomous Community Construction Technology to Achieve Service Assurance in ADCS", IEICE Transactions on Information and Systems, Vol. E91-D, No. 9, pp. 2259–2266, 2008.

[39] K. Mori, "Introduction on Autonomous Decentralized Systems", Morikita Publishing, 2006 (in Japanese).

Part 2
ADS applications in intelligent infrastructure

Overview

Along with sophistication of society, the values among the people have been diversified and their transition has been also faster. This trend affects the concepts of social infrastructure. To meet variety of needs multifunctionalities of social infrastructure systems and collaboration among the systems are advancing, and the system tends to be complicated. And dependence on the system is also increasing. This requires higher availability for social infrastructure systems and it is becoming more difficult to suspend the system long time for maintenance or expansion.

Autonomous decentralized system is known as a technology which realizes the property of online maintenance, online expansion and fault tolerant in complex large-scale system. Therefore, autonomous decentralized techniques have come to be utilized and have been put into practical use in various fields. Factory automation, public facility management, building facility management, steel production management, power equipment management, etc. are the examples.

In this part, the applications of autonomous decentralized system to control system of important social infrastructure which has extremely high mission criticality are shown. In Chapters 2–4, the applications for railway control system and variety of assurance technologies are described. In Chapter 5, the application for air traffic surveillance and its fault tolerant technology is described. In Chapter 6, the application for manufacturing system is described.

Chapter 2

Autonomous decentralized safety critical system

Takashi Kunifuji[1]

Abstract

Along with sophistication of society, multifunctionalities of social infrastructure systems and collaboration among the systems are advancing, and the system tends to be complicated. And dependence on the system is also increasing. This trend is no exception for safety critical systems like railway control systems. Conventionally, in safety-critical system, it has been considered that system integrity is the most significant requirement to ensure its safety. Since frequent system expansion involves a risk of impairing the integrity of the system, it is considered desirable to avoid expanding the safety system as much as possible, and extension of the system during operation has been prohibited. For that reason, the construction period for system expansion had been becoming long, and a technology to drastically shorten the construction period was required. In this chapter, as one of the solutions to the issue mentioned above, autonomous decentralized railway control system which has developed and introduced in JR East and its system-construction technology is described.

2.1 Introduction

Railway which is a safe and stable transportation plays central role in transportation of inner or inter-city thanks for its bulk and high-speed transport characteristics. So, it has become an important social infrastructure. And now, needs for a railway is changing in accordance with changing of social environment. During the period when the economy is significantly growing, traffic of people and logistics became active, railway was required to increase transportation capacity, shorten trip time, and expand operating hours. Especially in Japan, railway's role in public transportation is very significant from its terrain characteristics. Tokyo, the capital of Japan, has been developing railways for a long time, and it is one of the cities with a

[1]East Japan Railway Company, Tokyo, Japan

high dependency on railway in public transportation. It is also evident that the safety and stability of railway transport is trusted. Therefore, railway operators have focused on improving the safety and reliability of railway control systems. On the other hand, in sophisticated society as seen in developed countries, safety and stability of railway transportation are regarded as given and further service improvement such as diversification of train types, shortening of total trip time, enhancement of information provision at transport disorder and so on are expected. And system expansion for improving transportation service is frequently performed mainly in urban railway lines.

However, since the expansion of the railway control system requires a very long construction period, it is difficult to catch up quickly to the needs of society. The reason for the long construction period is due to the short work time and the poor work efficiency.

Specifically, since the system expansion should be performed with the system in stopping, it should be carried out during few hours of train intervals at midnight. This means that the shorter the train interval, the shorter the time for system test, but it is impossible to expand train interval at the present. Therefore, to improve the work efficiency and shorten the construction period, it is necessary to develop a technology for safely performing expansion of the system during operation.

Autonomous decentralized system (ADS) is known as a technology which realizes online expansion in complex large scale system. East Japan Railway Company (JR-East) decided to adopt an ADS when renewing the railway control system in the Tokyo Metropolitan Area [1–3]. In detail, as described in Chapter 4, the autonomous decentralized railway-control system has been introduced to the 23 lines of the Tokyo metropolitan area over 25 years. And the system controls about 350 stations and manages the train traffic of railway lines over 1,000 km in total. Introduction of the system has been performed step by step. In the line, but it was difficult to complete the construction without the excellent online extensibility of the ADS.

By the way, the railway control system is a huge complex system, and the transport operation control system is only a part of it. From now on, autonomous decentralized technology is indispensable for railway control system to improve constructability of whole system to continue further evolution while catching up social needs, but autonomous decentralized technology is required to continue its own innovation. JR East has been proactively applying ADS to railway control system to improve constructability, especially to safety-related systems. For example, following the transportation-management system described above, we introduced an autonomous distributed system into the train control system. In addition, this paper describes application to signal control system.

2.2 Railway control system

2.2.1 System structure

Figure 2.1 schematically illustrates a typical setup of the railway control system. Railway control system mainly consists of traffic control system, route control

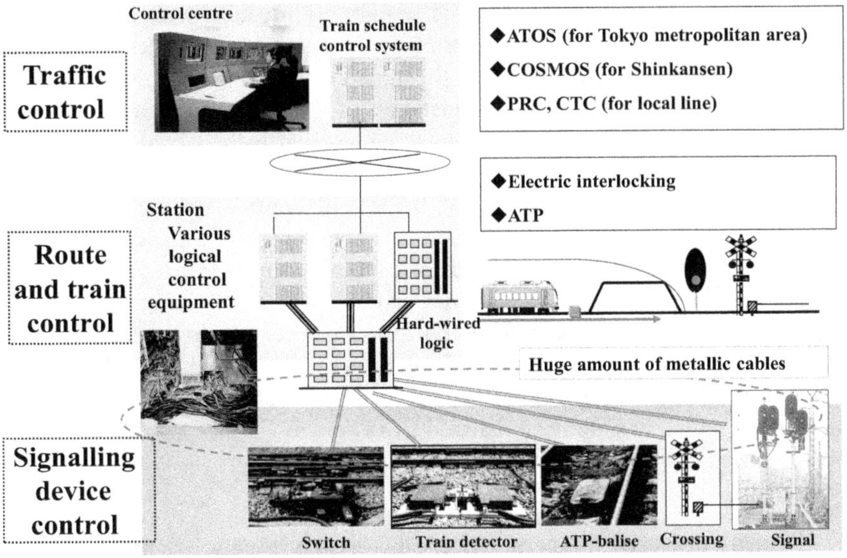

Figure 2.1 A typical railway control system

system, and signal control system, and they are configured hierarchically. The role of each system is described below.

1. Traffic control system
   To keep stable train operation, the system understands the status of train traffic, and when transport disorder has occurred, it reschedules the train schedule and applies it.
2. Route control system
   Route control system sets up the route of station to lead a train into the station yard or depart a train from there. When setting up the route, the system ensures safety by not setting up conflict route at a time.
3. Signal control system
   Signal control system controls signalling devices electrically according to route control system's command.

2.2.2 Issues in conventional railway control system

As mentioned above, the conventional railway control system is a hierarchical system, so if suspending of operation caused by error, maintenance, or construction has occurred in upper layer, it affects lower layer broadly. Therefore, even with small system expansion, it is often necessary to suspend train operation over the entire line. It is not suitable for the train operation system in the Tokyo metropolitan area where the train is operated at high density and the system expansion is frequently performed. So, JR East introduced an ADS to modernize the railway control system for the Tokyo metropolitan area, but only the signal control system was left as it is.

This situation is, largely due to the system architecture with safety as a top priority. Specifically, as shown in Figure 2.1, the control logic is centralized in the signal-control equipment installed in the machine room of the station, and the signalling device as the actuator is installed at wayside. These are electrically controlled through a large amount of metal cables, that is the signal control system and the signalling device ensure high integrity as an integrated system.

As described above, the safety of the signal control system is ensured by wiring integrity, and care must be exercised for wiring change work in system expansion. The work of installing a lot of cables, wiring and checking their connections is a work that requires extreme caution, and since human errors during wiring work sometimes cause serious transport disorder, signal control system not dependent on wiring was required. In developing a new signal control system, we decided to incorporate the idea of an ADS and to realize online expansion and online testing which has not been realized from the viewpoint of safety until now to improve constructability.

2.3 Signal control system utilized x-by-wire technology

2.3.1 Motivation for system change

As the amount of construction of the signal control system increased, securing construction quality was a serious problem. However, the quality assurance of the wiring work must rely only on human's attention, and it has been difficult to show effective solution other than tightening up the management. Under such circumstances, on 28 September 2003, a very serious transport disorder occurred in the Chūō Line, Tokyo's aorta, affecting 180,000 people. The cause was only a few wiring mistakes, but there was pressure on the process due to the congestion of construction. In response to this, JR East began developing a signal control system that does not rely on wiring technology.

Although the direct cause of the transport disorder of the Chūō Line is the wiring mistake as described above, there is a current situation that it takes time and effort to construct the signal control system as a factor behind it. For example, in the aging-replacement work of the signal control system, as a preparation for the test, the wiring is transferred from the old system side to the new system side, and a test (connection test) for checking each wiring to the on-site signalling equipment is performed. After completion of the test, wire back to the old system side and conduct a confirmation test (soundness check) on whether adverse effects due to work are not occurring. Since there is no time to do only the connection test of several signal equipment overnight, this will be continued for several months in the switching work of the large station. When replacing a aged signal control system by new one, huge amount of wiring work is required because an average of about five pairs of core wires are used for signal light on the wayside, and it was also a hotbed of human errors.

From this, it is considered that reducing the wiring between the controller in machine room and the field signalling device and not having the hard-wired logic greatly contributes to the improvement of the workability and the prevention of human error. It can also be prevented by connecting the controller in machine room and the field signalling device with network and shifting from electric control to information control as its main concept. In addition, from the viewpoint of improving the availability, efficiency of system maintenance and system expansion, and fault tolerance, we considered that online maintainability and online expandability should also be required. From the above, it is decided that the new signal control system is directed to autonomous distributed system.

2.3.2 System configuration

2.3.2.1 Autonomous decentralized signal control system [4–6]

Based on the concepts mentioned above, JR-East has developed a new railway signalling system that utilized autonomous decentralized technology. Figure 2.2 schematically illustrates the new signalling system which called network signalling system. In the network signalling system, an optical network connects the logical controller in the machine room and the I/O controller built in the field signalling device. This is to increase noise resistance, improve reliability, and protect from lightning. To facilitate procurement of network equipment, IP was adopted as a communication protocol. The system consists of a central control unit (logic controller: LC) and signal devices (field controller: FC) connected with optical cables. Both the LC and the FC are duplex.

This system is also an ADS. The LC is an autonomous safety-related equipment located at the signal house. The LC generates the signal control information (such as aspect for signal light and operation for switching devices) and translates it into the IP-formatted command data, which is broadcasted to FCs through optical fibre network as a data field (DF). The FC is also an autonomous safety-related

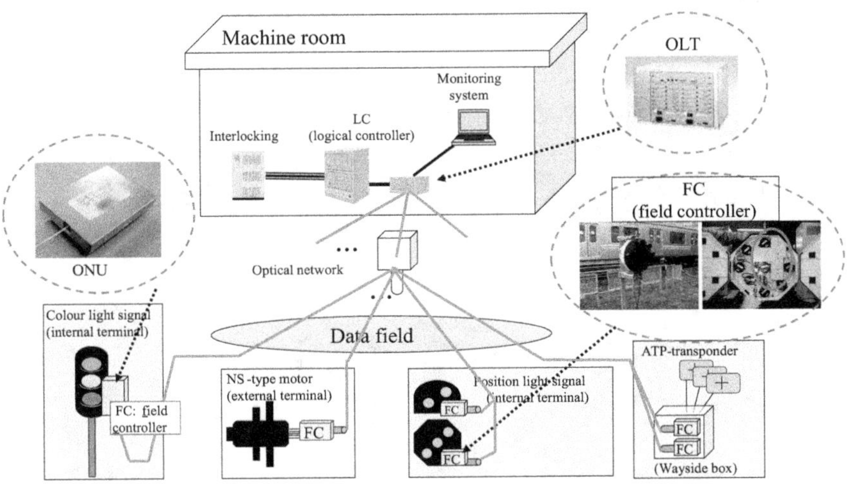

Figure 2.2 Configuration of a new signalling system

equipment located at the wayside. The FC controls the signal device electrically based on the extracted data from the DF. The FC also translates the obtained information from the signal device into the IP-formatted feedback data, which is transmitted to the DF. We have developed two types of the FC: one is equipped in signal device itself (Figure 2.3(a) is one of examples) and the other one is installed in a wayside case, and we use copper wires from the FC to the signal devices (Figure 2.3(b) is one of examples).

2.3.2.2 Method of control

The networking and ADS technology drastically changed the control method of the field devices. Figure 2.4 shows the comparison of conventional electrical control method and new one controlled by data. In a new method, the control data is broadcasted to the network that is DF, and the device receives the data selectively from the DF and operates its aspect or manipulation according to the data.

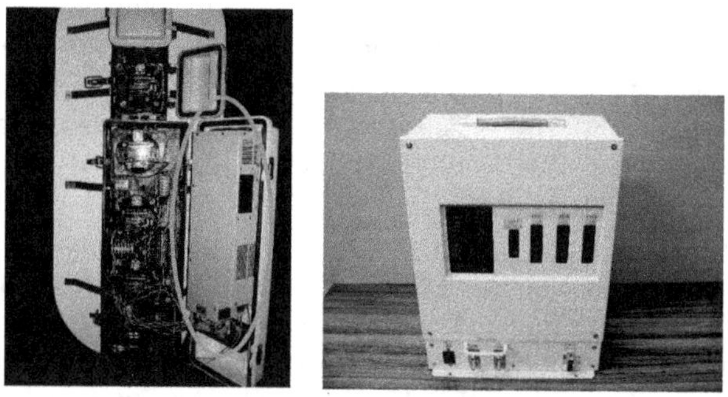

(a) (b)

Figure 2.3 Two types of FC: (a) built-in type and (b) in-case type

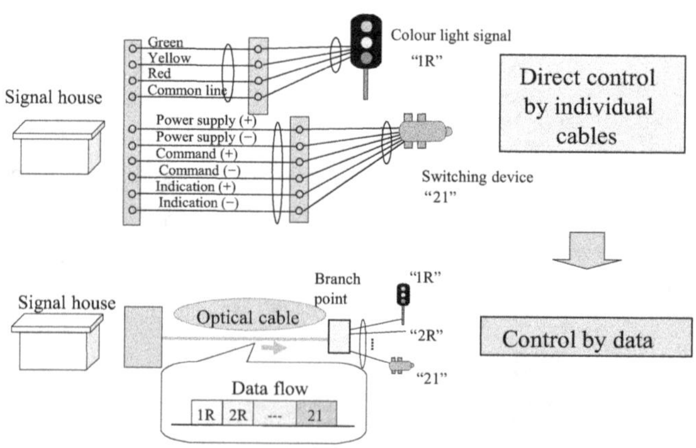

Figure 2.4 Two methods of signal devices controlling

2.3.3 Functions for shorten construction period

2.3.3.1 Issues in current constructing procedure

When current signal control system is replaced by new system, many tests should have been performed before cutover of the new system. The test procedure is as follows.

1. Transferring of wiring
   Transfer the wiring from the current system to the field signalling device to the wiring from the new system.
2. Functional test of the new system
   Confirm the functionality of the new system by controlling the signalling device.
3. Restoring of the wiring
   Restore the wiring from the new system to the field signalling device to the wiring from the current system.
4. Integrity test
   Confirm the functionality of the current system by controlling the signalling device.

In procedure (1) and (3), many of wiring work has performed. It takes long time and has a risk of human error.

In addition, although these are tasks incidental to the examination, they cannot be performed unless it is a train intersection, which is pressing down the essential test time. Reducing these and increasing the in test time will shorten the construction period and improve the construction quality.

2.3.3.2 Assurance technology [7]

Degradation of workability in the conventional system is due to physically separating the control signal between the current system and new system under testing. The necessity of transferring a large amount of wiring for its separation has spurred a decline in workability. This can be easily solved by giving attributes to data in data-driven system. Also, troublesome transfer of a large amount of wiring can be solved by sharing a transmission path with data for train operation and data for test.

By changing from electrical control to data control, it became possible to coexist various operation modes at a time, and a function for improving workability using the data was developed. Coexistence of heterogeneity and adaptability to change is called assurance property.

Assurance technology utilizing data-driven technology has been developed. In current system, signal devices are controlled by electric power. In power-driven system, the information to control the device and the medium to transmit the information are conjoined. So, when the device has been replaced, the test of the logic is needed. And online maintenance is highly restricted because electric power cannot have multiple modes at a time while in data-driven system information and medium are separated. So, it does not need to test the logic when device has replaced. And operation mode of the FC is included in control data, so multiple modes such as train operation and maintenance could be co-existing (Figure 2.5).

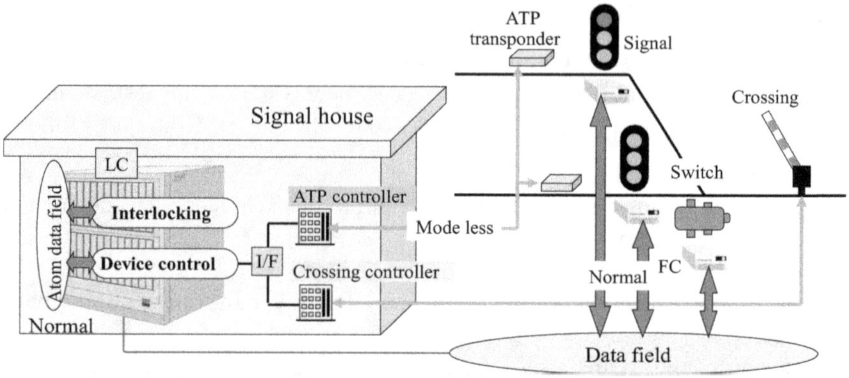

Figure 2.5 State of the system before expansion

In many cases of signalling construction, the functions of LC are expanded step-by-step. In conventional technology, when functions are expanded, the whole system should be set offline. In the network signalling system, each function (e.g. interlock, device control, etc.) is configured as autonomous decentralized software module. And the functions are able to be expanded or tested at online without disturbing other functions. The procedure of online expansion is described below.

Figure 2.5 represents the initial state. In this state, the functions of interlocking and device control are performed by the new system, while the functions of automatic train protection (ATP) and railway crossing control are performed by the external system.

2.3.3.3 Online test applied assurance technology [8,9]

Continuity check

When the signalling device is newly installed or exchanged, the test for confirming the operation by connecting the new signalling device to the logical controller is required. This test is called continuity check. Conventionally, since the different condition is assigned to each core wire which transmitted the control condition by using plural core wires from the LC to the signalling device, it should be confirmed that all the core wires are correctly connected. For this confirmation, it is necessary to operate other signalling devices interlocked with the signalling device under test to a predetermined state, which requires time and labour for testing and has a wide range of influence.

In the case of data-driven system, since the control pattern is determined by the value of the data, if the control pattern corresponding to each data is confirmed in the factory, it is only necessary to check whether the data reaches or does not to the FC at the field.

When testing whether the data reaches to the FC or does not, it is not required to interlock the signalling device under test with other signalling devices. Therefore, only the device to be tested is set to the test mode, and the test can be performed separately from the logical controller. At this time, other signalling devices which are originally interlocked with the device under tested may receive the value on the safe side from the device under test.

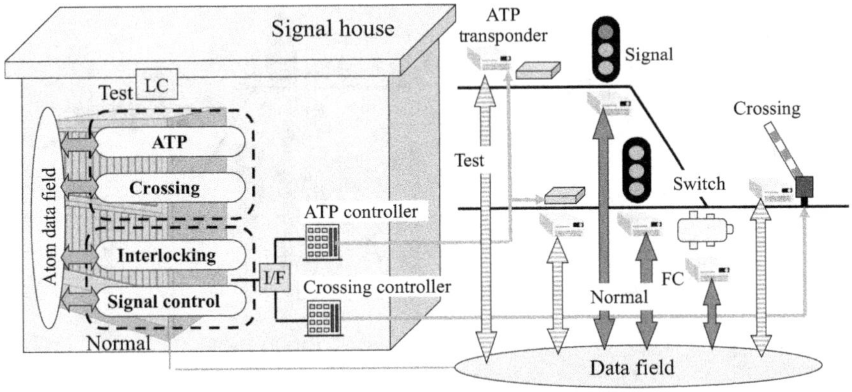

Figure 2.6 Online testing of newly added functions

Table 2.1 Relationship between the function mode and the I/O message mode

	Function mode			
	Normal		Test	
Message mode	Input	Output	Input	Output
	Normal only	Normal		
	Normal and test	Test	Any mode	Test
	Test only			

Figure 2.6 represents the state of online testing for the functions of ATP and railway crossing control, which are added to the new system.

During this state, the test-mode is set to both functions of ATP and railway-crossing control. The functions of interlocking and device control remain normal mode. The relationship between the function mode and the I/O message mode is shown in Table 2.1.

Figure 2.7 represents the final state. In this state, all functions are performed by the new system.

Monitoring test

On the other hand, when updating the logical controller, especially when it is replaced with a new mechanism, it is required to perform a long-term test as to whether the signalling device can be controlled equally with the current logical controller. In the conventional test method, not only does it require hand over conditions using a large amount of cables between the current and new logical controller, it takes a very long time to prepare for the test, and the cable is unnecessary after switching to the new system. So, it will be removed, but human errors often occurred at that time.

In the case of data-driven method, not only a large amount of conditions can be transmitted with one optical fibre but also the confirmation of the test result is

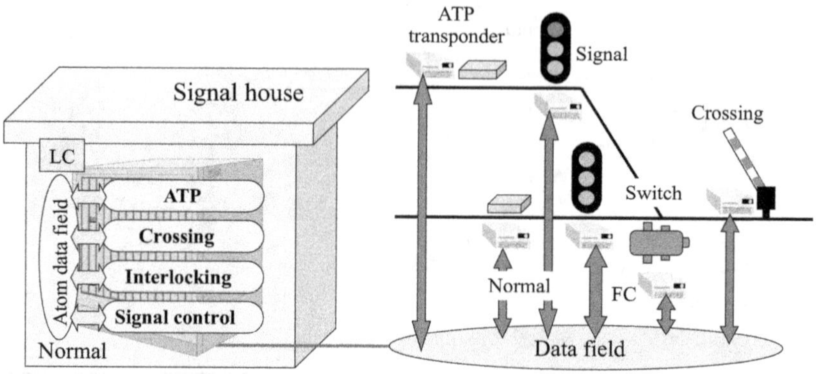

Figure 2.7 Adding of the functions to the LC

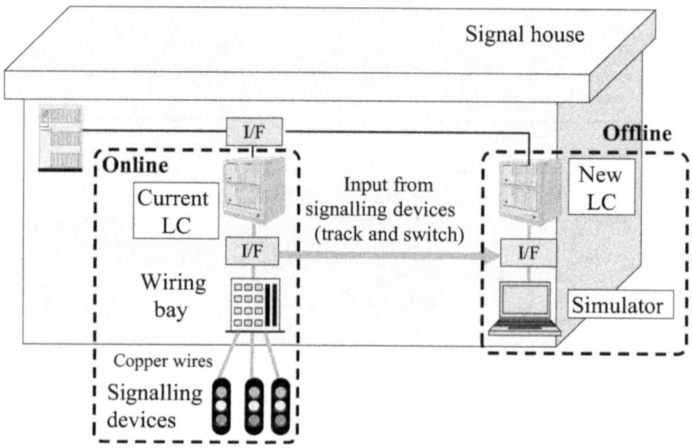

Figure 2.8 Monitoring test with conventional technology

performed only by matching data, so it can be processed by a computer, so it is efficient and free from human errors.

Monitoring test is one of the tests which is required in system-replacement construction. In this test, the processing results of current LC and new LC are compared. The comparison items are state of the control and timing. Figure 2.8 shows the method of the monitoring test with conventional technology. In this method, new LC is tested at offline with field signalling device simulator. The issues of this test method are listed below:

1. Large amount of cables and equipment only for the test is temporarily required.
2. Accuracy of the test is not so high because the simulator cannot replicate actual disturbance in field devices such as delay or chattering.

Figure 2.9 shows the method of monitoring test in the network signalling system.

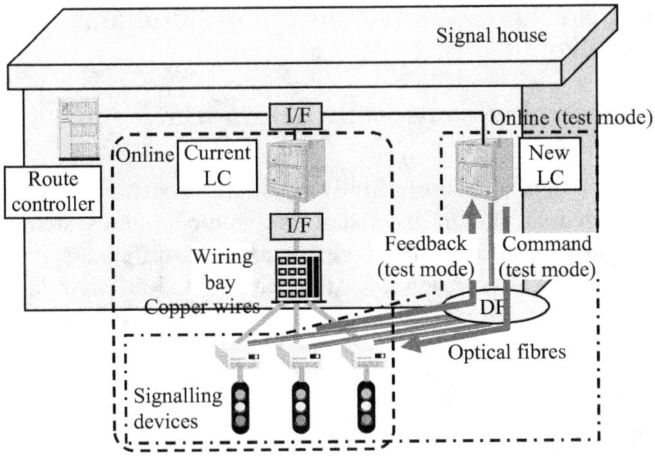

Figure 2.9 Monitoring test in the network signalling system

Table 2.2 Decision table of FC's behaviour according to the mode

FC's mode					
Normal			**Test**		
Message mode		Device control	Message mode		Device control
Input	Feedback		Input	Feedback	
Normal	Normal	Yes	Normal	No	No
Test	Test	No	Test	Test	Yes

In this method, both current LC and new LC are at online, and FCs accept commands from both LCs, and as an FC has intelligence, so it is able to decide the works autonomously by the combination of input message's mode and own mode. The details are shown in Table 2.2.

In this way, the monitoring test of new LC at online is realized without disturbing the control by current LC. And the merits of this method are listed below:

1. No temporal cables or equipment only for the test are required.
2. Accuracy of the test is high because this test is using real FCs and signalling devices.

As application of online-expansion technology, online monitoring in replacement from current signalling system to the network signalling system and online test in step-by-step function expanding of network signalling system.

2.4 System configuration technology of autonomous decentralized railway control system

2.4.1 Architecture of autonomous decentralized railway control system

Figure 2.10 shows an architecture of railway control system which applied autonomous decentralized signal control system. In this model, the system is composed of three horizontal DFs and a vertical DF. In each layer, different service is realized from other layers. The DF of each layer is separated and different layer of the DF are interconnected by gate way (GW). And that is also an autonomous decentralized subsystem.

2.4.1.1 Traffic control DF

This DF is for exchanging information related to train operation. The traffic-management subsystem receives train traffic condition of entire line section via this DF and adjusts the operation plan so that stable train operation is maintained. The changed operation plan broadcasted to the DF, and the route control subsystem set upped at each station receives it.

In this DF, the information should be updated every 2 s, and if the deadline miss occurs, the stability of train operation will be decayed.

2.4.1.2 Logic DF

This DF is for exchanging information related to route configuration for each train at the station. The route control subsystem requests route configuration to the interlocking subsystem based on the train-operation plan. And the interlocking

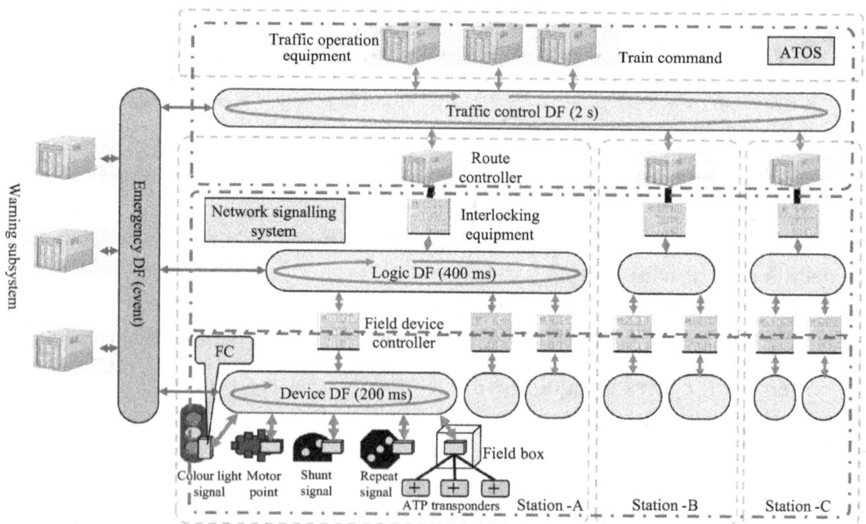

Figure 2.10 System architecture of autonomous decentralized railway control system

subsystem checks collisions between routes. As a result of the collision check, the route setting request judged to be configurable is broadcast to this DF. Field device control subsystem extracts the route configuration request broadcast from the route control subsystem via the DF.

In this DF, the information should be updated every 400 ms, and if the deadline miss has occurred, the safety is decayed.

2.4.1.3 Device DF

This DF is for exchanging information related to field signalling device control. The field device control subsystem broadcast control information for each signalling device to this DF based on route configuration request. The signalling device control subsystem named FC extracts the control information and control the signalling device electrically. And the FC broadcasts controlled status as feedback.

In this DF, the information should be updated every 200 ms, and if the deadline miss has occurred, the safety is decayed.

2.4.1.4 Emergency DF

This DF is for exchanging information related to emergency such as heavy rain, strong wind, earthquake, and so on. Variety of disaster prevention systems is connected to this DF, and they broadcast emergency information to stop the train when the safety of train operation is threatened. This DF is connected to all horizontal DFs.

All subsystems autonomously behave on the safe side according to the extracted information. Due to the difference in the update cycle of DF, the timing of the emergency information at each level is different, but the subsystem integrates the emergency information between DFs, thereby matching the states.

2.4.2 Heterogeneous real-time autonomously integrating system

As described above, a system that integrates ADSs with different real-time properties is called heterogeneous real-time autonomously integrating system (HRTAIS).

2.4.2.1 Definition of heterogeneous real-time

Heterogeneous real-time is defined by two properties of deadline and service level (Figure 2.11).

1. Deadline
   Deadline means a time limit which is defined for each processing. Generally, when the time of processing is over, it is called deadline miss. If deadline miss occurs, the system will fail down.
2. Service level
   Service level means quality of service which is secured by real-time processing such as safety, stability, and so on.

2.4.2.2 Generic system architecture

Generic system architecture of HRTAIS is shown in Figure 2.12. HRTAIS consists of several heterogeneous real-time DFs (RT-DFs) and many autonomous real-time subsystems (RT-SSs). Emergency DF (EMDF) is one of the event-DF and connected

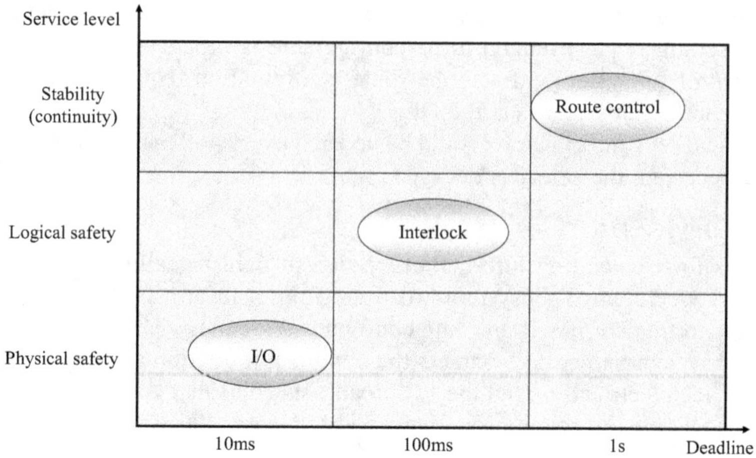

Figure 2.11 Heterogeneous real-time property

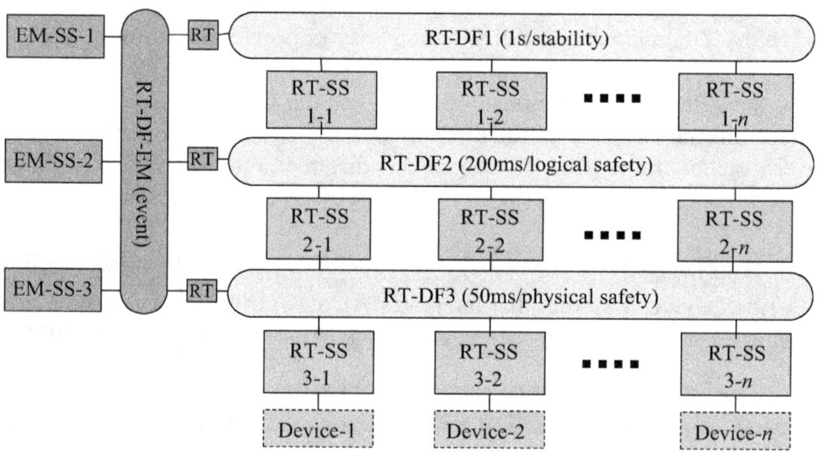

Figure 2.12 Generic system architecture of HRTAIS

to all DF's via router. Emergency subsystems (EM-SS) are only connected to the EM-DF. EM-SS activate an alarm, and this alarm is broadcasted to all the subsystems via each RT-DF.

Real-time data field

In HRTAIS, all kinds of the data which flows in the DFs have real-time properties that are defined in section two, and all the data in same DF should have same real-time property. This kind of DF is called RT-DF. To ensure keeping of time restriction, the DF should be refreshed within properly time. The refreshing method should be selected according to the service level by which real-time is guaranteed.

1. Refreshing by event
   In this method, the data is refreshed only when the processing result has changed. In the subsystem that uses data concerned, when there is no reception from the DF, the latest value is used for processing. When the delivery confirmation is done to confirm the reception in time by the subsystem that uses it when there is a change in the processing result for this method, and it is not possible to receive it, it is necessary to send it again. Therefore, when time limits are shorter than overheads according to re-transmission, it is not possible to apply. Moreover, the technology that the transmission line judges whether data disappeared on the way whether data is sent on the receiving side is newly needed.
2. Refreshing by cyclic
   In this method, the data is refreshed periodically regardless of the presence of the change in the processing result. Latest information is expected to be regularly received, and in the subsystem that uses data concerned, when the time-out is generated, a pre-set value is used and it processes it. It should be noted that this method causes high load to the transmission system. Moreover, it is necessary to shorten the transmission cycle to the extent in which it can follow to the change frequency of the transmission data.

Real-time subsystem
Architecture of the RT-SS is shown in Figure 2.13. The subsystem has two heterogeneous autonomous real-time control processors (RT-ACP). Each RT-ACP is connected to different RT-DF; subsystems are connected to maximum of two heterogeneous RT-DFs. The data is shared by real-time applications (RT-AP) via atom DF (ADF) in the subsystem.

RT-ACP is periodically driven and driving cycle is the same as another subsystem's RT-ACP which is connected to the same RT-DF. Hence, RT-ACPs in the same RT-SS is driven asynchronously. Generally, in ADS, subsystems are driven by data, that is when all required data is received, the process is driven. In periodically driven RT-ACP, there is no guarantee that all required data has been received within the period. So, RT-AP adopts alternative value and starts processing when data loss has occurred. This alternative value is called default.

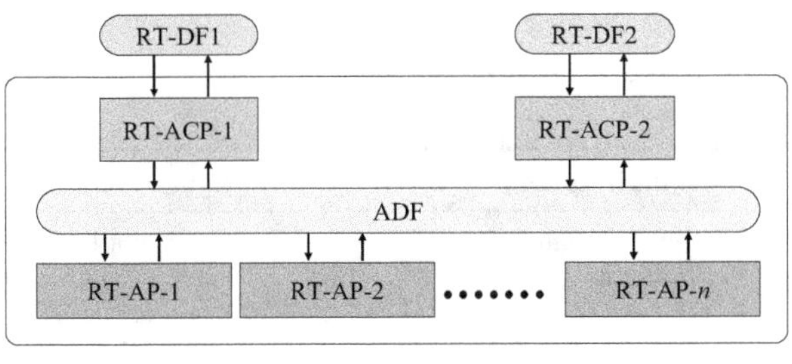

Figure 2.13 Architecture of real-time subsystem

Variation of default value and its role

Objective of adopting default value is to keep service level of applications as high as possible. To do this, it is desirable that default value shall be autonomously specified at input according to the application's requirements. Default value is classified into four types: previous value, application specified value, no data, and flag.

Previous value: This value is the same as what is used at previous processing cycle. This contributes to keep availability of the system.

Application-specified value: This value is constant value which is specified by application.

NULL value: This value means that the data is not for use.

Flag: This means an indicator that a deadline miss has occurred, and the data is not for use.

Variation of default values for input and output are summarized in Table 2.3.

2.4.2.3 Heterogeneous real-time autonomous transparentizing technology

In this section, the heterogeneous real-time autonomous transparentizing technology which filters volume and integrity of the data between heterogeneous RT-DFs is described.

1. Filtering of time-limit
   In the RT-DF, the data is broadcasted cyclically by RT-SSs. When the data is exchanged between heterogeneous RT-DFs, volume of the data per processing cycle is different. To deal with the data transparently between the heterogeneous RT-SS's transparentizing of data volume is required.

   In the case of transmitting the data from short-period RT-DF to a long-period one, typical data should be extracted from short-period one. In the case of transmitting the data from long-period RT-DF to short-period one, lack of the data from the long-period one should be complemented. The former filtering is named integrated filtering, and the latter filtering is named divided filtering. Filtering method and typical application of each filtering type is shown in Table 2.4.
2. Detailed filtering procedure
   Figure 2.14 shows the model of data exchange between heterogeneous RT-DFs. In this model, the data is exchanged from RT-DF1 to RT-DF2. The order of filtering procedure is mentioned below.

Table 2.3 Variation of default value

Type	Default value	Usage
Input	Previous value	Keep availability
	Application specific	Keep service level
	NULL	Keep accuracy
Output	Flag	Notify the default to the receiver
	NULL	Notify the default to the receiver

Table 2.4 Filtering method and typical application

Type	Adopted data	Typical application
Integrated	Latest data	Control system to control system (control data)
	Selected with condition	Control system to information system (alarm data)
	All data	Control system to information system (journal data)
Divided	Latest data Copied one when lack of data	Information system to control system (instruction data)
	Fragmented data	Information system to control system (instruction data)

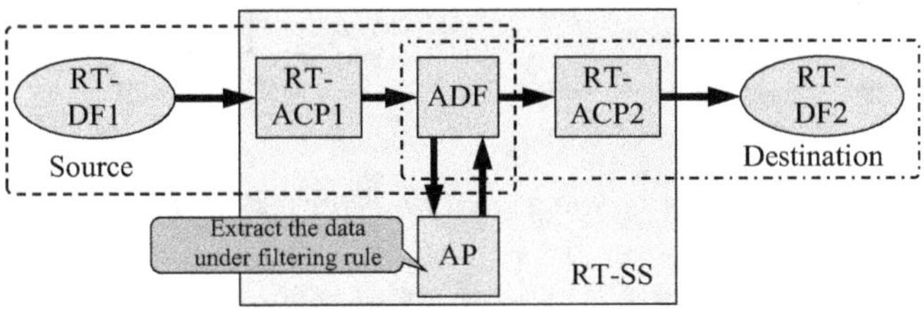

Figure 2.14 Data-exchange model between heterogeneous RT-DFs

(i) The data received from RT-DF1 is inputted into AP via ADF.
(ii) Extracting of data is performed autonomously by AP according to the filtering rule.
(iii) The extracted data is broadcasted to the RT-DF2 via ADF.

Detailed filtering procedure with this model is described below.

3. Example of integrated filtering
   Figure 2.15 shows the example of integrated filtering. In this case, the period of RT-DF2 is three times as long as that of RT-DF1. All data received from RT-DF1 are stored in ADF and packed per cycle. At the timing of transmitting to RT-DF2, latest package of the data is broadcasted to RT-DF2 by one packet.
4. Example of divided filtering
   An issue of divided filtering is separation of no receive and data loss by error. Figure 2.16 shows an example of divided filtering. In this case, the period of RT-DF1 is one-third of RT-DF2. Therefore, at the timing of transmitting to RT-DF2, a case that no data has been received from RT-DF1 is considerable. When this situation has occurred, previous data may be adopted. On the other hand, in the case of data loss at receiving from RT-DF1, the default is set alternatively. This is performed under responsibility of autonomous real-time communication of RT-ACP1 in RT-SS. Therefore, data loss and no data timing are distinguished.

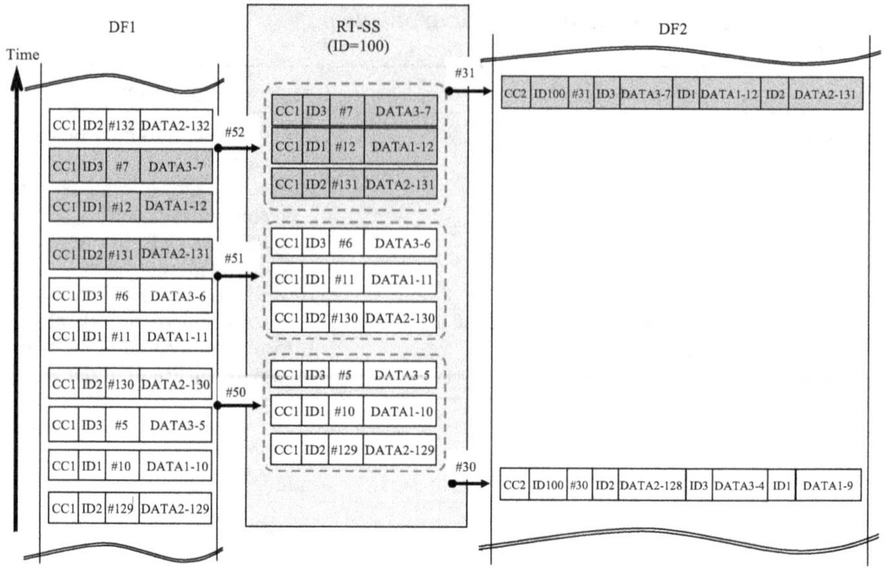

Figure 2.15 Integrated filtering (latest data)

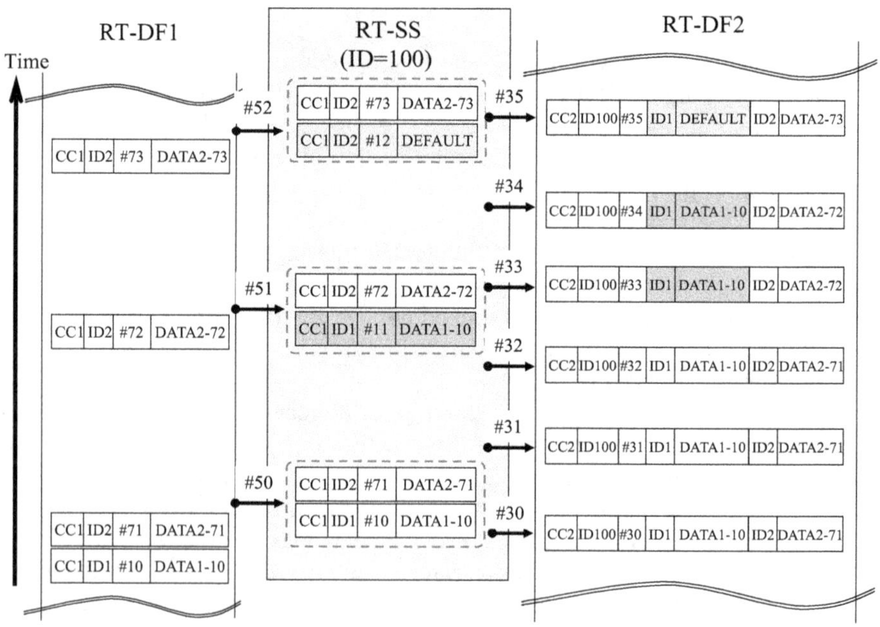

Figure 2.16 Divided filtering (complement)

2.4.3 Safety technologies in autonomous decentralized system

2.4.3.1 Autonomous safety

In autonomous decentralized safety-related system, it is required to have characteristics of autonomous controllability, autonomous coordinability, and autonomous safety to achieve online property. Autonomous safety is defined as the characteristic that if any subsystem fails, is repaired and/or is newly added, the other subsystems can keep safety. The requirements to realize autonomous safety is that each sub system autonomously processes fault detection and safety control in real-time.

2.4.3.2 Autonomous fault detecting and safe-side control

Figure 2.17 shows autonomous fault-detecting model. In this model, built-in tester (BIT) is prepared as an application of RT-SS. And external tester (EXT) is prepared as one of RT-SSs of a system. There two types of testers that detect the fault from different viewpoint. BIT is an autonomous application which connected to ADF. It processes self-checking of RT-SS and healthy checking of applications in the RT-SS. When a fault is latent in BIT, it is impossible to detect fault in RT-SS. To solve this issue, EXT is utilized (Table 2.5). EXT is one of an autonomous RT-SS which is connected to the DF set up in same DF. It notifies occurrence of the dangerous fault to all the subsystems in the DF. Subsystems receive the notification and replace the input from the subsystems in dangerous fault mode to the default value autonomously.

2.4.3.3 Fault-detection mechanism on content code communication

Table 2.6 shows fault-detecting method in content code communication classified for error type. To make general-purpose network equipment usable, error detection

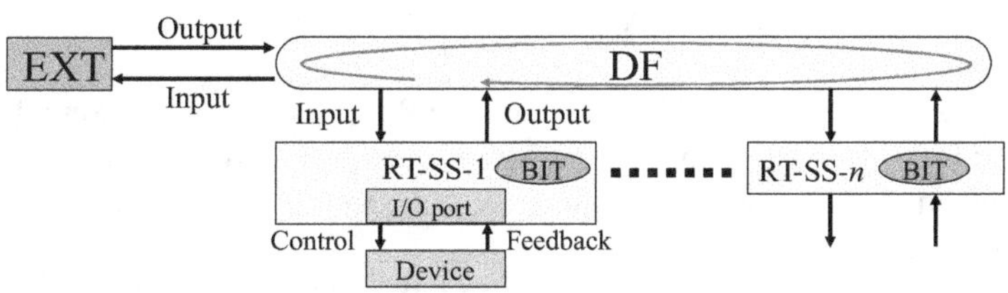

Figure 2.17 Autonomous fault detecting and safe-side control model

Table 2.5 Definition of safety fault and dangerous fault

Fault detection	Preventing fault propagation	Failure mode of RT-SS	Performed by
Detected	Enable	Safety failure	BIT (built-in tester)
Detected	Disable	Dangerous failure	EXT (external tester)
Missed	–		

Table 2.6 Fault detecting method in content code communication

Error type	Example	Caused by (example)	Detected by (example)
Data error	Bit error (random)	Noise	CRC
	Bit error (systematic)	Hardware error on communication device	Reverting data
Time error	Delay	Congestion	Time out
	Missing data	Hardware error on communication device	
	Disorder	Congestion, store and forward	Sequential no.
Delivery error	Misdirection	Systematic error (software bug, system configuration)	Test when construction

is performed in the subsystem. And based on the characteristics of ADS, the following conditions should be applied:

1. All types of error should be detected by SS autonomously.
2. The error should be detected by error-detecting code.
3. Error-detecting code should be included in the message.

2.4.3.4 Safety control

The role of safety control is to suppress the dangerous side output to the outside due to an abnormality locally generated in a certain subsystem and to prevent the local abnormality from being propagated to the entire subsystem and thus to other subsystems. It consists of safety control of the subsystem itself and safety control at the time of communication abnormality detection.

Safety control in subsystem itself

1. Hardware error
   Hardware error is detected by BIT, and the RT-SS will be shut down by BIT. In other RT-SS, the data that was expected to be received from the failed RT-SS is replaced with the default value.
2. Software error
   Software error is detected by BIT, and the autonomous application which caused the error will be halted by BIT. The data expected to be submitted by the failed application is replaced by default value. If error has occurred in ACP or BIT, RT-SS should be shut down by EXT.

Safety control in communication error

Communication error is detected by ACP, and the data will be replaced with safe-side value.

1. Detecting method for the delay caused by asynchronous system. In autonomous decentralized system all subsystems working asynchronously, so we developed a method for detecting delay among autonomous sub systems (Figure 2.18).

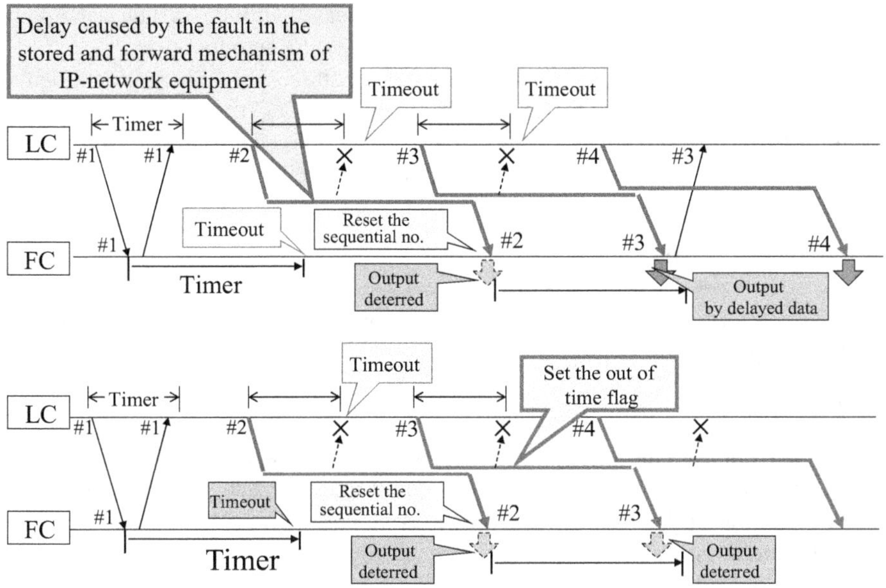

Figure 2.18 Autonomous transmission delay detection in asynchronous system

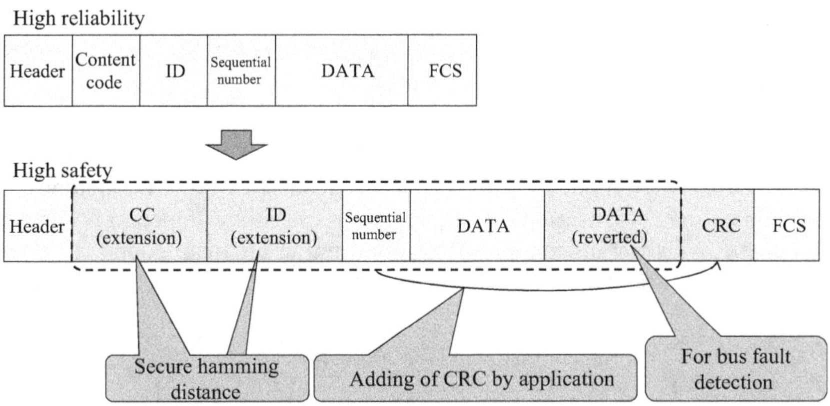

Figure 2.19 Service level filtering (reliability to safety)

2.4.3.5 Heterogeneous service-level filtering

In case the service level which should be assured by system, such as reliability or safety, is different, integrity of input data becomes different. So, data exchange between heterogeneous RT-SS's filtering of service level is also required. Figure 2.19 shows the difference of data format according to the service level. In case service level is highly reliable, it may be sufficient that minimum level of redundant code is added to the data. However, in case service level is highly safe, addition of diversified redundant code and extension of code system to increase hamming distance should be required. This filtering is performed by AP in the RT-SSs autonomously.

To perform these filtering functions mentioned above, RT-SSs should recognize the real-time requirements of each RT-DF. In the ADS, these requirements are recognized as common knowledge, and each subsystem distinguishes the requirements according to the CC of the data. So, filtering is able to be performed autonomously.

2.5 Future study

In this section, future railway control application applied heterogeneous real-time autonomously integrated architecture is proposed.

2.5.1 Expansion for future railway control system

As the extension from the Network Signalling System which described in Section 2.3, the next generation railway system is proposed. Figure 2.20 shows the model of the logical controller of the Network Signalling System. Figure 2.21 shows the model of the next generation railway control system. In the next generation railway control system, the logical DF and the device control DF are integrated into one logical DF.

Also, the autonomous subsystem is installed on the train, the route control DF is provided to adjust the route-control sequence by inter-train communication, the route control DF is configured on the radio-communication space, and the logical DF is connected by radio communication. Further, a command DF for sharing information among the respective command systems is provided and is coupled to the route control DF by radio communication.

2.5.2 Example of flexible route control

In this section, as an application of a heterogeneous real-time autonomous integrated architecture, a realization case of flexible route control function is described.

Figure 2.22 shows an example of configuring an alternate course by flexibly avoiding failures of field devices. In this case, it is not possible to configure the route

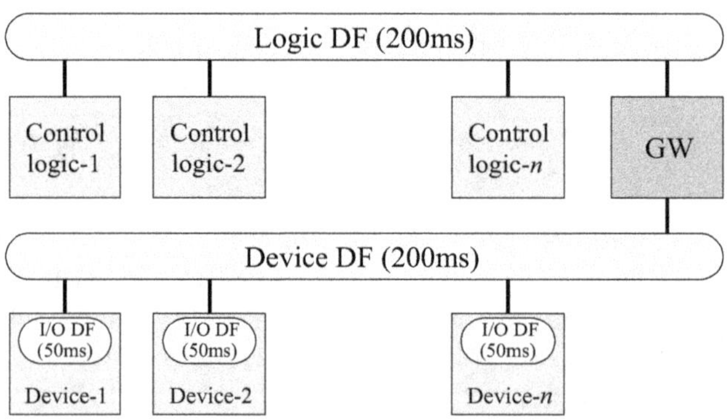

Figure 2.20 System architecture of current logical controller

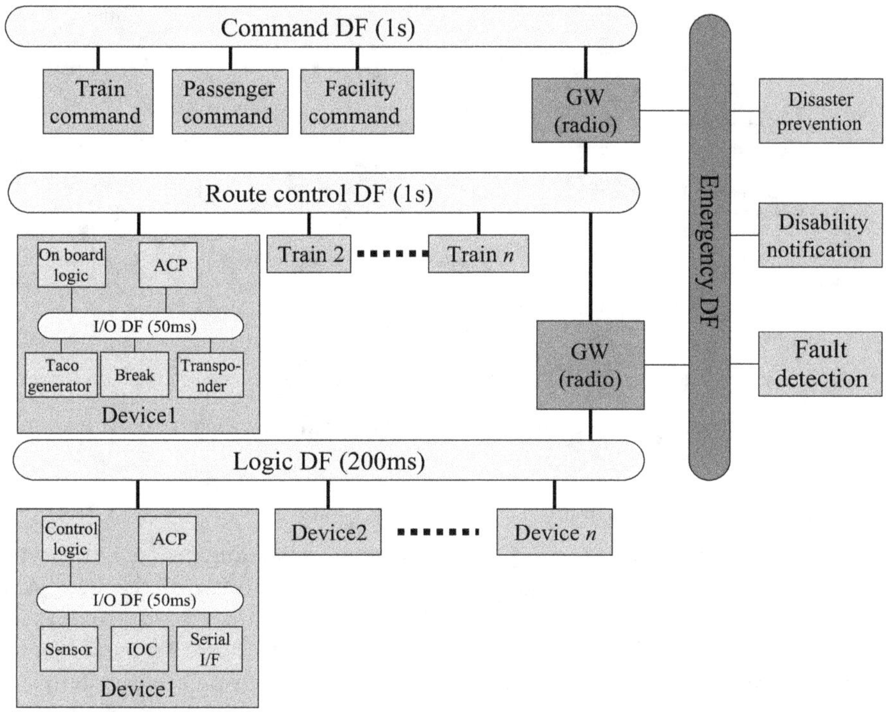

Figure 2.21 Architecture of future railway control system

entering into the track no. 2, which is originally scheduled for the A train due to the failure of the point machine no. 11. In the conventional system, as a result of requesting the route configuration towards the track no. 2, a response that the configuration is impossible is returned only, and the configuration of the alternative course has to be decided by a train commander. The track layout shown in this example, track nos. 1, 3, 4, and 5, can be considered as alternate routes, but the tracks which cannot be used because they contradict the operating conditions are also included.

Here, in the proposed architecture, we show a case where the top level autonomously judges contradiction avoidance by integrating data to the upper level via the subsystem.

1. The I/O control module in the device subsystem detects the failure of point machine no. 11 and broadcasts the information to the I/O DF which is in the ADF and further to the logic DF.
2. The control logic module in the device subsystem extracts the failure information of point machine no. 11 from the ADF and the logic DF, integrates it with the route information of each control logic, and uses the failure information of point machine no. 11 which is on the route towards the track nos. 1 and 2 is generated and broadcasts to the route control DF via GW.

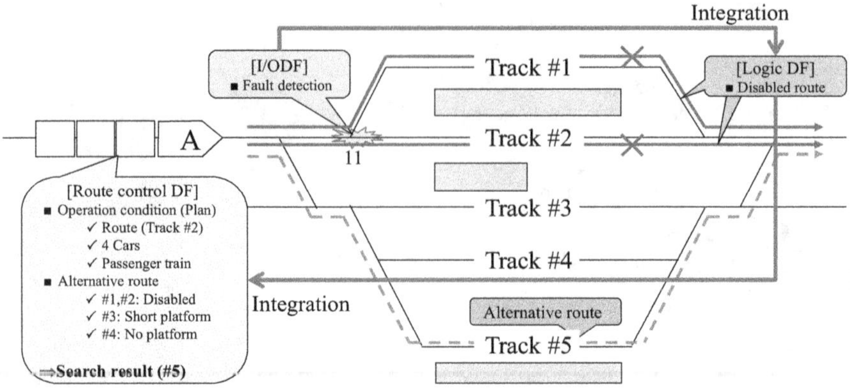

Figure 2.22 Flexible route control on autonomous decentralized railway control system

3. The train control subsystem on the train A integrates failure information of the route towards track nos. 1 and 2, and the operation plan of the train A is obtained by on-board logic module and contradictions that track no. 3 does not have satisfied length, and track no. 4 does not have plat home for passenger are recognized. As a result track no. 5 is autonomously selected as an alternative route and route configuring request is broadcasted to the logic DF via the GW.
4. Each device subsystem autonomously extracts the route configuration request and controls the field-signalling device. As a result, the route towards track no. 5 is configured.

As shown in this example, a flexible route configuration which was impossible in conventional system has become possible by utilizing HRTIS.

2.6 Conclusion

In this chapter, the needs for social infrastructure systems diversify as society advances. In order to respond to this need, the system has been increasingly dependent on social infrastructure in people's social life by repeating integration and expansion, and as a result, it is not even possible to stop the system for maintenance or system expansion. The problem was raised as an issue.

In a mission-critical system such as a social-infrastructure system, the ADS is known as a technology to realize maintenance or expansion during the system is operating. In this paper, the railway control system in the Tokyo metropolitan area, Japan is introduced as a case of application to the railway signal control system, which requires the highest safety. In addition, as a generalized model of autonomous decentralized railway control system, the architecture of a heterogeneous real-time autonomous integrated system and its system construction technology are shown.

With this technology, it is easier to construct a system of systems in which various real-time control systems cooperate, and we believe that the social infrastructure system will contribute to appropriate development and development of social needs appropriately.

References

[1] F. Kitahara and K. Kera, "Widely Distributed Train Traffic Control System", Journal of SICE32 No. 7, 1993.

[2] F. Kitahara, T. Iwamoto, K. Kikuchi, K. Fujiwara, H. Kawashima, and H. Yamamoto, "Widely-Distributed Train-Traffic Computer Control System and Its Step-by-Step Construction", Proc. of ISADS 95. Second International Symposium on Autonomous Decentralized Systems, pp. 93–102, 1995.

[3] F. Kitahara, K. Kamijou, Y. Kakurai, K. Bekki, K. Kera, and K. Kawano, "Phased-in Construction Method of ATOS" Proc. of ISADS 99. Fourth International Symposium on Autonomous Decentralized Systems. Integration of Heterogeneous Systems, pp. 415–424, 1999.

[4] Y. Hirano, T. Kato, T. Kunifuji, T. Hattori, and T. Kato, "Development of Railway Signalling System Based on Network Technology", Proc. of 2005 IEEE International Conference on Systems, Man and Cybernetics, Vol. 2, pp. 1353–1358, 2005.

[5] R. Ishima, Y. Fukuta, M. Matsumoto, N. Shimizu, H. Soutome, and M. Mori, "A New Signalling System for Automatic Block Signal between Stations Controlling through an IP Network", WCRR, 2008.

[6] J. Nishiyama, H. Sugahara, T. Okada, T. Kunifuji, Y. Fukuta, and M. Matsumoto, "A Signal Control System by Optical LAN and Design Simplification", Proc. of IEEE International Conference on Systems, Man and Cybernetics, pp. 1711–1716, 2007.

[7] K. Kera, K. Bekki, K. Fujiwara, F. Kitahara, and K. Kamijo, "Assurance System Technologies for Large Scale Transport Operation Control System", Technical Report of IEICE. FTS99-29, 1999.

[8] T. Kunifuji, G. Kogure, H. Sugahara, and M. Matsumoto, "A Novel Railway Signal Control System Based on the Internet and Assurance Technologies", IEICE Transactions Vol. E91-D, No. 9, pp. 2293–2299, 2008.

[9] T. Kunifuji, T. Miura, G. Kogure, H. Sugahara, and M. Matsumoto, "A Novel Railway Signal Control System Based on the Internet Technology and an Assurance Technology", Proc. of 2008 The 28th International Conference on Distributed Computing Systems Workshops, pp. 581–586, 2008.

Chapter 3
Train control system

Masayuki Matsumoto[1]

Abstract

Control systems have been put into practical use in all the fields of the industrial world. In a conventional system, a central system controls devices directly. This is called a central control system. However, this system configuration has a limitation to meet variety of needs and situations that change from hour to hour. Therefore, autonomous decentralized techniques have come to utilization and have been put into practical use in various fields. Factory automation, public facility management, building facility management, steel production management, power equipment management, etc. are the examples. Train control systems have also been utilizing autonomous decentralized techniques. As a result, these systems have contributed to improve not only safety but also service, such as transport capacity increases and ride comfort improvement. In this chapter, the background of realization of these systems, transition of technology, and assurance techniques are described.

3.1 Introduction

Many accidents occurred in the history of Japanese railways. Innovation in railway systems has taken place by studying lessons of those accidents. The railway operators could not fully understand the needs of operation and technical needs without daily operation. Moreover, they could not find problems without it. The Japanese railway system has achieved the highest level of safe and stable operation in the world. What is an operator's view point concretely? How have railway operators achieved innovation in operations and technology from an operator's viewpoint? These are described below.

3.2 Safety and stability of a railway operation

Railway operation between Shinbashi and Yokohama, launched in 1872, was the origin of the railway in Japan.

[1]Matsumoto Signalling System Consultant, Japan

The government had owned and operated the principal railway line during their first century, with nationalization under the "Railway Nationalization Law" in 1906 and subsequent transformation into a public corporation in 1949. Japanese National Railway (JNR) was privatized in 1987, followed by complete privatization of the three companies of Honshu by revision of the JR Company Law in 2001. As a result, their railway operation was released from governmental fetters. Each company has its own original management strategy and began to tackle various new businesses which benefited from improvements in the appeal of the railway itself and the infrastructure of the railway.

Mass transportation, high-speed travel, and punctual service are features of the railway.

Therefore, when an accident occurs, there are many victims and the impact on the society is very big. The first railroad accident in Japan was a derailment and rollover accident in the Shinbashi station yard in 1874 (2 years after establishment of the railway). The following accidents are representative examples with many casualties: a train collision (105 dead) on the Hachiko line in 1945, a train derailment and rollover (184 dead) on the Hachiko line in 1947, a train fire (106 dead) at Sakuragicho station in 1951, a train derailment and collision (160 dead) at Mikawashima station in 1962, and a train derailment and collision (161 dead) at Tsurumi station in 1963.

The investment for safety in JR East occupies more than half of all the investment, and it is also increasing every year. As a result, the number of accidents itself is decreasing.

Accidents, such as the derailment and rollover (107 dead) on the Fukuchiyama Line in JR West in April 2005 and the limited express derailment (5 dead) on the Uetsu Line of JR East in December 2005, have still occurred recently. In a railway business, the guarantee of safety is the foundation of management strategy. There have been few examples of cases in which the extensive improvement of a signal control system or a train control system has been advanced as a countermeasure against these big accidents (Figure 3.1).

On the other hand, neither stable transportation nor customer service must be neglected just because the safety guarantee is predominant. Mitigation of congestion in the rush hour in urban areas or improvements of the degree of riding comfort are examples. It is also necessary to consider competition, not only with the railway business of other companies but also with other means of transport, as will be mentioned later. That is the differentiation to be made after guaranteeing safety and stable transportation.

For that purpose, maintenance and improvement of the structure for present safety and stable transportation are needed first. Now, there are 180 railway accidents per year including very small accidents in Japan. So, not only strict keeping of the present structure of safety but also development of what has higher reliability is required.

The railway has come to be, from the length of its history, one of "the mature industries." However, the share of cars and airplanes has increased, and passengers have left the railway. The various efforts by railway business, the environmental

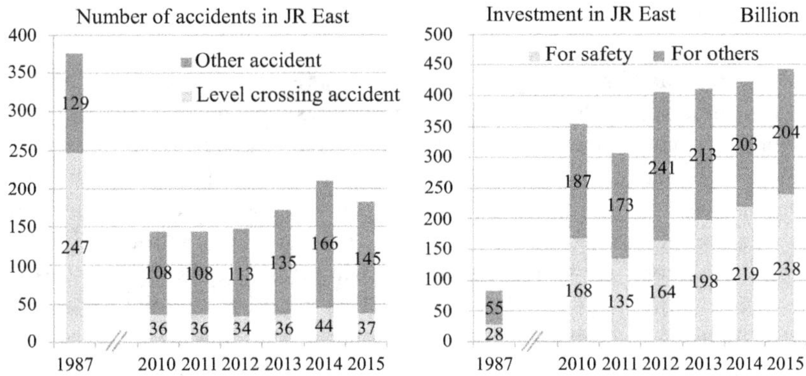

Figure 3.1 Accident and safety investment in JR East

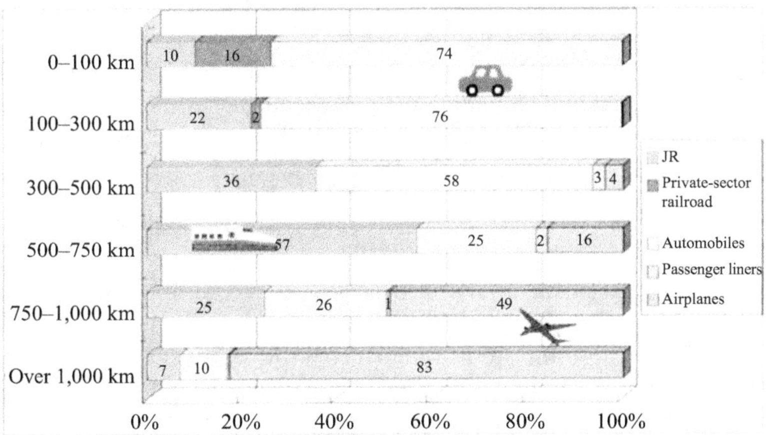

Figure 3.2 The transportation ratio according to the distance

problem, and other factors have affected this "modal shift" and brought recovery of the railways, but never to the situation of railway predominance. In transportation services, users of cars and airplanes increased in number. On the other hand, users of railways decreased in recent years.

The history of a railway is long, and it has been supposed to be one of "the mature industries." Various efforts, environmental problems, modal shifts, etc. were brought by railway services. A recovery of a railway is also seen these days.

As shown in Figure 3.2, a railway is never in the situation of predominance. It is especially a big theme how customers can be increased in number in short-distance transportation and long-distance transportation. Furthermore, a decrease in the birthrate of whole Japan and reduction of the number of workers make the whole traffic reduce. If such tendency continues, railway business will decline. It is a big problem how to get customers to use the railway.

3.3 Development of train control system

The earliest railways had no signal systems, so station employees had to use hand gestures to indicate to train drivers whether they should stop or go on. Even so, there were still many train collisions because drivers missed or ignored the hand signals due to human error. The race was on to develop foolproof signaling system that could prevent train accidents. However, it was more like an arms race because as soon as a digital automatic train control (ATC) system was introduced to prevent one type of accident, another accident was caused by another type of error that required yet another digital ATC system. This race has culminated in today's almost accident-proof train control systems, but even so, new train control systems are still being developed with the focus on improving operation efficiency and passenger services quality.

This section describes development and improvement of train control systems to prevent train accidents, with a focus on the train operation requirement and related technologies.

Since it was quickly realized that directing trains using signals (hand, semaphore, lamp, etc.) alone could not stop train accidents, cab warning devices indicating to the train driver that the train was approaching a stop signal soon came into common use, but even so, train accidents ascribable to driver oversight occurred from time to time.

3.3.1 Outline of development history

The worst railway accident in JNR's history occurred in 1962 at Mikawashima Station, killing 160 passengers and injuring 296. A down freight train in the Mikawashima Station yard on the Joban Line entered the safety siding when the driver missed the stop signal. The locomotive and one freight wagon derailed, blocking the main track of the down line. A following down train hit the freight train and derailed blocking the main up line. Six minutes later, an up passenger train collided with the previously derailed cars, derailing its first four cars. This disaster prompted the decision to introduce a new automatic train stop (ATS), combining a cab warning system and ATS.

As a result, the frequency-shift type ATS was introduced on all Japanese National Railways (JNRs) lines to prevent train collision by April 1966. It was called an ATS-S (S type ATS). Trackside coils were installed at about 43,300 points on the ground and pickup coils were installed on about 13,000 cars (Figure 3.3).

ATS-S takes the driver partly out of the equation by adding the ability to stop the train automatically.

In this design, the ATS-S halts the train by automatically applying the emergency brakes if the driver overlooks or disregards a warning signal in the cab. But there was one clear drawback, once the driver acknowledges the in-cab warning (usually by pressing a confirm button), the automatic stop system is overridden and the train can proceed, so if the driver confirms the warning by mistake, a collision can still occur.

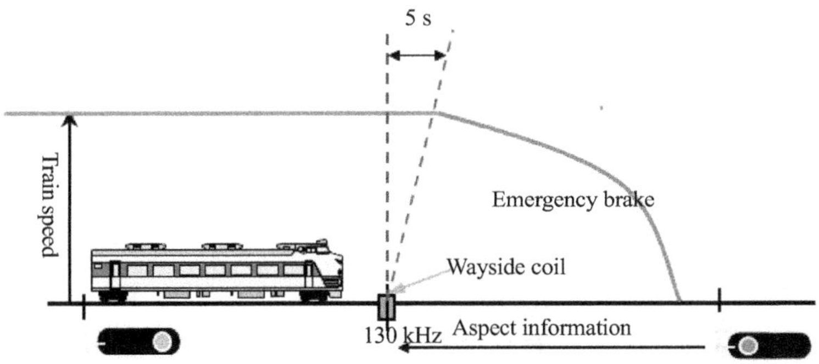

Figure 3.3 Outline of ATS-S

To solve this drawback, a new ATS-P (ATS-Pattern) system was developed to eliminate the weaknesses of the ATS-S. After a derailment at Hirano Station on the Kansai Line in December 1973, it was decided to add a speed-check ability to ATS-S because of the following three problems:

- No ability to check train speed.
- After train driver confirmed warning within 5 s, subsequent safe operation dependent entirely on him.
- Ground equipment transmitted information to on-board equipment only when train was at some specified point ahead of stop signal.

With ATS-P, the train speed is constantly checked against a speed pattern to bring the train to a halt at a stop signal. The ATS-P sends digital data about the distance to stop signals by transponder to on-board equipment which checks the train speed pattern against a reference model and automatically applies the train brakes if the train speed exceeds the pattern. Since the system reduces the number of braking stages, train headways can be shortened to improve train operation efficiency. The ATS-P system was put in practical use in early the 1980s (Figure 3.4).

On high-speed sections with trains operating at very short headways, failure to respond to any signal aspect for any reason can lead to a very serious collision within seconds. Therefore, another system that continuously displays permitted speeds on an in-cab monitor and automatically applies the service brakes was needed. The automatic train control (ATC) system was originally developed to support safe, super-high-speed railway transportation utilizing Japan's Shinkansen and introduced into conventional commuter lines to shorten the train interval (Figure 3.5) [1].

Today, this is called analog type ATC (analog ATC or ATC); it was introduced first in 1964 on the Tokaido Shinkansen where trains were running faster than 200 km/h because drivers could not safely recognize trackside signal aspects and could not respond quickly enough even when they saw the signals.

Sometimes slightly before the debut of analog ATC on the Tokaido Shinkansen, the Hibiya subway line in Tokyo introduced a trackside signal-based analog ATC system—the result of R&D into automatic train operation—to improve train safety

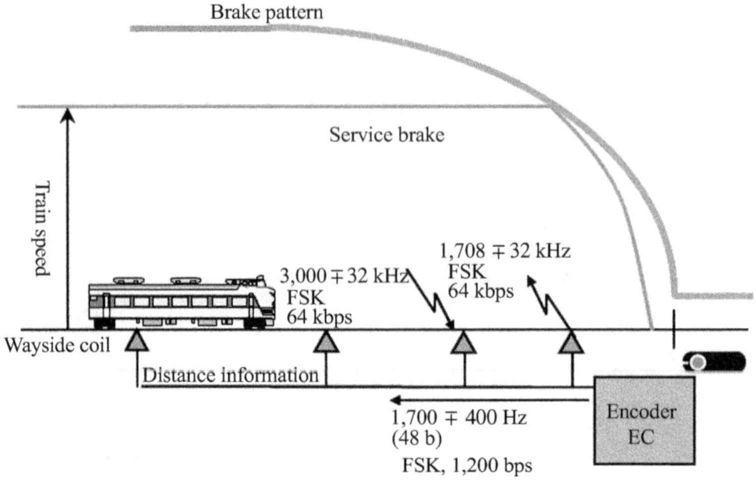

Figure 3.4 Outline of ATS-P

Figure 3.5 Tokaido Shinkansen opening ceremony

and operational efficiency. Then an in-cab signal-based analog ATC system to prevent driver mistakes was put into service on the Nagoya subway in 1965. This was soon followed by an installation on the Joban Line of JNR to support through operation with the Chiyoda subway line in 1971.

The analog ATC is a continuous control system that has improved the safety of train operation much more than ATS, which is an intermittent control system. This is largely because analog ATC assures safe train operation even if the driver misreads a cab signal. In analog ATC, the trackside equipment transmits an analog ATC signal matching the control train speed to the track circuit and the on-board equipment receiving the analog ATC signal displays the signal on the monitor and slows the train according to the signal.

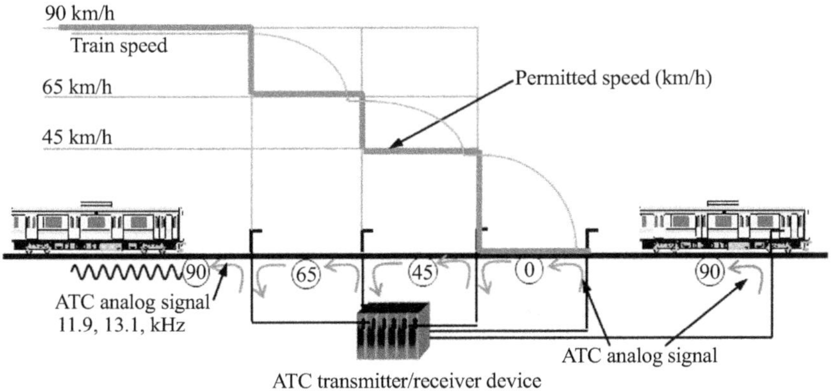

Figure 3.6 Outline of ATC

However, the analog ATC is relatively an old technology that was developed during the early stages of the Tokaido Shinkansen and has problem, including inability to cope with increasing numbers of trains.

The ideal train control system would determine the distance to the point where the train must stop—the only information needed to prevent a collision with the preceding train—and stop the train there safely. In this case, the basic required system information is the current exact train position and exact point where the train must stop.

Reconsideration of the purpose of conventional train control systems have led to new decisions on the basic functions of a new train control system where the trackside equipment transmits the point where the train must stop to the on-board equipment, which determines the current position of the train and calculates the distance to the stop point. The system applies the train brakes as necessary at curves and on down grades.

Then, the D-ATC (Digital-ATC) system (digital and decentralized ATC) had been developed using autonomous decentralized system (ADS) technology. In the D-ATC system, the device on each train autonomously calculates the permitted speed of each train. This system was introduced to the Keihin-Tohoku Line as D-ATC in 2003 and to the Shinkansen as DS-ATC (Digital Communication & Control for Shinkansen-ATC) in 2002 (Figure 3.6) [1–3].

Existing train control systems have been in long use and boast high degrees of safety. At the same time, they have contributed to better track utilization. However, some of the technologies, such as detection of train position using track circuits and transmission of data to the on-board equipment via the rail, are very old.

There has been recent remarkable progress in mobile communications and computer technologies which makes possible a new train control system where train itself recognizes its location and communicates with other trains through radio transmission and which can be used for train control systems to reduce the amount of ground equipment and cut costs. Therefore, a new radio-based train control

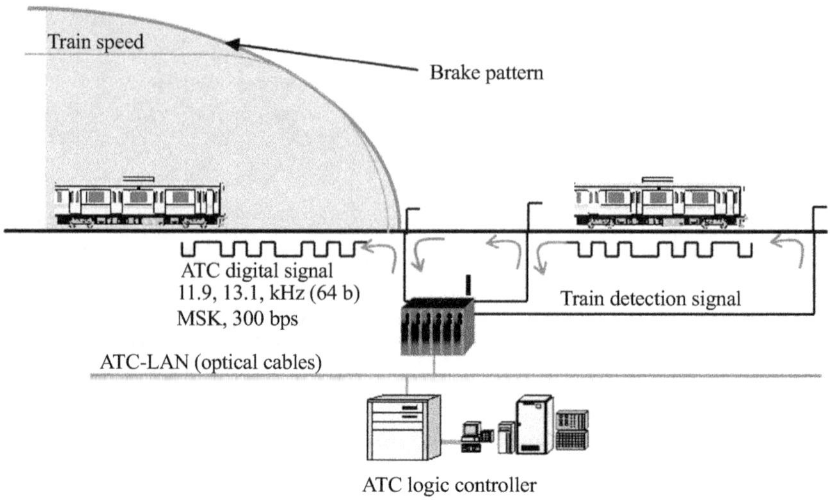

Figure 3.7 Outline of D-ATC

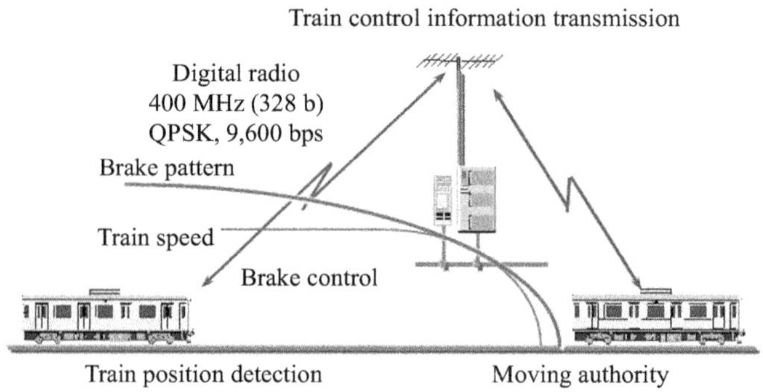

Figure 3.8 Outline of ATACS

system called ATACS (Advanced Train Administration and Communications System) using digital radio, computer and information communication technologies (ICT) had been developed. This adopts a new framework for trackside and on-board equipment functions (Figure 3.7) [1].

ATACS is the new train control system using information technology and ADS technology. ATACS system was put into practical use in 2011 at Senseki Line in Tohoku district (Figure 3.8).

These train control systems are classified into five levels from "Generation 0" to "Generation 4."

"Generation 0" system is ATS-S, "Generation 1" system is ATS-P, "Generation 2" system is ATC, "Generation 3" system is D-ATC, and "Generation 4" system is ATACS see Table 3.1. Progress of the train control system is shown in Figure 3.9.

Table 3.1 Comparison of train control system

<table>
<tr><th></th><th>ATS</th><th>ATS-P</th><th>ATC</th><th>D-ATC</th><th>ATACS</th></tr>
<tr><td>Signaling system</td><td colspan="2">Way side signal</td><td colspan="3">Cab signal</td></tr>
<tr><td>Signal</td><td colspan="3">Analog signal</td><td colspan="2">Digital signal</td></tr>
<tr><td>Block system</td><td colspan="4">Fixed block</td><td>Moving block</td></tr>
<tr><td>Train detection or position recognition</td><td colspan="4">By track circuit</td><td>On-board</td></tr>
<tr><td>Transmission</td><td colspan="4">Transponder (Balise+Loopcoil)</td><td>Radio</td></tr>
<tr><td>Control method</td><td colspan="2">Point control</td><td colspan="3">Continuous control</td></tr>
<tr><td>Generation</td><td>0</td><td>I</td><td>II</td><td>III</td><td>IV</td></tr>
</table>

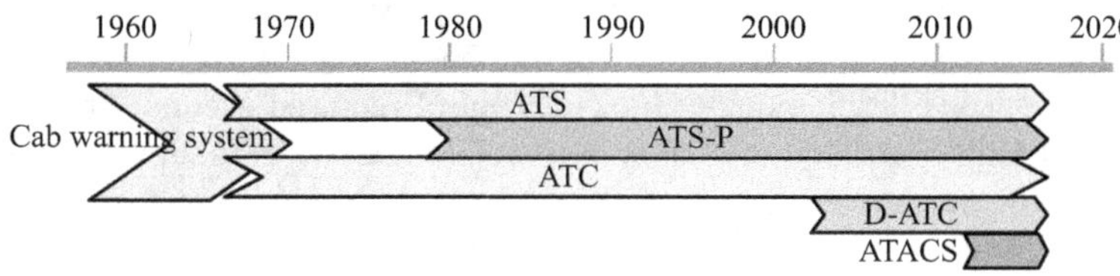

Figure 3.9 Progress of train control system

3.3.2 Automatic train control (ATC) system

3.3.2.1 Analog ATC

Figure 3.6 illustrates train speed control of the analog ATC system. In the analog ATC system, a signal device that indicates permitted speed is located in the driver's cab on trains, and it continuously receives information on the permitted speed transmitted from ground equipment.

The central ATC logic device sends ATC signals to track circuits. The analog ATC signal is information on the permitted speed, and at the same time, it is used as a train detection (TD) signal. The logic device can determine the section on which a train is present by monitoring the level of received ATC signal power because the wheels of the train short the track circuits. Boundaries of the track circuits and the pattern of permitted speed are designed in order to maintain train headway, which is necessary for train traffic control.

In the analog ATC system, the central ATC logic device takes charge of most of the train interval control, and on-board equipment only controls the braking of the train according to instructions from the central device. The analog ATC improves security and traffic efficiency but some problems still remain as listed below.

1. The braking control pattern is not smooth at the boundary of a section, therefore, the time of a braking operation tends to increase, driving operability is bad, and the train interval cannot be shortened.

2. The improvement of rolling stock performance does not improve the efficiency of transportation because the headway mainly depends on the permitted speed and the boundary point of the section.
3. Most equipment is placed in the central part of the system, and many cables are laid from the central device to the track circuits, which is costly.

3.3.2.2 Digital ATC (D-ATC)

The Yamanote Line and Keihin Tohoku Line—JR East's important arteries in metropolitan Tokyo—must maintain stable and high-density transport. The analog ATC systems came into use in 1981 and had become obsolescent after 20 years, making renewal important. In renewing the old systems, rather than simply replace the old ATC with new ATC systems of the same type, JR East decided to undertake a comprehensive review and develop a system based on an entirely new concept compatible with the basic purpose of ATC (controlling the headway between trains safely).

Interval control is one approach to train control. As long as the distance between preceding and succeeding trains is known, it is possible to secure safety by controlling the train speeds. Implementing interval control required accurate train position detection and high-speed communication between train and ground. Clearly such an on-board intelligence system is very different from the analog ATC.

What is digital ATC? The only data required to stop a given train at the rear of a block with a stop signal aspect is the point where train must stop and the distance to that point.

Based on this concept, JR East decided to develop a new on-board digital ATC in which the ground equipment transmits only the stop point as digital information to the on-board equipment, which recognizes the train position and continuously calculates the distance between the train position and the stop point and applies the brakes at the right moment, taking into account relevant factors such as curves and grades.

The characteristics of the digital ATC system are as follows:

1. On-board equipment with sufficient flexibility improves train deceleration and permits headways reduction without modifications to ground equipment. Continuous braking control and recognition of train position by the on-board equipment reduced headway and allowed high-density operation. High-density operations can lessen train congestion through single-phase braking control and on-board train position recognition.
2. Comparatively compact and economical ground equipment were realized by using general-purpose ICT devices and distributed system configuration.
3. Brakes are engaged smoothly and riding comfort is improved by conveying line service conditions to crews and finely conducting ATC brake control.
4. System informs a driver of route conditions, thereby improving train maneuverability.
5. If deceleration performance of rolling stock is improved, the system is flexible enough to shorten the interval between trains without changing ground facilities.

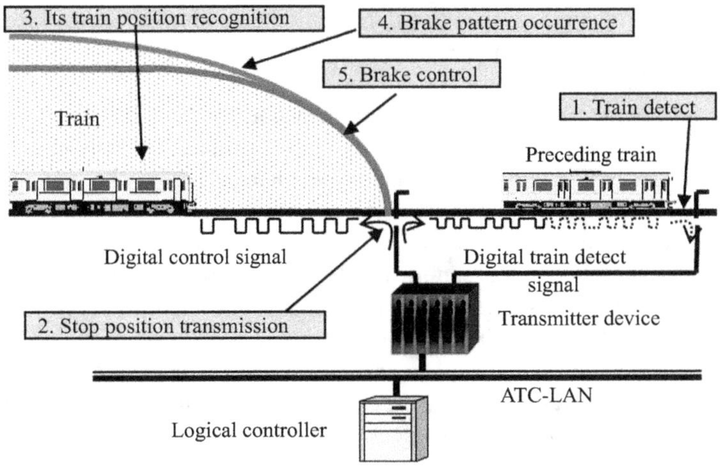

Figure 3.10 Operating principle of D-ATC

The new ATC system is less costly than the analog ATC system and has reduced headways of 150–120 s with stop times of 50 s and a deceleration of 3 km/h/s. In addition, the new system has much lower construction costs. The digital ATC functions are outlined below (Figure 3.10).

1. Train detection: A TD signal is transmitted constantly to the track circuit. As soon as a train enters the track circuit, the ATC logic unit detects the train.

2. Stop point transmission: Based on the train position information supplied by each track circuit and the route setting information supplied by the interlocking device, the ATC logic unit calculates the section permitting train entry and transmits it as a digital ATC signal to the on-board equipment via the track circuit.

The ground equipment transmits an minimum shift keying (MSK) modulated, 80-b ATC signal (64 b for text and 16 b for cyclic redundancy check (CRC)) using an high-level data link control (HDLC) compatible transmission protocol to the on-board equipment, which performs a CRC check on the received ATC text. It also validates the serial number, content, etc. of the ATC text and invalid ATC texts are not used in control of train operation.

3. Own position recognition and 4. Barking pattern: While the train is running, the on-board equipment constantly determines the train position (specific position in specific track circuit) by counting tachometer pulses from an axle generator. The train receiving the ATC signal retrieves the relationship between the stop point information given by the signal and the train position from the on-board database calculates the distance to the stop point from that relationship and generates a braking pattern.

In the digital ATC system, the on-board equipment must find the train position accurately, using a tachometer generator. Even if the tachometer generator error is on the large side, the data is processed on the safe side. In addition, to prevent accumulated error, correction-position transponders are installed between stations at intervals of about 1 km. The transponder also performs distance correction, while the train is running at a low speed or during slide detection to ensure safe position detection. If the train position becomes unknown, the system immediately applies the emergency brakes.

5. *Brake control:* The on-board equipment checks the train speed against the braking pattern and applies the brakes if the train speed exceeds the pattern. Immediately after the brakes are applied and immediately before the train stops, the on-board equipment also applies weak braking to minimize the shock of the braking action. Also, before the on-board equipment applies the brakes, it indicates that the train is approaching the brake pattern by flashing a lamp. Furthermore, a monitor displays the condition of the route ahead to improve the train maneuverability.

Reliability and testing: The digital ATC system is configured to secure both higher safety and reliability; the ground equipment is triplex and the on-board equipment is dual-duplex, lowering the critical failure rate of the entire system by 10%–12% or less, compared to the electronic interlocking system.

For about 6 months from October 1998 before putting the system into practical use, comprehensive final testing (32 night tests and about 180 days of day tests) was carried out with new ATC ground equipment installed between Minami Urawa and Omiya on the Keihin Tohoku Line and the new on-board equipment mounted on a Series 209 train set. Since the component technologies of the new system had already been validated by the 1995 on-track testing, the emphasis was on confirming and evaluating system functions, performance, constants, etc., required during revenue operation from the system startup to changing driver's cab, shunting, etc.

JR East planned to introduce digital ATC called DS-ATC, with "S" standing for Shinkansen as the control system for the soon-to-be-opened Morioka–Hachinohe section. Testing started near Koriyama Station in 1999 with good results. In 2002, JR East decided to introduce D-ATC on its narrow-gauge railways and DS-ATC on its Shinkansen. Both projects went smoothly with the DS-ATC system introduced on the Hachinohe section of the Tohoku Shinkansen in December 2002, and the D-ATC system introduced on the Minami Urawa–Tsurumi section of the Keihin Tohoku Line in December 2003. Then, in November 2005, the DS-ATC system was introduced on the Furukawa–Morioka section of the Tohoku Shinkansen. The DS-ATC system had introduced on the Joetsu Shinkansen in 2008, and the D-ATC system had introduced on the Yamanote Line in 2006, and on remaining sections of the Keihin Tohoku Line in 2007.

3.3.3 Radio-based train control system

JR East had developed a train control system using radio communication.

Existing train control uses the track circuit to detect the train position and displays the appropriate signal according to the detected train position.

Conventional train control systems using track circuits for TD require huge investments in equipment and maintenance. There are several reasons: various ground facilities must be installed on and around the track; train positions cannot be detected very accurately; many signal cables are required to connect ground facilities; etc. In addition, because train control is implemented by block, conventional systems

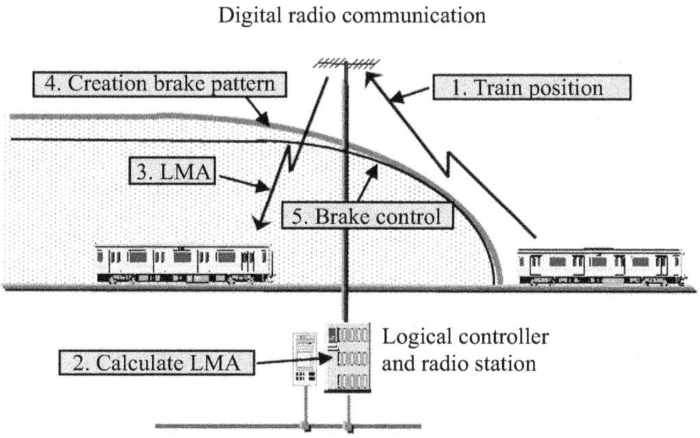

Figure 3.11 Operating principle of ATACS

cannot effectively support changes in transportation mode, such as development of new, high-performance vehicles.

However, by utilizing ICT as the core and redesigning the distribution of functions between the ground and on-board equipment, it is possible to configure an ideal control system in which individual trains exchange train position information with each another to control train intervals.

The recent great progress in ICT has now made it possible to build a train control system called ATACS, in which each train recognizes its position and exchanges data with other trains by radio (Figure 3.11). It is an innovative new train control system that uses ICT and autonomous distribution technology.

This new train control system utilizing ICT has the following purposes:

Cutting costs: Reducing the number of ground facilities cuts construction and maintenance costs. In addition, the costs of large-scale improvement and signal installation works during renewal of train control systems can be cut.

Facilitating system renewal: Since the system does not depend on ground facilities, it can easily support changes in transportation mode, such as higher train speed and shorter headway.

Train interval control: The train speed is controlled by calculating the permissible speed from the limit movement authority (LMA) of the train, instead of directly indicating the speed signal as in the present system. The base procedure for train interval control is as follows:

First, (1) the train position is transmitted by radio to the ground controller. Next, (2) the ground controller calculates the point to which that train can be permitted to move (LMA) on the basis of data on the train position and route of each train in the control area.

(3) The LMA is transmitted to the following train. Further, the on-board computer determines the permissible distance of running based on the LMA and the present train position. With braking performance and track conditions (gradient, curve, speed limit, etc.) taken into consideration, (4) the on-board computer creates

a brake pattern which can cause the train to stop at the limit of this permissible distance of running, and (5) applies the brake control.

Improving safety and reliability: The system does not depend on the train crew for control. It improves safety by positively preventing entry of trains into a closed section and by providing a protective pattern for brake control at level crossings. In addition, the fewer ground facilities reduce problems, contributing to improved transportation reliability. ATACS has many functions; the main functions—train interval control and level crossing control—are described here.

Testing: Toward practical application an ATACS system configuration and specifications were prototyped with future practical application in mind. Seven radio stations were installed on the 17-km Aobadori–Higashi Shiogama section of the Senseki Line and on-board ATACS was installed in each of 18 train sets. The test section is divided into several control areas, each of which has control equipment and a radio station. The control equipment has a number of functions, including TD, train interval control, switch control, level crossing control, and track-work safety management. The radio stations exchange information with the on-board computer and are installed at appropriate intervals according to radio area; some are linked to the control equipment. The on-board computer controls the train brakes according to data received from the ground control equipment. It also transmits train position information to the control equipment.

Level crossing control: ATACS controls level crossings by exchanging data between the ground and on-board equipment. To adjust the warning time of an approaching train, the system uses the train speed and train performance to estimate the time when the train will arrive at the crossing and transmits it to the ground control equipment to actuate the crossing signal. Then, the system transmits the information to the train and recalculates the brake curve. If the crossing signal is not actuated, the system stops the train before the crossing.

Other functions: The other functions include track work safety management, switch control, bidirectional distance control, and temporary speed regulation.

The distributed ground equipment are linked by a network. Each on-board system performs autonomous brake control according to LMA data. The ground equipment consists of a traffic control system and a train control system, and each device is configured, so that it works independently and even the failure of one device does not affect any other devices.

If one radio station goes down, the neighboring station automatically backs it up, keeping the system functioning.

First, the on-board equipment determines the accurate train position. The initial train position is obtained when the train passes the trackside coil installed at the boundary between train departure section and train entry section. Next, the on-board equipment continues recognizing the train position using the cumulative running distance calculated from the train speed. The distance is corrected each time the train passes each trackside coil installed at appropriate intervals. The train position is allocated as a unique number to the associated control equipment, and data about train positions in virtual sections (divisions of control area), related sections, etc. is processed by the ground and on-board computers. Radio stations are installed at intervals of about 3 km (varies according to reach of radio wave). To avoid mutual interference between adjacent radio stations, four different frequencies are used so stations in individual control areas can select a suitable frequency. Each station communicates with trains in its area every 1 s. Since transmission errors can occur, space diversity and Reed–Solomon error correction are used to improve transmission quality.

Day running tests (cumulative distance of more than 1 million km) and night control running tests (28 times) were performed from October 2003 to February 2005. Since the test results showed no functional or safety problems with the prototype system, R&D has started into practical application.

JR East's faith in the concept was finally realized 7 months later. On October 10, 2011, commercial operation of ATACS began on the Senseki Line, an urban commuter line in the Sendai area, with all operating trains switching from conventional train control to the new system.

3.4 ADS technology

In the present signaling system, because all trains are controlled in a uniform way, a higher performance train can't run according to its higher performance. Consequently, higher performance will be achieved only when all trains are changed to higher performance trains.

In D-ATC, the ground system is decentralized, and each device is connected by optical network. Each on-board system does brake control autonomously by using the stopping point given in transmitted information by the ground controller and own position. The ground system is divided into a logical controller and transmitter devices. As a result, it prevents a failure of some equipment from affecting the whole system.

In D-ATC, when the on-board system receives the stop point from logical controller via track circuit, it generates the brake pattern by using the on-board data base. The brake pattern has geographical conditions such as gradient and curve. The on-board controller controls the brakes based on this pattern. Therefore, if a new high-performance train is introduced even when the ground system is not changed, this train can run according to its performance. This means it has autonomy. As a result, it becomes extremely easy to speed up and to increase traffic volume of the line. And flexibility of the system improves.

The following are definitions for system modeling and testing concerning the analog and digital ATC systems by using ADS technology [4,5].

3.4.1 Functions of analog ATC system and definition of testing

Figure 3.12 is a model for the analog ATC system. S_a and T_1-T_n are the analog ATC ground system and trains, respectively, and V_{j-1} is permitted speed for train T_j.

As shown in Figure 3.12, ground system S_a acquires the position of train T_j. Then, the permitted speed information V_{j-1} for the following train is created based on T_j's position information. The created information is transmitted to T_{j-1}, the train behind. T_{j-1} controls the brakes based on the received information [6,7].

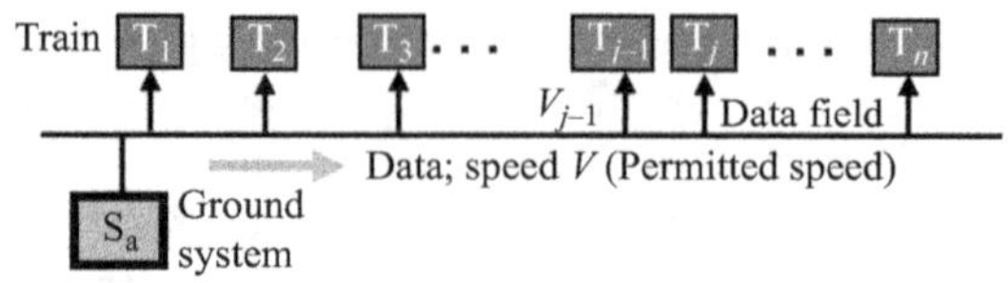

Figure 3.12 Model for the analog ATC system

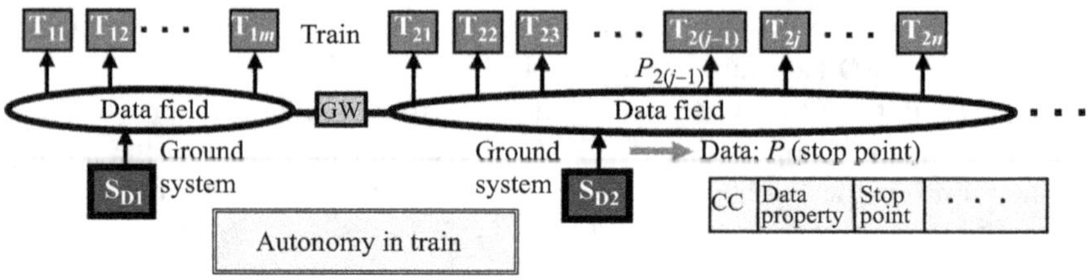

Figure 3.13 Model for the D-ATC system

This system will be tested regarding the following items: (1) whether the train position is detected appropriately; (2) whether the speed data are created appropriately; and (3) whether the brakes are controlled appropriately. The testing is defined as follows:

1. S_a detects the position of T_j.
2. S_a creates V_{j-1} according to the position of T_j and transmits it.
3. T_{j-1} controls its speed to be equal to or lower than V_{j-1}.

3.4.2 Functions of the D-ATC system and definition of testing

Figure 3.13 is a model for the D-ATC system. S_{Dj} and T_{ij} are the D-ATC ground system and trains, respectively, and the connection with the neighboring data field takes place at GW (gateway) [6,7].

As is shown in Figure 3.13, S_{D1} and S_{D2} are ground system connected with a GW. The GW has a role which prevents mixing of the ATC signal from an adjoining equipment room. System S_{D2} first obtains the position of train T_{2j}. Then, it creates the information for the following train according to the T_{2j} train position information. The created information $P_{2(j-1)}$ is then transmitted to the data field. When there are data about it, train $T_{2(j-1)}$ obtains these data. It then generates a pattern based on the obtained information and carries out the brake control.

This system will be tested, like the analog ATC system, regarding the following items: (1) whether the train position is detected appropriately; (2) whether the data for the following train are created appropriately based on the detected data; and (3) whether the brakes are controlled appropriately. The testing is defined as follows:

1. S_{D2} detects the position of train T_{2j}.
2. S_{D2} creates $P_{2(j-1)}$ according to $T_{2(j-1)}$ and transmits it.
3. $T_{2(j-1)}$ is controlled according to $P_{2(j-1)}$.

3.5 Assurance technology

Assurance technology is a technology to secure stable operation of systems and is the general term for technology utilizing a system that considers the two elements of heterogeneity and adaptability [8,9].

The system is configured to satisfy a large number of needs, including safety and reliability. The needs levels vary depending on systems, and when linking systems with different needs levels in a network, it is necessary to allow the coexistence of different needs levels. This is called heterogeneity. On the other hand, the environment which systems are placed in is constantly changing, including the systems themselves. These changes include system malfunction, initial system settings (startup), system repairs, system changes, and such. The requirement to cope with these factors is adaptability.

Continuation of train operation under different systems and during a replacement work period, smooth system change, and securing of safety and maintainability are issues of heterogeneity and adaptability. It is expected that assurance technology can resolve these issues.

Due to the advancement of railway signaling technologies, railway safety has been remarkably improved. With this improvement, what our society expects of railways has reached a higher level and become more diversified, with heightened requirements for stable transportation together with secure safety. For this reason, we need to assure continuation of train operation as much as possible in ordinary operation and under many other circumstances.

We previously depended on fault-tolerance technology to minimize effects in a case of system-down. However, system-down is not the only circumstance in which ordinary operation is not possible. In other words, in cases of system upgrade to accommodate improvements at stations, speed-up of trains or an increase in the number of trains to meet user needs or for the replacement of aged systems, stable operations of systems are required. We will indicate examples of these below.

Upon the introduction of new systems, since they are large in scale, it is not possible to upgrade the whole system at once. In other words, as a requirement, new systems and active systems need to coexist and either system should not hinder train operation. This presupposition needs to be taken into account when designing systems. These different systems include not only new systems and active systems but also various systems for different needs. Under these systems, train operation should not be interrupted, and these systems are required to operate in harmony and should not interfere with each other.

3.5.1 Modeling of system replacement

So far, models have been created for the analog ATC and the D-ATC systems. In this section, a model for system replacement is created based on these system models.

When the new ground system of a section has been installed, functioning of TD is tested first. Then data transmission to the on-board system and brake control are tested. The sections and trains of which these tests have been finished

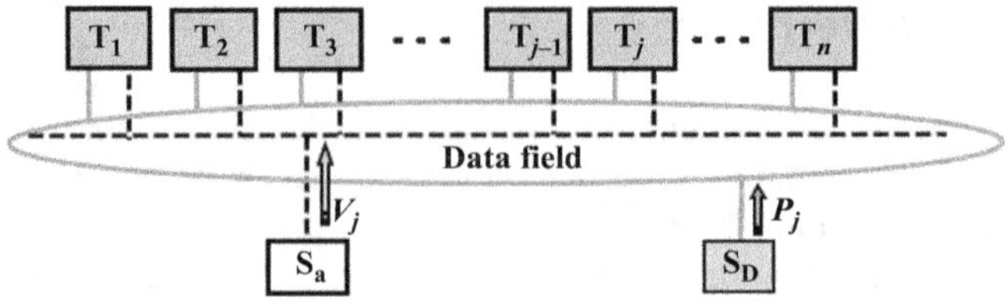

Figure 3.14 On-board integrated system

will be put into service one by one. Therefore, the system replacement takes place step by step.

In creating a model for system replacement, integrated on-board systems are considered. The integrated on-board system can access both data fields of the digital ATC and analog ATC systems. This is because the analog ATC and D-ATC adopt different signal frequencies which separate the data fields.

Since the on-board system is integrated with the analog ATC and digital ATC, both signals can be read. Therefore, system S_D can be built without considering coordination with S_a. When construction of S_D is finished, S_a and S_D coexist. On the other hand, on-board systems will be replaced with integrated ones successively, and all the trains are to have ultimately integrated systems. Train T_j with an integrated system can access both types of data fields and test the digital ATC system while executing control with the analog ATC system. After the coexisting period, S_a will be removed. Then, the system will be the entirely new one (Figure 3.14).

3.5.2 Testing assurance

In this clause, a testing assurance is defined and the total test time based on modeling of the system replacement is described.

3.5.2.1 Definition of testing assurance

Heretofore, when replacing systems, the current system was stopped temporarily, the digital ATC system was tested before its actual replacement, and then the current system was switched back on-line after completion of the test. This temporary stoppage of the current system lowered the assurance of the system. However, assurance can be improved by conducting on-line tests of the digital ATC system that allow the current system to remain on-line (On-line test: the test during train service operation).

Under the on-line test environment, the current system is either operated or stopped. Therefore, the influence to the current system cannot simply be evaluated by stop time alone but also depends on the on-line testing time. The test consists of N test items. For each of the test items, the system is tested n_i times. And it takes t_i for each test. At this time, the total test time in each test item is $n_i \times t_i$.

The sum of all the test items as the total test time is then defined as follows:

$$\text{Total test time} = \sum_{i=1}^{N} n_i \cdot t_i \tag{3.1}$$

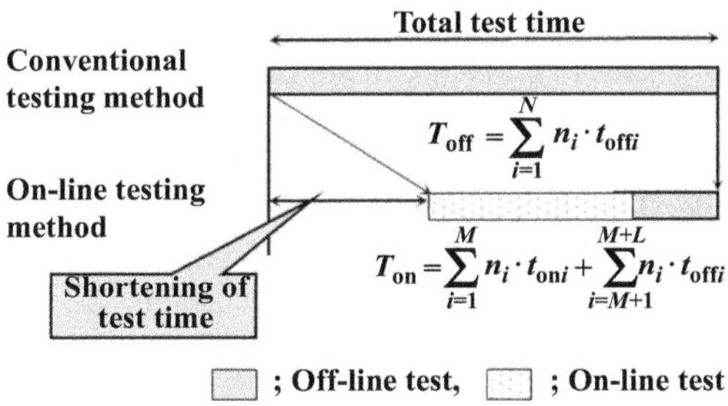

Figure 3.15 Testing assurance

Based on the above, the change in the off-line test is to be combined by introduction of the on-line test. The total test time of each test item when all systems are stopped for test is T_{off}. The total test time when the on-line tests are introduced is T_{on}. It is thought that the off-line test obstructs the operation of the current system, while the on-line test does not obstruct the operation but influences the testing work. Therefore, we have to shorten the total test time. The ratio of the total test time when on-line tests are applied to the total test time when all tests are done off-line is defined as non-assurance. This shows how the load by line tests can be decreased, and the smaller this value is, the higher the assurance. Therefore, the assurance of the test is defined in the equation below and Figure 3.15.

[Definition]

$$\textbf{Testing Assurance} = 1 - \frac{T_{on}}{T_{off}} \tag{3.2}$$

In this way, as the definition also shows, the testing assurance can be numerically valued by conducting on-line tests.

3.5.2.2 Number of tests

Considering the total test time of replacement of D-ATC, there are three kinds of tests for the D-ATC: "train position detection," "data creation and transmission," and "braking control." We set each testing period as $t_1 - t_3$, then the total test time of each test is described below.

1. **Train position detection test**
   To test train position detection, the data field is divided into several sections and a test is made whether the train position can be detected or reorganized accurately in each section. Only one test is done in each section. When replacing the analog ATC system with the digital ATC system, the number of tests to be conducted is the number of sections (S). So total test time is $S \times t_{on1}$ in the case of on-line test and $S \times t_{off1}$ in the case of off-line test.
2. **Data creation and transmission test**
   The data creation and transmission test checks whether the data for the stop position are created correctly and transmitted with regard to the position of a

train in a certain section. The test is conducted the same number of times as the number of train position detection tests. Therefore, the total test time to be conducted is $S \times t_{on2}$ for on line test and is $S \times t_{off2}$ for off line test.

3. **Braking control test**
   This tests whether the train controls braking correctly according to the stop position data. For the integrated system, it is sufficient to send correct control data to each section; therefore, the number of tests is S. Therefore, the total test time is $S \times t_{off3}$ same for on line test and off line test.

Therefore, the testing assurance for replacement is as follows:

$$\text{Testing assurance for system replacement} = 1 - \frac{t_{on1} + t_{on2} + t_{off3}}{t_{off1} + t_{off2} + t_{off3}} \tag{3.3}$$

3.5.3 Application of assurance technology to D-ATC system

3.5.3.1 Continuation of train operation during replacement of systems

Replacement of systems can be split into three stages in correlation with progress made in replacement work; during construction, testing period, and after actual use commences.

1. **During construction**
   During the construction period, while operating active systems, digital ATC systems will be introduced. It is necessary to undertake construction without interfering with train operation by affecting active systems during the construction period till digital ATC systems start to operate.
2. **Testing period**
   To test digital ATC systems, if we stop the operation of active systems, which means stopping train operation, this will interfere with stable transportation. Then a request arises to test the systems while operating trains. In other words, a design that allows for coexistence of digital ATC and analog ATC systems is required.
3. **After use commences**
   Even after a digital ATC system has begun to be used, removal of the current system continues, and both systems will be present together for a while until the current systems are all removed. Therefore, steps that prevent current systems from influencing digital ATC systems are necessary.

3.5.3.2 Smooth change of systems

To upgrade a whole system at once is impossible from a construction point of view. It is necessary to expand a section with a digital ATC system step-by-step and to replace the system in a short period of time.

Although installation of D-ATC will proceed steadily, the actual startup of use will not be as soon as construction is completed. Rather than that, exchange to the digital ATC system will be done in set sections and must be done in one night. Consequently, it is essential to have facilities that enable exchange to D-ATC done

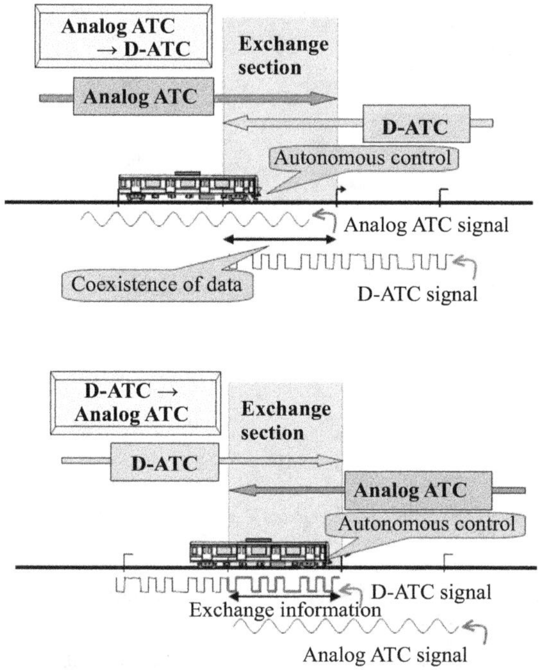

Figure 3.16 Continuous train operation

with minimal operations since the exchange to starting D-ATC use must be done in a limited time as described. Specifically, common data fields are set up, data is used that adds up to whether the D-ATC text is valid or invalid and the on-board equipment that receives this data selects which system to operate. As a result, it is possible to check the transmission of digital ATC signals while using the analog ATC control and to switch systems in a short time just by changing the valid/invalid information of the digital ATC signal (Figure 3.16) [10,11].

1. **Staged expansion of a section with digital ATC system**
   As construction progresses, sections with the digital ATC system gradually increase. There are boundaries between sections covered by the digital ATC system and those by old systems. It is necessary to assure that trains operate across these boundaries without hindrance. For this reason, at the boundaries between sections where digital ATC systems are already introduced and where they are not yet introduced, switching of systems is required and smooth continuation of operations between them needs to be assured.
2. **Rapid replacement**
   Even when the operation of the digital ATC system is started, there must be no influence on the operation of trains that run from early in the daytime to late at night. Therefore, the replacement by the digital ATC system has to be conducted in a short train interval from after the last train to the first train.

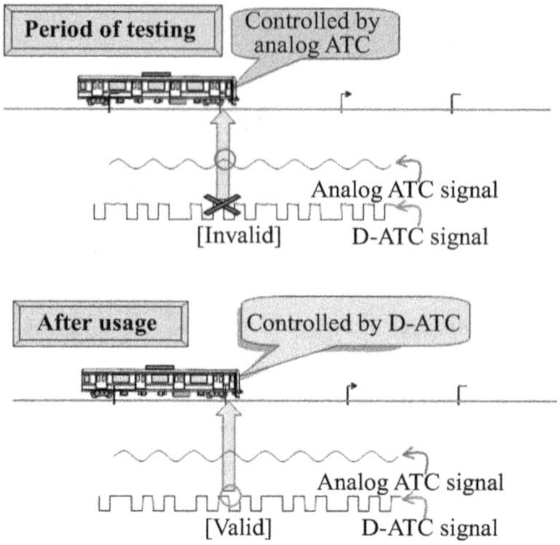

Figure 3.17 Rapid system change

To accomplish this, it is necessary to establish a method for quick replacement with minimal operations. In addition, in the future, a technique to operate analog ATC and digital ATC systems together and then sequentially switch to digital ATC systems will be required (Figure 3.17) [10,11].

3.5.3.3 Coexistence with heterogeneous systems

Railway-signaling-related systems, there are the traffic management system, train control system, signal control system, and others. These systems are not only required to operate in harmony without interfering with each other but also to coexist with other systems surrounding these systems such as business management system, maintenance work management system, and rolling stock system.

Trains that operate over sections with D-ATC and sections with the analog ATC or ATS must switch their systems at those borders. The current method employed is switching both the ground and on-board systems after the train has stopped. This time, the data fields are made to coexist in switching sections so that the train that receives data automatically switches the mode of the on-board equipment in correlation with the data. Train operations can be continued by employing this method of having heterogeneous data fields coexist in the same area.

While analog ATC is a system for preventing the collision of trains, ATOS is a system for conducting appropriate route control in interlocked station yards, preventing train derailing, and other functions. The coexistence of D-ATC and ATOS is sought for interlocked station yards in which ATOS is installed. In other words, there is a need to have two systems with differing data fields and needs which coexist. By placing a converter, mutual data exchanges between systems with different needs are executed to absorb the needs differences to achieve system coexistence (Figure 3.18) [12].

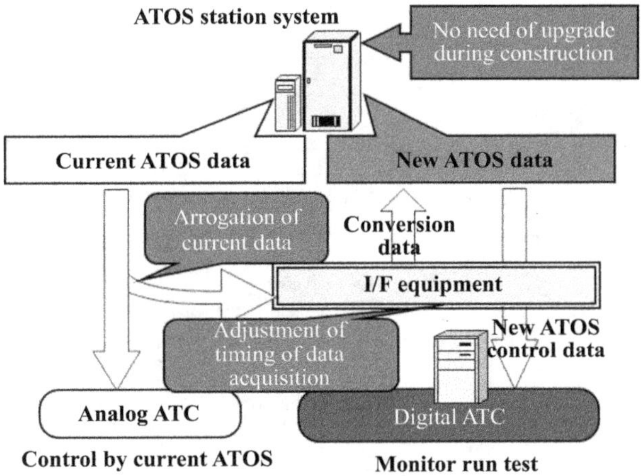

Figure 3.18 Coexistence of D-ATC and analog ATC systems with heterogeneous system

3.5.3.4 Securing safety and maintainability

At an occurrence of system failure in any of the failure modes, it will be no problem from a safety point of view if all trains on the system are stopped. However, this hinders continuation of train operation. To secure the continuation of train operation as much as possible without impairing safety, it is necessary to design system configuration which can reintegrate to ordinary operation as soon as possible during recovery from such failures, securing the continuation of train operation even at failure occurrences in a range which does not hinder safety.

3.6 A chain of a concept, technology, and a system

JR East (including former JNR) has problems which must be conquered, such as accident prevention through train control, improvement in passenger service, and cost reduction. The railway operator has so far developed and introduced various systems. Needless to say, the development of principal base technology of those systems is a precondition.

The following technologies are the key technologies.

Induction radio technology for a cab warning device, braking technology for ATS, track circuit transmission technology for ATC, automatic control technology for ATS-P, autonomous decentralized techniques and assurance technology for D-ATC and autonomous decentralized techniques and wireless control for ATACS.

There are three important points in construction of such systems and technology.

1) First, it is important to construct the image of the system that will satisfy needs and to discern the optimal technology for achieving it.
2) Second, the life of a huge infrastructure system is inevitably long, and it is requires a large amount of investment to install a digital ATC system. The train

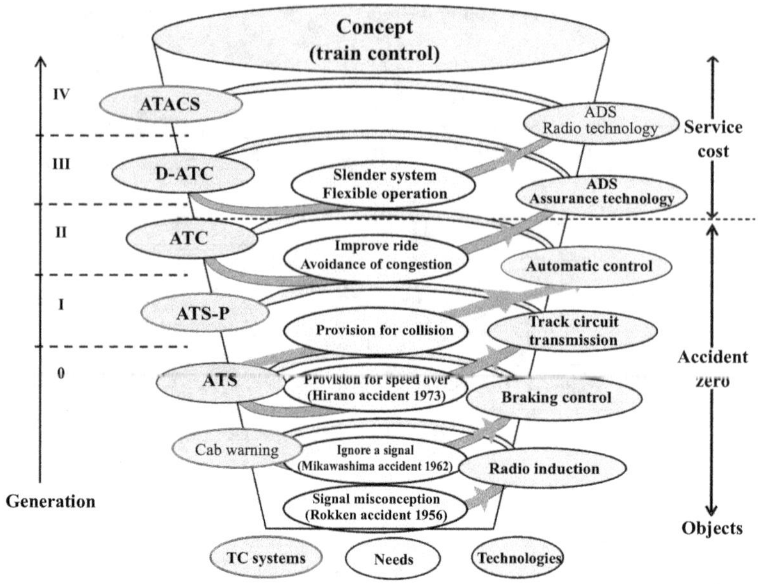

Figure 3.19 The chain of technologies, needs, and train control systems

control system connects with various heterogeneous systems. The important point is how to adjust these elements so that they do not produce inconvenience.

3) Third, it is railway operator, JR East, who must image the system, design, and develop it, and strengthen and evolve the whole system. It is indispensable to have close cooperation with railway equipment suppliers and system vendors by this leadership.

Therefore, a train control system can be said to be an aggregate of the dynamism of evolution, not a replacement or an accumulation. In other words, the construction process of these systems can be called a chain of technology and a system. The aspect of this chain is shown in Figure 3.19.

As shown in the figure, in the JNR age before 1987, needs for train control were mainly accident prevention. A prime objective was preventing the recurrence of following accidents: the Rokken station accident in 1956 (signal misconception), the Mikawashima accident in 1962 (red signal violation), and the Hirano Station accident in 1973 (speed excess and derailment).

A "cab warning device" was created by development of "inductive radio technology," "ATS" was created by development of "braking technology," "ATC" was created by development of "track circuit transmission technology," and "ATS-P" has been created by development of "automatic control technology." As the development of countermeasures against accidents was completed during the JR age, after privatization, customer needs evolved from countermeasure against an accident to include factors such as the degree of comfort and flexibility of operation. Corresponding to this trend, "autonomous decentralization" and "assurance technology" were developed, and "D-ATC" was installed in the metropolitan area. Subsequently, "wireless control technology" was also developed and the operation of the "ATACS" began.

When building a system for a railway, there is no need to mention that securing safety is a basic premise. However, in today's society, the need for securing stable operation is increasing and how securing stable transportation can be achieved is important. For this reason, under any circumstance, reliable operation of systems is required. With the recently developed railway signaling system, by using assurance technology how this need was solved was described, giving an actual example. In addition, in improvement of systems or coexistence with existing systems, flexible correspondence is possible. To sustain stable operation, system configuration and maintenance was made to be easy to handle. Furthermore, the technical possibility of replacing a system in stages with different degrees of life was discussed.

Assurance technology is expected to function as an important infrastructure in the field of signaling systems in the future. Moving toward the actual construction of a digital ATC system, further studies will be conducted and the assurance properties will be heightened even more.

References

[1] M. Matsumoto, "Learning from Past Railway Accidents—Progress of Train Control", JRTR, vol. 43/44, pp. 86–98, Mar. 2006.

[2] M. Matsumoto, A. Hosokawa, S. Kitamura, D. Watanabe, A. Kawabata, "Development of the Autonomous Decentralized Train Control System", IEICE Trans. Commun., Vol. E84-D, No. 10, pp. 1333–1340, Oct. 2001.

[3] M. Matsumoto, A. Hosokawa, S. Kitamura, D. Watanabe, A. Kawabata, "The New ATC System with an Autonomous Speed Control with On-board Equipment", ISADS2001, Dallas, TX, USA, pp. 235–238, Mar. 2001.

[4] K. Mori, "Autonomous Decentralized Software Structure and Its Applications", IEEE Fall Joint Comput., pp. 1056–1063, Nov. 1986.

[5] K. Mori, "Autonomous Decentralized Systems Concept, Data Field Architecture and Future Trends", ISADS1993, Kawasaki, Japan, pp. 28–34, Apr. 1993.

[6] M. Matsumoto, T. Tsurumaki, D. Watanabe, K. Mori, "Modeling of Train Control System and a Proposed Method of Assurance Evaluation", ADSN2002, Vienna, Austria, pp. 89–94, Jul. 2002.

[7] M. Matsumoto, "Assurance Technology for Coexisting Test and Operation Subsystems and Its Application to Train control System", ISADS 2003, pp. 259–265, 11 Apr. 2003.

[8] K. Mori, "Application in Rapidly Changing Environments", IEEE Comput., Vol. 31, No. 4, pp. 42–44, Apr. 1998.

[9] K. Mori, "High Assurance Application System and Their Technologies Under Situations", HASE1997, pp. 168–174, Aug. 1997.

[10] M. Matsumoto, S. Kitamura, M. Sato, "High Assurance Technologies for Autonomous Decentralized Train Control System", HASE2001, Boca Raton, FL, USA, pp. 220–227, Oct. 2001.
[11] M. Matsumoto, T. Tsurumaki, S. Kitamura, D. Watanabe, K. Mori, "Assurance Evaluation Technology of Train Control System for Achieving Expansion during Operation", HASE2002, Tokyo, Japan, pp. 71–78, Oct. 2002.
[12] M. Matsumoto, K. Bekki, "Application of Assurance Technology for Railway Signaling System", ISADS2007, Sedona, AZ, USA, pp. 69–76, Mar. 2007.

Chapter 4

ATOS (autonomous decentralized transport operation control system)

Kazuo Kera[1], Satoru Hori[2], and Takashi Kunifuji[3]

Abstract

In traffic-management system (TMS) for existing lines of a metropolitan area, there was a limit in the control of a large station in the centralized TMS based on the conventional CTC (centralized traffic control). In the case of TMS, all information is gathered and judged based on train schedule, namely train diagram and train-tracking data, in the center, and is controlled via CTC. For this reason, it is difficult to control all stations in a metropolitan area, which have many large stations, making introduction of the system difficult. Development to resolve this big issue was done in the Tokyo autonomous decentralized transport operation control system (ATOS). It is a total system that carries out transport operation control and signal control of the existing 24 lines in the Tokyo metropolitan area. It is the largest system in the world, covering complex and high-density operations of existing lines (minimum headway of about 2 min). Its features are as follows: (1) Including the electronic interlocking system into the transport operation control system, constituting an ATOS that arranges the diagram and route-control function at the station. This makes automatic control of line with complex large stations possible. (2) Even when the diagram is disturbed by any accident, it is a system that can quickly recover the diagram by the operation adjustment function that the operator can operate with the diagram-based operation*. (3) When the concept of new interlocking logic which integrates maintenance work management function was considered, the management of occupancy information of track was adopted to interlocking logic by the method of exclusively controlling this track occupancy information. (4) By adopting autonomous decentralized system (ADS), online testing of the system and step by step system construction of the system became

[1]Railway System Consultant, Kera Consarutant Office, Japan
[2]Hitachi, Ltd, Japan
[3]East Japan Railway Company, Japan

*The train schedule diagram is displayed on the screen and the operation method in which the diagram is directly changeable by mouse operation.

possible, it became easier to construct a large scale system step by step and to partially replace. In this way, ATOS contributes greatly to the system construction of large-scale and complicated line sections.

4.1 Introduction [1–3]

Traffic management on railway is to ensure safe and stable transportation, that is, to manage the quality of transportation. The traffic is managed according to a predetermined timetable, but the timetable is often disordered due to natural disasters, accidents, or other causes. Hence, train-rescheduling tasks are carried out to restore the timetable to the original plan. Train-rescheduling task means that collecting various situations related to the train operation, creating a new operation plan based on the situation, controlling the station signals, providing passenger information, and maintenance work management are carried out according to the new plan. Since many operations related to traffic management were carried out manually, eradication of human error and improvement of the efficiency had been a challenge.

In the Japanese case, systemization of traffic management has been promoted since the 1970s. The typical system is centralized traffic control (CTC), and it is still being introduced from the local lines to the main lines, even in the case of the Shinkansen throughout Japan. But it is difficult to apply it to high-density lines in the Tokyo metropolitan area because of the performance of computers and transmission speed of communication lines. In this reason, systemization of the traffic management in the Tokyo metropolitan area was not realized even in the late 1980s.

After privatization, each JR, which inherited Japanese National Railway, started its own improvement of the traffic-management system (TMS) to realize transport service optimized for the region. In particular, in JR East, the modernization of the traffic-management task in the Tokyo metropolitan area was an urgent management issue, and JR East started development of the brand new TMS named ATOS in the early 1990s.

In the development of ATOS, JR East aimed for a comprehensive TMS that improves the efficiency in its management and customer service by incorporating all operations related to transport operation into the system, including those that have been performed by the manual due to technical difficulties.

Specifically, those are modernization of command task, improving efficiency of traffic-management task at the station, and improving safety and efficiency in maintenance of work-management task.

System architecture for the new TMS is not centralized system based on the conventional CTC's technology, but autonomous decentralized system (ADS). In this system, the traffic-management function based on information technology and the signal control function based on control technology are integrated into one intelligent station-control system. The data field (DF) architecture that is the core concept of the ADS enables one to integrate these heterogeneous functions.

As a result, information is shared between the central train command system and the station control system, enabling safe, efficient, and flexible information control.

Below, the basic concept of transport operation control system and, as concrete example, autonomous decentralized Tokyo area transport operation control system are described.

In Section 4.2, outline of ATOS is mentioned, which includes the system concept and outline functions. In Section 4.3, advancement of ADS is mentioned, which includes important expansion of online testing function. In Section 4.4, the step-by-step system construction technology for large-scale transport operation control system is mentioned. Finally, this part is summarized in Section 4.5.

4.2 Outline of ATOS (autonomous decentralized transport operation control system) [4,6]

4.2.1 Issue of transport operation control system

When considering the introduction of a transport operation control system to the existing urban area intercity line, the issues are organized as shown in Figure 4.1.

In terms of safety, a signal system such as automatic train protection has been introduced, but in the metropolitan area, the route control according to the diagram and the operation adjustment to recover the diagram are needed when the diagram is disturbed by the operation management from the viewpoint of improving the efficiency of operations. It is also necessary to change the dispatching task, and crew-directive system from the dispatching method via the station staff to the direct dispatching method of "dispatcher–driver (crew)" and to further improve the

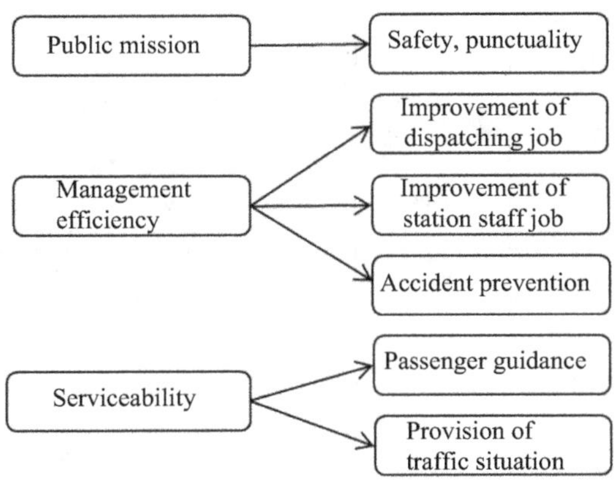

Figure 4.1 Issue of transport operation control system

efficiency of the station operation is there. The signal operation at the station is not only important for the traffic control for main line but also for the shunting and the systematic safety management of the maintenance work. Services to passengers from the perspective of bridges are also necessary to provide more detailed passenger information or to provide accurate operational information at the time of occurrence of an accident.

From the above, in constructing the transport management and operation control system, it is an important task to control comprehensive transport management information considering the dispatching system, efficiency of peripheral operation of the station, from the system of the train route control.

4.2.2 Concept of transport operation control system [4,6]

4.2.2.1 System architecture (centralized and decentralized systems)

The system architecture of the transport operation control system is described. Figure 4.2 shows the architecture of the centralized and the decentralized systems. Historically, the transport management operation control system has been developed as a centralized system of hierarchy structure based on CTC, on which a route-control system and an information-processing system are introduced. Although this structure is good for mainly route-control function, when considering the expansion to the transport management operation control system, it is necessary to control the information of station level and to make the station intelligent. There is a limit in CTC's slow communication line. Therefore, the ATOS is introduced, which carries out route control and other transport management tasks at the station level autonomously.

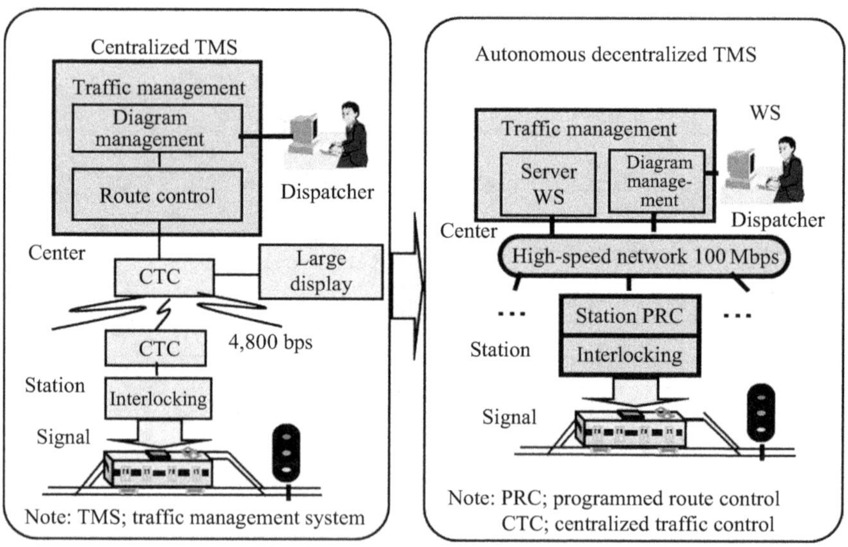

Figure 4.2 Centralized system and autonomous decentralized system

4.2.2.2 System architecture and characteristics

The system architecture of the transport operation control system uses an ADS, whose concept is as follows:

1. Place route control at the station level and communicate all of stations by a high-speed network.
2. By adopting a high-speed network (100 Mbps), total line information can be shared between operation center dispatchers and station staffs and many traffic-management relevant staffs.
3. For the man–machine communication of the operation center, dispatchers WS (work station) and personal computers (PCs) are used. And for servers and other equipment, real-time control server (RCS) and general-purpose controllers are adopted as constituent equipment of the system.
4. Reliability availability safety (RAS) data collection and software maintenance of the entire system is carried out from the center as centralized management.
5. By adopting an ADS, it is configured so that it can operate both in the station system alone and in the entire line section. In addition, the failure of the station system does not affect the operation of the whole system.
6. Using the online test function that is a feature of the ADS, the system can be tested at the system construction stage in parallel with online operation. (Refer to Section 4.3.1 for details.) In addition, the system can be constructed step by step from the part. It also enables partial replacement of the system. (Refer to Section 4.4 for details.)

4.2.3 Overview of autonomous decentralized Tokyo area transport operation control system

4.2.3.1 System overview

East Japan Railway Company decided to introduce a transport operation control system in the main line section of the Tokyo metropolitan area in order to modernize the transport management operation work in that area. In the plan at the project start, the target route is a 19-line section including Chuo line and Yamanote line. The total line length is about 1,050 km, targeted station number is about 270. It is far beyond the transport operation control system and scale which had been introduced in the conventional line section. Figure 4.3 shows the target lines at the project start and it is expanded to 24-line section in 2018.

4.2.3.2 System target

The purpose of this system is as follows:

1. Improvement of transport management work: For this reason, the conventional TMS was the main line control system, whereas the system aims to improve the shunting work in stations and yard and maintenance work management, which is a burden of station staffs and dispatchers in operation center.
2. Improve service to railway customers: For this reason, it connects with the passenger information system, provides detailed guidance broadcasting and display, accurately provides information on train operation (train schedule,

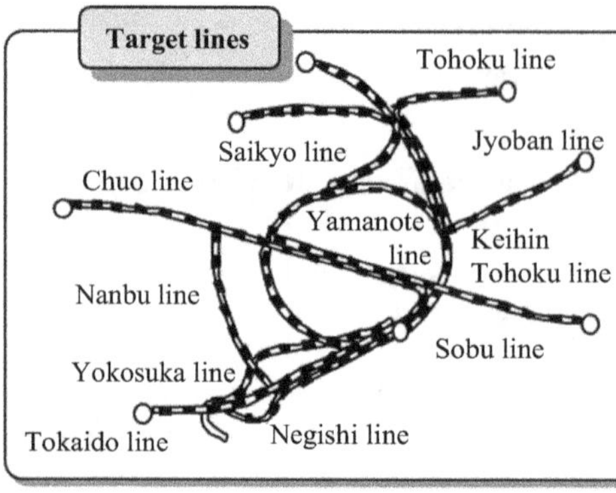

Target lines	19 lines
Number of stations	270 stations
Turnout stations	150 stations
Total line length	1,050 km
Min. headway	2 min

Figure 4.3 Target lines

train delay information) to relevant departments such as stations and drivers districts through the information terminal. It is for improving service.

3. Ensuring the safety of maintenance work and improving efficiency of maintenance work management: Safety measures to prevent train from entering the maintenance work area. Increase efficiency of maintenance work planning task and maintenance work management procedure, also efficient inquiry of operation information.
4. Increase efficiency of transport operation control work: Improvement of man–machine communication and efficient operation of dispatcher's work in operation control center (OCC) using WS.
5. Step-by-step construction of large-scale system: Being able to construct a large-scale system step by step by making full use of the online testing function of the ADS.

4.2.3.3 System configuration and characteristics [6]

System configuration concept

ADS is constructed with the station systems as the core. Also, at the construction stage of the system, it is impossible to build a system of about 300 stations at once; therefore, step-by-step construction way has to be adopted. For this reason, the station system alone is made operable, and further the system can be operated as a whole system.

The configuration of this system is shown in Figure 4.4. The station PRC (programed route-control) system is located at each station and is responsible for route control, shunting control, and passenger information system. The line section management system of the OCC and each station are connected by an optical high-speed network of 100 Mbps, and the operation monitoring data are transmitted from the station PRC to OCC in real time, so that the operation condition can be fully grasped on the dispatcher's WS screen. When the diagram is disturbed, a recovery diagram is made on WS by the dispatcher in OCC, then the recovery diagram is transmitted to PRC system at each station, changes the diagram, and carries out the route control.

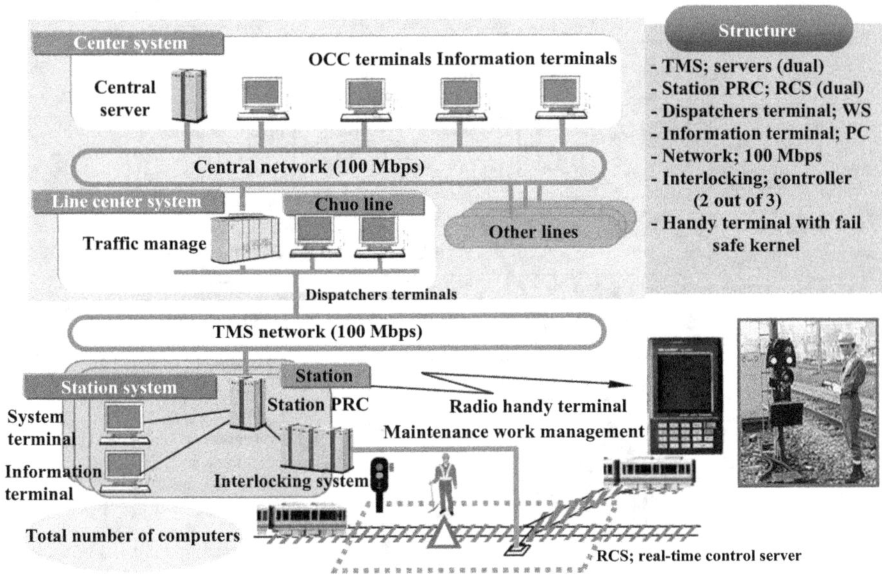

Figure 4.4 ATOS system configuration

System monitoring and remote maintenance [7]

As it is distributed over a wide area, management of the operation state of each station, collection of RAS data at the occurrence of system failure, and management of distributed software are major issues. Figure 4.5 shows the management method of large-scale distributed software.

1. System monitoring: The system-monitoring method regularly collects the operation status of all the devices to OCC monitoring system. It is also possible to start and stop the normal system by operating from the system-monitoring terminal.
2. Remote maintenance of software: Regarding the maintenance of distributed components in a wide area, it is possible to collect RAS data at the system failure to show the data to system-monitoring terminal in OCC through the network. Software maintenance can also be remotely loaded from the central development support system through the network.

4.2.3.4 System main function and characteristics

New type electronic interlocking system [4]

The interlocking system is a core equipment for ensuring the safety of train operation. Although this history was originally realized by relay logic, the electronic interlocking device formed dedicated devices by using electronics next, and dedicated logic was constructed. Since it was dedicated on the other hand, it had difficulties in terms of functionality extension and maintainability. Therefore, a new electronic interlocking system is developed, based on a triple system (2 out of 3) of a general-purpose controller. In addition to the normal interlocking function, it is possible to set the maintenance working area and set the route of the maintenance car. Also, the

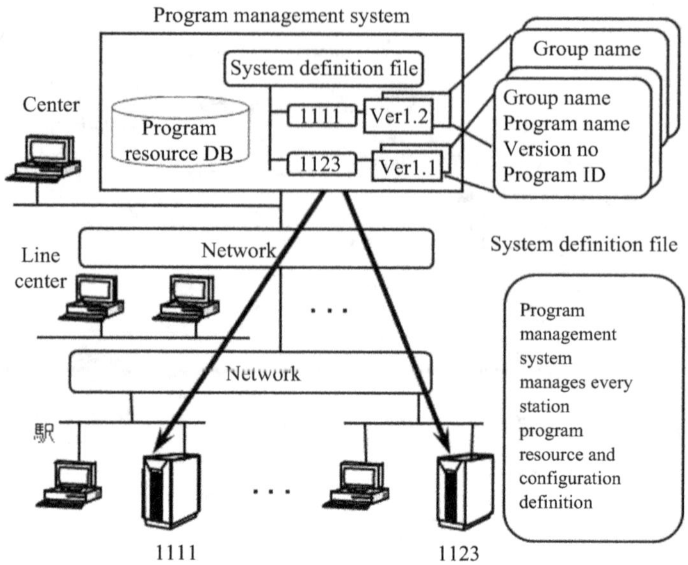

Figure 4.5 Large-scale distributed software management

interlocking logic is designed to store the interlocking table as it is in the controller and interprets it and executes it. Therefore, maintainability of interlocking logic has dramatically improved. Above all, the wonderful thing about this new type of electronic interlocking system is that it is organizing interlocking logic using software, which enabled us to incorporate the logic of maintenance work management. The interlocking system is a key to safety, and ensuring the dangerous side failure rate is a critical issue for this new electronic interlocking system. This new type interlocking system is developed at the time of development in 1994 and the dangerous side failure rate is 10^{-12}/h or less. This is the level that clears the current safety integrity level 4.

Control of large stations

On the conventional line in the Tokyo metropolitan area, there are many line sections with a rolling stock depot and large stations where the number of signal is about 500. In these large stations, the TMS only displays the position of the trains or control status of the signals in the stations, furthermore train operation such as signal control or train rescheduling are carried out by manual.

The reason for this is that the conventional TMS is a centralized system based on the CTC, concentrating information-related train operation at the train command center; train command arranges the timetable based on the traffic situation which changes in real time and controls the signals of the station. It was impossible to control a large station using a computer with low throughput and a communication line with small capacity.

However, ATOS adopts an ADS, a distributed PRC is installed at the station, and a timetable is held so that the signal can be controlled. This makes it possible to control large stations. As a result, even in the case of a conventional line in the

Tokyo metropolitan area, automatic route control can be performed consistently for all the lines. This is also an advantage of introducing an ADS.

Maintenance work management [4]

In order to secure the safety of train operation and to prevent accidents of workers during maintenance work such as rail exchange and track maintenance, it should have to prevent trains from entering the area of the maintenance work place and during the time period from the start to the end of work. In the past, because it relied on communication by human system, efficiency was not good, and accidents sometimes occurred. For this reason, in this system, the maintenance worker has a wireless (or wired) handy terminal whose maintenance work plan has been loaded in advance and directly inputs the work scope, work start, and end time to the maintenance work management system. On the other hand, the maintenance work management system systematically checks it and interlocks not to let trains enter the corresponding place and time zone.

Designation of the working range is made possible for each track circuit. It is also possible to set the route of maintenance cars using this wireless handy terminal.

With these, maintenance work management system which had traditionally relied on the human system is realized. And it dramatically improved its efficiency and safety of the maintenance work management. Regarding safety, confining logical judgment into the new type electronic interlocking system with high safety design, securing data reliability by terminals and communication paths by protocols, and ensuring safety as a maintenance work management system were taken care of. The dangerous failure rate of this system is less than 10^{-10}/h.

With these, we have succeeded in automating the task of maintenance work management that had traditionally relied on the human system, and dramatically improving its efficiency.

Maintenance work management system is shown in Figure 4.6.

Passenger information

At concourse and platform of each station, departure guidance displays and audio broadcast system are installed. Make detailed guidance corresponding to the type of train (limited express, ordinary train, etc.). Delay indication when train delay

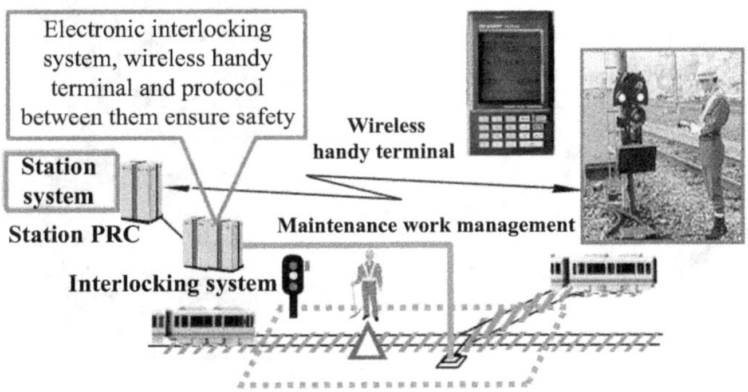

Figure 4.6 Maintenance work management system

occurs and message display input by human are also made available. In addition, it has a terminal that displays train traffic status at each station. These contribute to improvement of passenger information service.

Central command system

Manage the operation of each line section at the central command center. Instead of conventional CTC's large-sized display board, it displays operation by WS and displays execution schedule and operation record in train diagram chart format on WS. Train operation also has greatly improved operability by operating "streak" on the chart on WS as it is.

Here, from the viewpoint of integration of information and control, consider the recovery method when the transport disorder has occurred. TMS has a timetable as a plan, and signal control is carried out according to the timetable. When the transport disorder used to occur in the past, since changing procedure of the timetable was not easy, station personnel used to control the signal manually. As a result of this operation, the timetable did not match with the control state of the signal, and it was supposed to prolong the influence of transport disorder.

Therefore, it is easy to change the diagram by operating the "line" on the chart of the WS screen as it is, so when the transport disorder occurs, first the timetable is changed and as a result, signal control follows it. By controlling the information, by actually controlling the signal, information can be quickly communicated between the commands in the change command center and information can be transmitted to the relevant department even when the transport disorder is occurring. The ATOS makes it possible to deliver and be consistent with the train traffic information among train command, stations, train crew depots, and maintenance depots systematically. For the train commander, the operation efficiency was greatly improved by shifting from complicated timetable adjustment operation to directive operation mainly based on judgment. Figure 4.7 shows the screen of operation adjustment and operation monitoring.

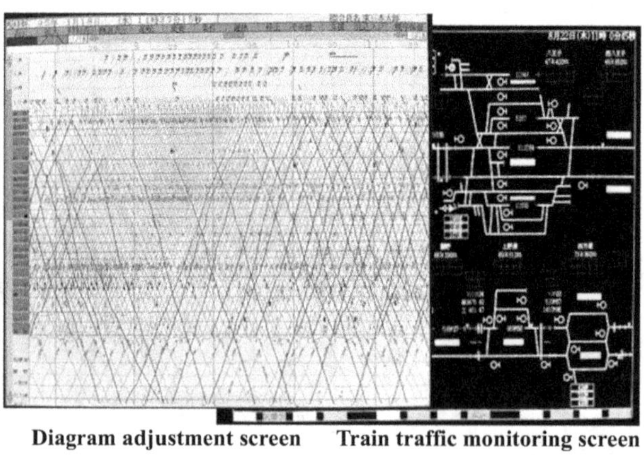

Figure 4.7 Diagram adjustment and train traffic monitoring

4.3 Advancement of ADS technology

4.3.1 Online testing

In order to test new system with existing train TMS running, online testing of ADS was applied. On testing, it takes a lot of time to make test data and decide generation timing of test data. In addition, it is very difficult on testing of large-scale system because relation of function modules is complicated. Thus, online testing can greatly reduce the load of test preparation when testing large-scale train TMS.

The system expands in two parallel phases: station-computer subsystem addition and application programs addition or modification in station-computer subsystems.

On subsystem addition, a new station-computer subsystem needs to be added to already operating station-computer subsystems without stopping the existing subsystems' operation and be tested using online data. Only after the operation of the added subsystem is verified, the subsystem starts actual operation. In general, online testing is essential in mission-critical real-time computer-control systems.

On application programs addition or modification, new application programs are added to application programs installed in station-computer subsystems, or installed application programs are modified. In these cases, tests using online data are also necessary.

4.3.1.1 Online testing for subsystem expansion

In the following, the online test techniques for the expansion of subsystems and that of the application programs are respectively presented.

First, online testing for subsystems expansion is described. At certain points in the step-by-step construction of widely distributed train-traffic computer-control systems, a newly constructed station-computer subsystem is added to the DF of a train-line network. The subsystem needs to be tested using online data as a final check as to whether it normally works. At this time, online data are sent from the subsystems in online mode and test data are sent from the subsystems in test mode in the train-line network. Likewise, when a newly added computer of a station subsystem is tested, both online data and test data flow in the local Ethernet.

In general, online data are more important than test data for the purposes of train-traffic control systems. Test data, however, is also important to improve the reliability of station computer subsystems. Hence, it is required that these two kinds of data coexist without disturbing each other. In order to permit this, each DF in the three kinds of networks (two Fiber-Distributed Data Interface (FDDI) and one Ethernet) is logically divided into online DF and test DF. Only online data flow in the online DF and only test data flow in the test DF.

A computer in online mode or test mode connects to both the online DF and the test DF and distinguishes online data and test data by their content code (CC) (Table 4.1). Each computer works as follows (Figure 4.8):

1. A computer in online mode
   - (i) receives and processes online data,
   - (ii) receives test data but does not process it, and
   - (iii) broadcasts online data.

Table 4.1 Relation of data and mode

Data	Online mode	Test mode
Online data	• Receives data • Processes data	• Receives data • Processes data
Test data	• Receives data • Does not process data	• Receives data • Processes data

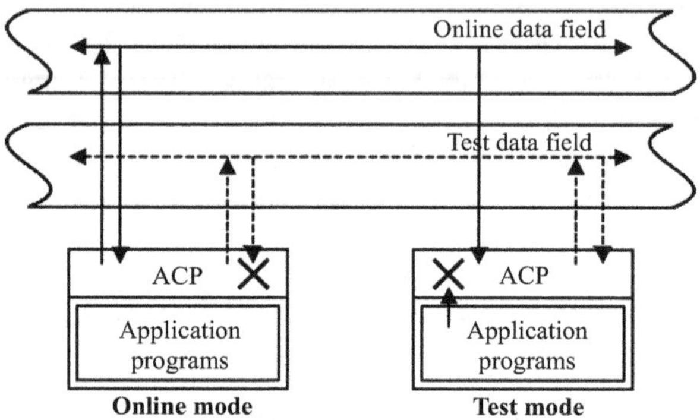

Figure 4.8 Separation of online data and test data

2. A computer in test mode
   (i) receives and processes online data,
   (ii) receives and processes test data, and
   (iii) broadcasts test data.

4.3.1.2 Online testing for application software expansion

Next, online testing for application program expansion is described. There are two diagnostic components to verify application software (AP) in ADS: a built-in tester (BIT) and an external tester (EXT). Online testing of AP is carried out using the BIT and the EXT.

The BIT is a testing management module installed in an autonomous controlled processor (ACP) of each subsystem. It attaches a test flag to test data and suppresses output according to the configuration.

The EXT is AP and connected to a DF. It receives messages broadcasted from APs and verifies APs by comparing the messages from several APs. The functions and installation location of the EXT for non-real-time management AP and real-time control AP are provided in Table 4.2.

4.3.1.3 Online testing for non-real-time management APs

An online testing technique using the EXT for non-real-time management APs and real-time control APs is described next.

Table 4.2 Functions and installation location of the EXT

Type	Functions	Installation location
EXT for non-real-time management AP	Checks whether the message is broken or not	In an operation room
EXT for real-time control AP	Checks the output-timing of data sent from APs	On site because it is essential to check critical timing of data sent from real-time APs

An online testing technique for non-real-time management APs is shown using an example of train-traffic rescheduling AP (Figure 4.9).

In the example, the train-traffic rescheduling management computer and the passenger information control computer are in test-mode and the train-traffic-management computer and RCS are already operating. The online testing technique for the train-traffic rescheduling AP in the train-traffic rescheduling management computer is as follows.

The EXT for non-real-time APs testing is installed in an operation room and is connected to the management subsystems by Ethernet. The CC of re-scheduled timetable data is registered in the CC in the ACP of the EXT.

1. An operator's inputs changed data of train-departure sequence on the terminal of the train-traffic-management computer. For example, the operator inputs the data that indicate that the departure sequence of train 1464M and that of train 4012H from Takao station on the Chuo line will be exchanged.
2. The train-traffic-management computer broadcasts the message that consists of the data as provided in Table 4.3.
3. The train-traffic rescheduling APs receive the message with the CC of rescheduled train-departure sequence data and change train-traffic scheduling table, which includes all train-traffic schedule data of the line of the day, according to the data in the message. The size of the timetable data is 4 MB. The changed data include the train number (e.g., 4012H), station names through which the train passes, the departure time, and the arrival time of the train of each station. The length of the changed data is a maximum of 8 kB.
4. The train-traffic rescheduling APs broadcasts the message that consists of the data as provided in Table 4.4 in order to transmit the rescheduled timetable data.

   It is necessary to check whether the message sent is broken or not because the message is large and is possibly broken.
5. The EXT for non-real-time management APs receives messages with CC of rescheduled timetable data if the CC is registered in the ACP of the EXT. The EXT receives two messages because the messages are also sent from the replicated train-traffic rescheduling AP. The EXT compares messages and checks whether messages are identical or not.
6. A timetable-display AP in the passenger information control computer in test mode needs to receive messages with CC of rescheduled timetable data. The ACP in the computer receives the messages broadcasted from the replicated

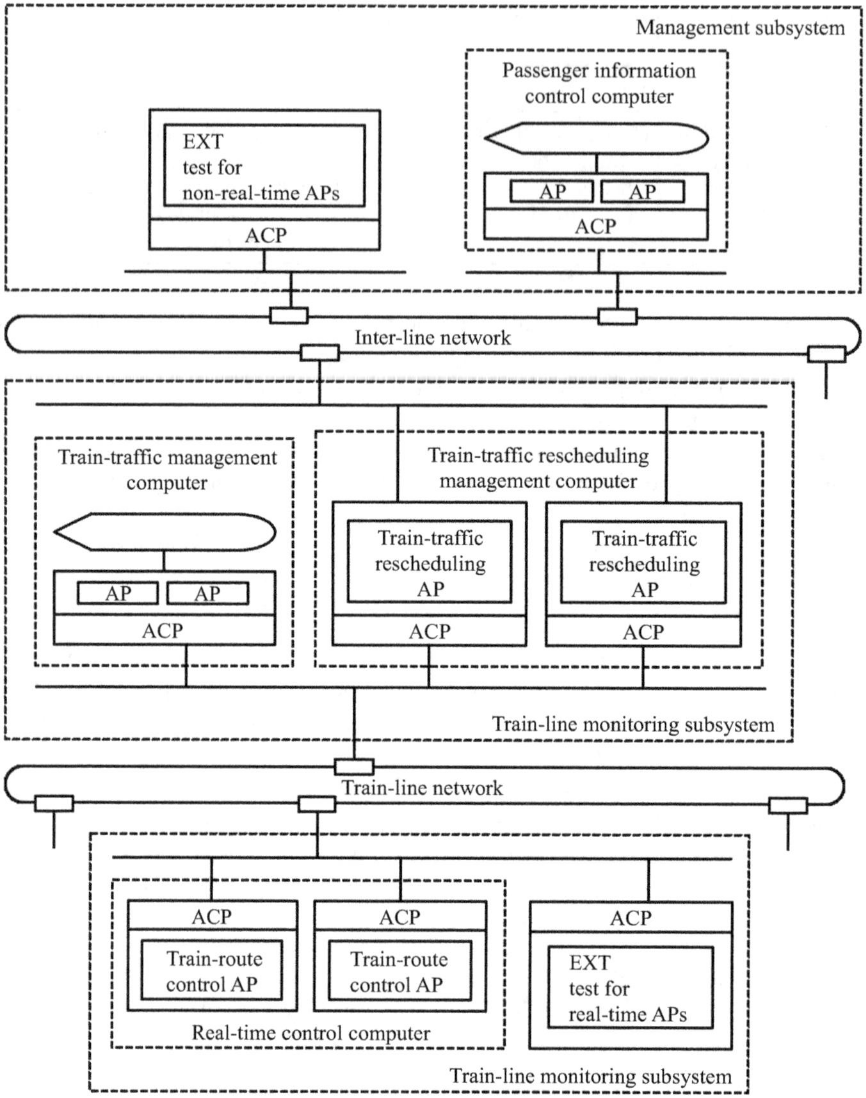

Figure 4.9 Detailed structure of the system

train-traffic rescheduling APs and compares these messages. If the messages are identical, the ACP sends the message to the timetable-display AP. If the messages differ, however, the ACP discards the messages. In this way, the timetable-display AP can receive the correct message. A train-route control AP in the RCS in online mode also needs to receive the message with CC of rescheduled timetable data. The ACP in the computer discards messages with a test flag because the computer is online. In this way, already operating RCSs are not disturbed.

Table 4.3 Message of rescheduled train-departure sequence

Item	Value
Content code	Code that indicates rescheduled train-departure sequence data
Flag	Online flag
Message sequence number	For example, #1 to the first message
Data	Changed data of train-departure sequence which include Takao station, train-number 1464M, and train-number 4012H

Table 4.4 Message of rescheduled timetable data

Item	Value
Content code	Code indicates rescheduled timetable data
Flag	Test flag because the train-traffic rescheduling management computer is in test mode
Message sequence number	For example, #1 to the first message
Data	Train number (e.g., 4012H), station names through which the train passes, the departure time, and the arrival time of the train of each station

By the above sequence from 1 to 6, the online testing can verify the consistency of non-real-time management APs without interrupting the system operation. It does not disturb already operating subsystems because ACPs in online computers receive only online data but reject test data.

4.3.1.4 Online testing for real-time control APs

Non-replicated APs testing

1. Online testing for train-tracking module: An online testing technique for real-time control APs is shown using an example of the train-tracking module and route-control module. The test system architecture for the train-tracking module is shown in Figure 4.10. Expanded subsystems (a real-time control computer, an electrical route-switching device, an I/O controller, and a train-tracking module) are connected to the existing electrical route-switching device and the expanded subsystems can get the real control data from existing sensors and electrical route-switching device. The timing specification is determined as follows:
   (i) The sensor sends state information of signals and devices (SD), and train-position (TP) in its control region. The synchronizing unit is connected to the sensors and gets SD and TP.
   (ii) The synchronizing unit is also connected to the existing electrical route-switching devices. It gets state information of the train-route-interlocking device (TRID) and sends SD, TP, and TRID to the expanded electrical route-switching devices.

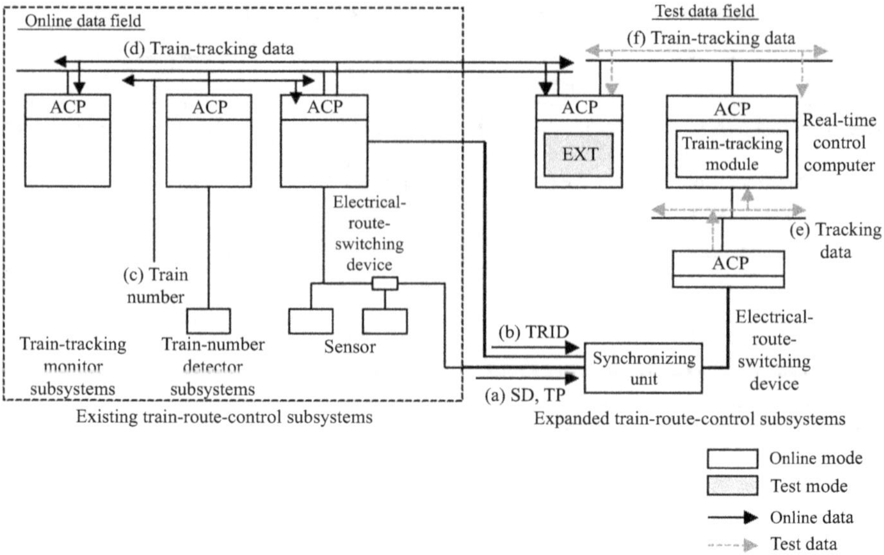

Figure 4.10 Online testing for train-tracking module

(iii) The train-number detector receives the train number and sends it to the train-tracking monitor.

(iv) The train-tracking data are broadcasted by the existing electrical route-switching device. The train-tracking monitor displays the state of SD and TP.

(v) The expanded electrical route-switching device is replicated as triplex-sequencers, and they are connected to the synchronizing unit to receive the real-time data, SD, TP, and TRIO. They broadcast the tracking data to the expanded real-time control computer. The timing specifications are evaluated to ensure that the data-read timing of the expanded triplex-devices allows all cyclic data, such as SD, TP, and TRID, to be received. They are evaluated by receiving the broadcasted train-tracking-data. If the data-read timing is too slow to catch the SD, TP, and TRID, the train-tracking-data may indicate inconsistent motion of trains, signals, and TRIDs. These sequence-checking methods are described in Figure 4.10.

(vi) The train-tracking module is installed in the expanded real-time control computer. It receives the tracking data and broadcasts the train-tracking data. The EXT receives the train-tracking data from the existing train-tracking monitor and the expanded train-tracking module. The EXT makes the sequences of the train-tracking data from the online data and the test data with the test flag. The sequences of the train-tracking data are consistent if the following conditions are satisfied.

   (a) The control sequences of signals and TRIDs of the same trains are identical in the online subsystem and in the test subsystem.

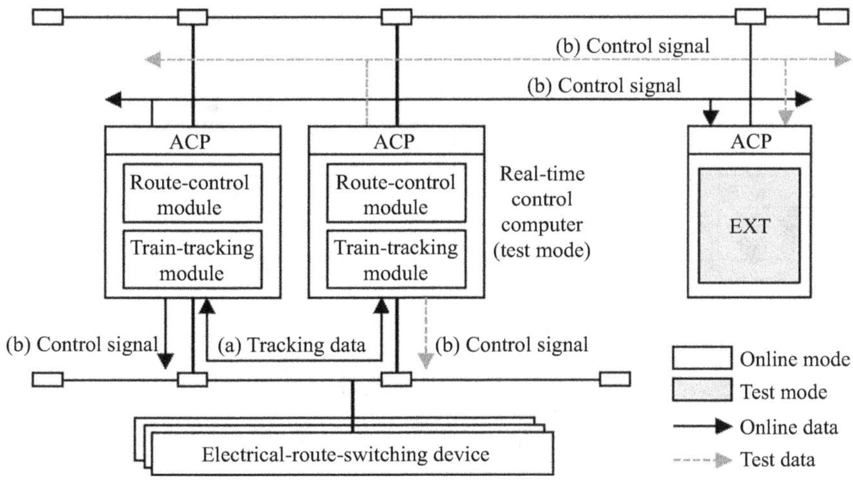

Figure 4.11 Online testing for route-control module

(b) The difference of the time when the real-time tracking data, which are sent every 600 ms, are sent from the existing train-tracking monitor and the time when the data are sent from the expanded train-tracking module is within a few milliseconds. The difference of the time of train-tracking data, which are sent every 2.4 s, is within 20 ms.

(c) The testing is executed for 1 week. After the testing finishes, the timing specifications of all expanded subsystems are fixed.

2. Online testing for route-control module: Online testing of the route-control module is basically evaluated by the following criteria (Figure 4.11):
   (i) Verify that the operation cycle is short enough that the route-control module can receive tracking-data. This is the most critical real-time performance indicator of control software. If the control software cannot receive signal or train state, serious accidents may occur. If the state is recognized less frequently than every 600 ms, when unpredictable signal state occurs due to operator's actions or the fail-safe logic, the module cannot calculate control output considering the signal state, which causes a control error. Thus, system engineers must check the performance of the state recognition.
   (ii) The interval of the control signal outputs of the route-control module is under 2.4 s.

Replicated APs testing

When important AP, such as the train-tracking module or the route-control modules, is replicated, the replicated AP is tested as follows:

1. Check consistency of the output data of replicated AP. The replicated AP must receive the same tracking data in 600 ms and send the same control signal in 2.4 s.

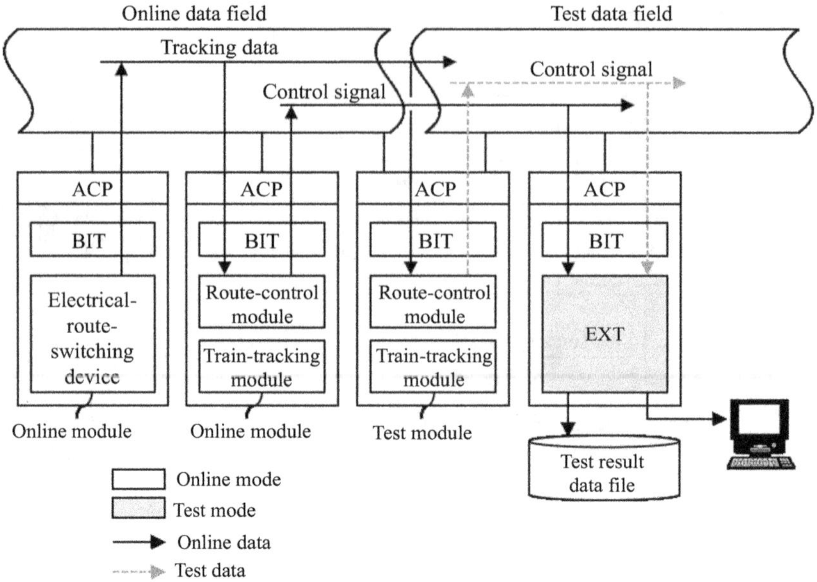

Figure 4.12 Data field isolation for online comparison testing

2. Compare the output of newer version control software and that of older version control software. When newer version software is implemented, the real-time performance must be compared with online operating software (the older version). The comparison criteria of real-time performance are the same that of (1). The newer version software is validated by confirming that the control sequence of each train calculated by the older version software and one calculated by the newer software are identical. The testing is executed in the following environment (Figure 4.12):
   (i) DF isolation: The test DF is isolated from the online DF. The testing equipment is connected to the test DF and testing is executed only in this DF. The two DFs are connected to the ACP that operates as a gateway. It passes the online data to the test DF so that online data can be used in the test DF and prevents the test data flowing into the online data filed. This isolation mechanism ensures the real-time performance of the online subsystems.
   (ii) Parallel execution of the online module and the test module: The BIT in the test module receives the online data from the test DF and executes the route-control module. The BIT also attaches the test flag to the data from the route-control module and sends it to the test DF. In the test DF, there is also online data from the online route-control module. The EXT connected to the test field monitors data from both the online module and the test module. With this testing mechanism, parallel testing can be executed during online operation.

4.3.2 Self-correction [6]

In order to make train TMS reliable, subsystem self-correction technique was proposed using fault tolerance technique of ADS.

In existing system, self-diagnosis technique was used. This technique enables subsystem to detect its failure when it behaves abnormally or receives erroneous data. With this technique, subsystem can detect its failure but cannot correct its behavior.

In ADS, subsystem is connected DF and can receive multiple data sent by other subsystems that have the same CC included in the data it sends. Using these data, subsystem cannot only detect its failure but also correct its data or restore related function module.

This section described the three extended features of ADS technology: function enhancement of DF for information control integration, online testing, and self-correction.

Enhanced DF and online testing were applied to already existing railway systems. Using these techniques, development time could be shortened and maintenance and renewal cost could be cut down. They can be applied to not only railway systems but also other widely distributed real-time systems and can contribute to development period and cost reduction.

4.4 Step-by-step system construction technology for large transport operation control system

4.4.1 Outline of large transport operation control system

As an example of the large transport operation control system, there is Tokyo metropolitan area ATOS, which consist of 19 train lines with 270 stations, and total length 1,050 km. That is a large system covering the entire railway network in Tokyo metropolitan area.

In the conventional way, this system had to be constructed as conventional centralized TMSs of the each line sections. For this reason, the partial expansion such as the functional enhancement, the terminal or computer version up had to be waited until next system replacement time. Originally, it should be a system that can be replaced, expanded functions, and increased performance as necessary, but since it adopts a centralized TMS on a line section basis, it had to be waited until large system replacement.

The system does not consist of a conventional centralized operation control system. It has intelligence on a station level. It consists of an ADS to enable each station to systematize its complex operations. The central system and the systems of each station are connected with an optical network of 100 Mbps. The hardware used consists of small general-purpose components (workstations, PCs, controllers, and small servers) to construct a system in an open system environment. The portion required to be reliable consists of a duplex or other redundant system.

Such a system must have an architecture designed with much consideration to reliability, expansion and maintainability. It is true that the reliability of the system and that of specific devices are both important. Yet more important is system assurance, including the methods to ensure system expansion, recover from a damage by an expansion failure (should such a failure occur), and prevent adverse effects on the entire system while performing online system expansion.

In order to expand or partially replace the system while the system is in operation, it is necessary to provide protection measures to avoid interference from the test system to the online system. Or it is difficult to construct a system at once or replace it, so it is necessary to construct a system in a step-by-step way. These are called system assurance technology.

Following is a description of how assurance technology is applied to ATOS, which is a representative railway transport operation control system introduced by East Japan Railway Company [8].

4.4.2 System construction issues and assurance

1. Coexistence of heterogeneous functions

   ATOS is a large-scale ADS, in which the information function is realized in a network-wide manner together with the route control. The conventional centralized TMS adopts a hierarchical structure, the control and information function has a hierarchy of computer systems, and communication lines are also divided. However, ATOS coexists with control and information function. By adopting such a configuration, although it is possible to make an efficient system configuration, it is conceivable that mutual influence is caused by coexistence of different functions such as control and information function. For example, when a disturbance of the diagram occurs, an inquiry from many terminals rush to raise the traffic peak of the information function, thereby affecting the control response.

   It is necessary a mechanism not to cause such a situation.

   At the system construction stage, it is necessary to conduct a test during online operation. This is also coexistence of two heterogeneous functions, online function and test function. Even if there is some interference from the test system to the online system at such a time, it must be able to protect it online.
2. Step-by-step system construction

   ATOS is a large-scale ADS incorporating not only the TMS but also the signal system such as the electronic interlocking system, and also the information service function. Construction of such a system cannot be constructed and operated at once because of the big size of the scale, so it is inevitable to construct a system step by step. The conventional TMS is a centralized system and has a hierarchical structure of "TMS–CTC–Interlocking system," each of which is separated as a system, the construction order first establishes interlocking device, CTC. As for the TMS, we have two systems of old and new, and use the new system only at the time of the test until the completion of the construction of the whole system, return to the old system at the end of the test, and switch to the new system all at once.

   However, the large-scale ADS such as ATOS has the route-control function of the electronic interlocking system and the TMS as a station system, and furthermore the information processing can be operated as a station system by including it as a station system. It has a constitution. Also, since the electronic interlocking system is integrated with the signaling equipment at the site, there is a need to start using the electronic interlocking system/station route-control function every time when changing the signaling system. This concept is to allow the system to be constructed and operated step by step during online operation, and it is a concept that it is possible to gradually grow the system step by step. ATOS is a large-scale growing system.

3. High assurance system

   As mentioned above, a large-scale ADS can be said to be a growing system in which heterogeneous needs coexist. Here, the heterogeneous needs indicate coexistence of different functions such as control and information service. In addition, system growth shows system expansion during online operation and step-by-step system construction. On the other hand, however, the expansion of such online systems is risking the reliability. Therefore, a mechanism and technology are needed to assure stable operation of the system when the system is expanded. Therefore, a high-assurance system has been proposed as a system to assure stable operation of growing systems coexisting with different needs [11]. Table 4.5 shows the issues in system construction.

4.4.3 Application of assurance technology [7,8]

Following is a description of the assurance technology for guaranteeing system operation in the process of step-by-step system construction.

1. Issues and assurance technology in system construction

   Figure 4.13 shows assurance technology as a set of solutions to the issues of system construction. To guarantee system operation in the construction process, there is a precondition consisting of a mechanism that tolerates an online system, test system, coexistence of a construction system, protection of an online system, or the coexistence of other different operation systems in order to ensure testing and construction while online. In addition, there is the optimization of a construction sequence for step-by-step system construction as an assurance technology. Following is an overview of the various technologies.

Table 4.5 Issues in system construction

No.	Issue	Description
1	Step-by-step system construction	Optimal construction sequence for step-by-step construction
2	Nonstop system expansion	Construction while online, and expansion by partial stoppage
3	Coexistence of online and system construction	Configuration of operation modes, and adaptation to transitions and changes to operation modes in the case of an expansion failure
4	Protection of online disturbance from construction system	Data protection from other systems on an online network

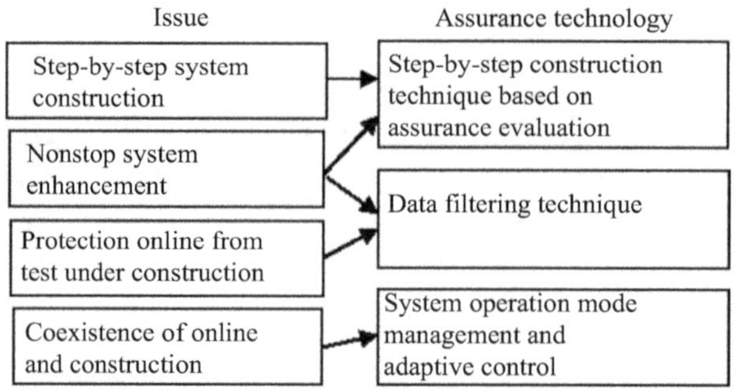

Figure 4.13 Issue of system construction and assurance technology

2. Management and adaptive control of heterogeneous operation modes
   (i) Configuration of heterogeneous operation modes

   For system construction in parallel while online, the system comes with such system modes as online, test, unconstructed ones. It also has three operation modes (station independent mode, neighbor-linked mode, and center-linked mode) according to the construction stage of the system. These operation modes are effective for step-by-step system construction, and also as a fault-tolerance function in the case of an anomaly.

   (ii) Management and adaptive control of operation modes

   In the construction stage of the system, the operation mode changes according to the construction status of the system. For example, when a system for one station is constructed and put into operation, it shifts to the station-independent mode. When the system construction for a neighboring station is complete, the system shifts to the neighbor-linked mode. Then, when the system construction for all stations and the central system is complete, the system shifts to the center-linked mode.

   On the other hand, when an expansion fails and degeneracy is necessary, the system shifts from the center-linked mode to the neighbor-linked mode or to the station-independent mode.

   When (in addition to the relevant station) one station switches to another operation mode, another station and/or device needs to change its operation mode.

   Figure 4.14 shows an overview of the operation mode control. In an ADS, all information is inputted and outputted through a DF. When the operation mode is changed, the system declares to the DF that its operation mode has changed as a change of operation mode. Other systems periodically monitor operation mode changes in those other systems. When they detect the operation modes of those other systems, they change their own operation modes adaptively.

   OMC (operation mode control) thus adaptively controls the operation mode according to changes in the operation modes of the other systems.

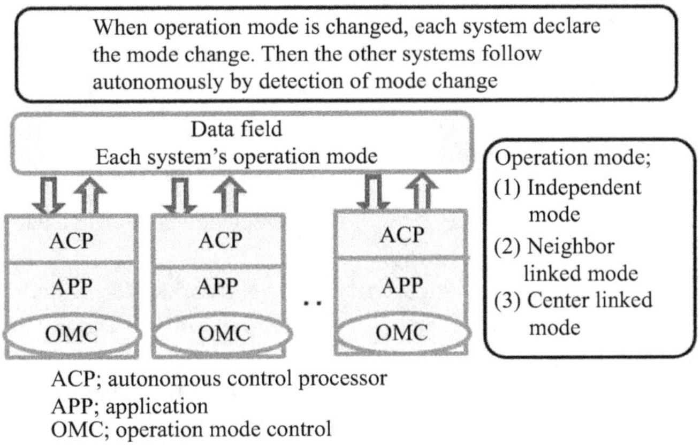

Figure 4.14 Operation mode control

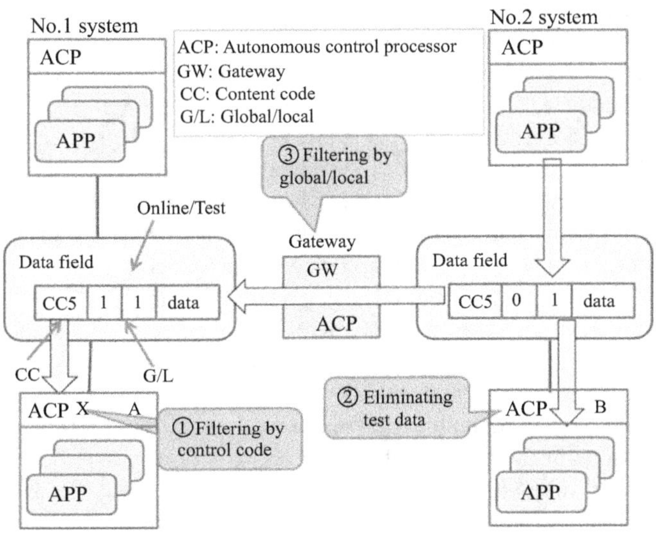

Figure 4.15 Data-filtering mechanism

3. Data-filtering technique [10]

   Since the system is constructed in parallel while online, various signals are emitted from a line being constructed to the online system. Whichever system may be constructed, the online system must prevent itself from being stopped due to such an event.

   A data-filtering mechanism is therefore provided as an assurance technology to protect the online system. The data-filtering mechanism consists of three mechanisms. Figure 4.15 shows how the online system is protected by the data-filtering mechanism.

(i) Data filtering by ACP

Each subsystem in the ADS uses the mechanism of selectively receiving only the necessary CCs, selecting only the necessary data, and rejecting other data.

(ii) Data filtering by online and test modes

A test flag is added to the data to make it possible to identify the test data, so that the online system can reject data provided with a test flag.

(iii) Data filtering with gateway

This mechanism is such that, when networks are interconnected, data filtering will be performed to ensure that only the data to be actually used is passed.

The three data-filtering mechanisms described above are used to prevent unnecessary data from going to the online system in quantitative or qualitative terms. Here, quantitative protection means reduction of traffic (in packets per second), while qualitative protection allows the passage of only the items of data with useful CCs without passing useless data.

4. Step-by-step system construction technology [8]

(i) Assurance of system construction

The best construction sequence in step-by-step system construction is to guarantee "function volume times available time" in a specified period (mission time) under "a stable operation of the system."

Assurance is defined as "the capability to adapt to heterogeneous modes changing over time and guarantee system operation" [10]. Consideration of heterogeneity and adaptability in system construction indicates that heterogeneity is "system construction when heterogeneous modes (such as online and system construction) coexist," while adaptability means "system construction in an environment where heterogeneous modes change (degeneracy and reconstruction due to expansion and expansion failure)."

In addition, assurance in system construction can be defined as "the capability to guarantee the maximum operation of the entire system functions in the process of system construction."

(ii) Concept of evaluation technology

The assurance (Af) of system construction was defined by means of an index of functional reliability to evaluate the maximization of "function volume times available time" in a specified period.

The Af of system construction can be defined as the average functional reliability including the entire range of the specified period (mission time) of functional reliability as shown in Figure 4.4. It can thus be expressed with formula (4.1). Functional reliability [9] is as shown in formula (4.2).

$$Af = \left(\frac{1}{TM}\right) \cdot \int_0^{TM} Rt(Xt)dt \tag{4.1}$$

where TM is the mission time for system construction and $Rt(Xt)$ is the functional reliability in the system status (Xt) at time (t).

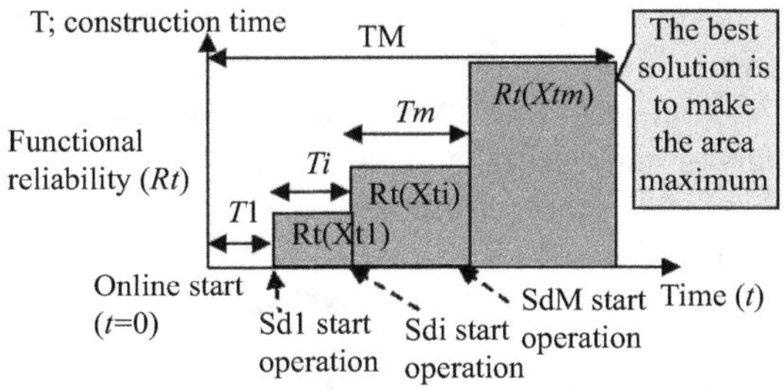

Figure 4.16 Definition of assurance evaluation

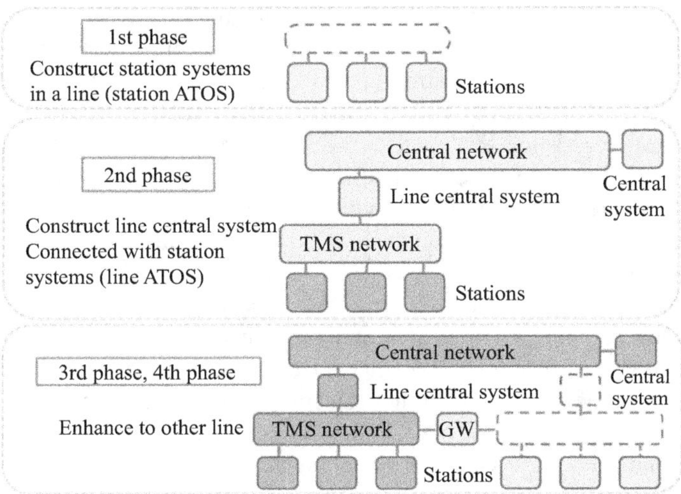

Figure 4.17 Step-by-step construction

$$Rt = \sum_{X} \Psi(S,X) \cdot P(X|t) \tag{4.2}$$

Here, as shown in Figure 4.16, the area maximization of "function times available time" means an increase in assurance.

4.4.4 Application results of the step-by-step construction technology

1. Concept of step-by-step construction of the entire system

   The Tokyo area transport operation control system employs step-by-step construction as a result of assurance evaluation. It also uses assurance technology described in Section 4.4. The concept of step-by-step construction is as described in Figure 4.17.

   Construction is roughly divided into phases 1–4.

(i) Phase 1: Construction of the station devices (making the station ATOS-compliant)

Each station starts operation by automatic route control in the station independent mode, after completion of construction. When the construction of neighboring station is completed, the system shifts to the neighbor-linked mode.

(ii) Phase 2: Construction of central system + connection of station system (making the line ATOS)

When all stations complete their station system constructions, the central system has the information regarding all stations, so that it is completely capable of monitoring the operation status of the entire line. At that time, the system shifts to the center-linked mode. At this stage, the line section central system manages the operation status of the diagram the whole line section and the train of the entire line section and can operate by connecting to all the stations.

(iii) Phase 3: Enhancing the range of operation (construct station devices in another rail line and connect through a gateway)

In this stage, the communication lines are extended. Similarly to Phase 1, the station devices will be built step by step. Since the device is to be connected by network with another rail line, which is already connected to the network, a gateway is provided to prevent data and other resources from affecting the online system during the construction. Different line sections are connected through a gateway.

(iv) Phase 4: Enhancing the range of operation (connection of central and station devices with another rail line through a gateway)

As in Phase 2, when all the station systems are prepared, they are connected to the central system, the next line section is completed, and it operates as a line section. In this way, the construction of all lines is completed.

2. Effects of step-by-step system construction
   (i) Effects for the system constructor side
      (a) In the case of step-by-step system construction, construction personnel planning can be developed in a more leveled manner than in the case of collective construction.
      (b) Few personnel repeatedly participate each construction. It is therefore easy to produce a feedback effect by these personnel through learning, thus making it possible to upgrade the quality of constructions one by one.
      (c) Step-by-step construction keeps the expansion risks lower than collective construction.
   (ii) Effects for the user's side
      (a) In the step-by-step construction, online operation begins after each step is completed. It therefore enjoys the effect of systematization better than in the collective construction where the system is put into operation only after the construction of the entire system is

completed. Therefore, when comparing the step-by-step construction and the collective construction, the step-by-step construction can be operated earlier from the part where the construction is completed without waiting for completion of the whole system, so the effect of the step-by-step system construction is preceded. For this reason, it can be said that the step-by-step construction is more effective. In order to evaluate the effect, we will compare the reduction of the operation and management human workload after systemization by the operation management system between the step-by-step construction and the collective construction.

(b) Figure 4.18 shows the trend of gradual construction of the Yamanote line and the Keihin-Tohoku line and the reduction of the human work volume divided into the central control room task and the station operation task. At the same time, this figure shows how the manipulated variable changed, assuming that collective construction was used and the overall online application was not executed until the central system was completely prepared. As shown in Figure 4.6, the traffic operation at the station is considered to be 38% lower in the step-by-step construction than the collective construction. In the case of the collective construction, since the system does not operate until the system construction of the entire line section is completed, the work does not decrease, but in the case of the step-by-step construction, if the station system is completed, it becomes possible to

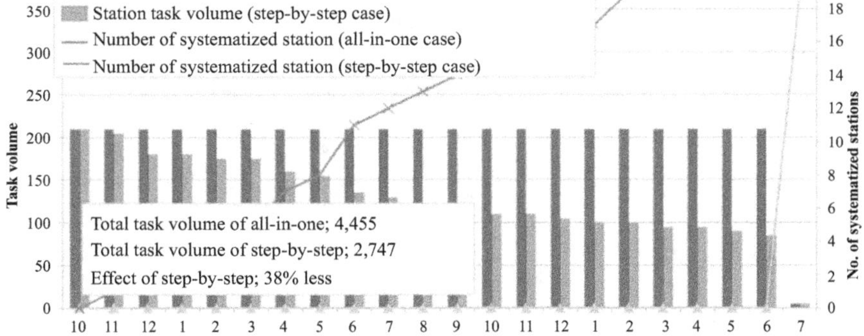

Figure 4.18 Implementation to Yamanote/Keihin-Tohoku line and the reduction of human traffic operation volume

operate from that part because the amount of work decreases with translation.

(c) The Tokyo area forms a network of railways. There is a case that four different lines stop at one station. In such a case, one-line station can be put into operation at a time in a single station. There is an example that human traffic operations came to be reduced by 77%.

3. Process and results of ATOS system construction

Figure 4.19 shows the construction process and results of the Tokyo areas transport operation control system. Figure 4.20 shows the construction results of Tokyo area railway network.

The Tokyo area transport operation control system (ATOS), which was picked up as an example of the large transport operation control system, has been expanding in order since the first station was put into operation in July 1993. In the entire line, the Chuo line was put into operation in December 1996. In July 1998, the Yamanote and Keihin-Tohoku lines were put into operation. The system was then expanded to the entire region of the Tokyo area. Step-by-step construction is thus continued while running online operations.

So far, a total of three system failures have occurred: one failure stemming from trouble with software management and two of them related to the network. The system in general has been running extremely steadily. No trouble has since then occurred as the result of drastic corrective actions.

This transport operation control system has been constructed step by step since the operation of the first station in July 1993. In the meantime, the system has undergone many partial updates. Taking the example of an area no. 2, which contains the Yamanote and Keihin-Tohoku lines, system installation (system exchange)

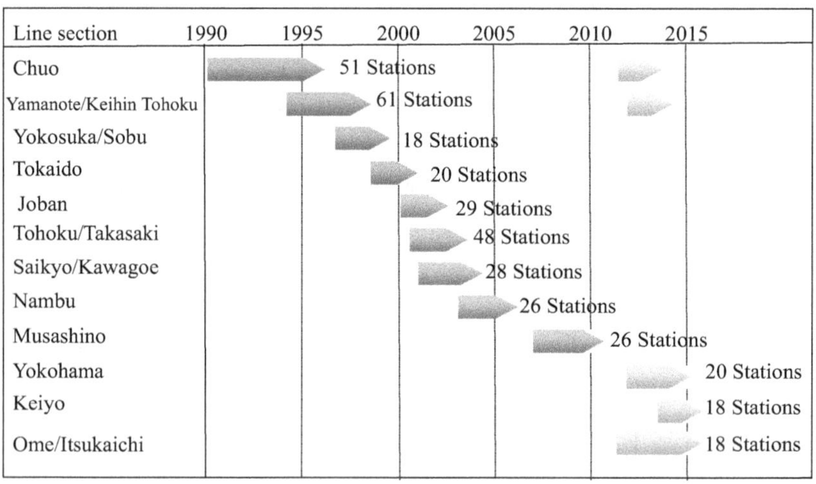

Figure 4.19 Actual results of ATOS construction

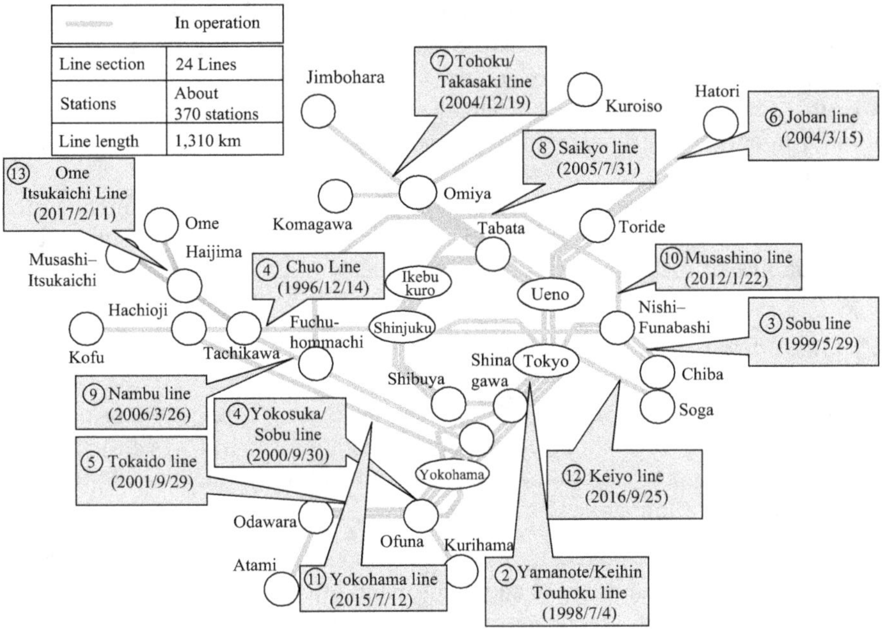

Figure 4.20 Tokyo area railway network (ATOS system construction results)

was conducted for a total of 173 times for 3 years for installing station devices, and 14 times for installing central devices. These installations are both prescheduled and conducted as the result of midway repairs or modifications. These system installations were conducted in an extremely short time during nighttime, but there occurred no trouble regarding system construction, Step-by-step construction has been conducted steadily. At ATOS, initially planned for the 19-line section, in March 2006 the original 19-line section began to be operated, and the line section expanded to the 22-line section. In ATOS, the replacement of the Chuo line of first introduced line has already begun. The replacement is carried out by a step-by-step way in autonomous station system unit and finally the entire Chuo line is replaced. Step-by-step construction technology has thus been contributing greatly to the construction of a large system and the stabilized operation of the system.

4.4.5 Summary

In Section 4.4, step-by-step system construction technology in a large-scale transportation operation control system has been discussed. The system consists of ADS, the minimum autonomous unit is constructed as a station system, a station system is constructed step by step, after all stations construction, the entire line section system is connected. Then line sections are expanded further, finally the whole system is completed. Such step-by-step construction has been made possible by assurance technology such as assurance evaluation, OMC, data-filtering technology to protect online system.

The effect of step-by-step system construction is that it can gradually enjoy the system introduction effect from before the completion of the whole system by sequentially shifting to online operation from the completed subsystem. This is particularly effective for a system where the system is large, and it takes a long time to build. In addition, it is also advantageous that leveling of system construction personnel is possible.

The step-by-step system construction is effective not only in the initial system construction but also in the case of replacing. In reality, ATOS is currently replacing Chuo line of first introduced line.

4.5 Conclusion

As an example of control and information integration application using ADS, ATOS (autonomous decentralized transport operation control system) that is a Tokyo area transport operation control system has been mentioned. Here, its characteristics are summarized as follows:

1. Development of the largest transport operation control system with high traffic density in the world by ADS and control and information integration.
    - (i) As the system architecture, ADS is adopted, and the minimum autonomous unit consists of a station system which is possible to operate independently or to operate in cooperation with the entire line section.
    - (ii) In the conventional centralized operation control system, if any traffic disturbance occurs, manual route control without diagram's update is done which causes to take a long time to recover from diagram disturbance. Also, the passenger information guidance control cannot follow up, and it remained largely disturbed. Therefore, when the traffic diagram is disturbed, it was made possible to easily modify the diagram by the screen of the dispatcher's terminal using diagram-based operation. Then dispatchers can carry out the route control or the passengers' guidance control according to modified diagram. As a result, ATOS with strong against diagram disturbance was realized even in high traffic density railway.
2. Development of maintenance work management system to integrate train operation and maintenance work information and to control interlocking system, for safety management and rationalization of maintenance work
    - (i) Electronic interlocking system by which maintenance work staff directly check the train traffic situation and maintenance work using wireless handy terminals without assistance of dispatchers or station staff, and carry out interlock to protect to invade maintenance work area, were developed.
    - (ii) As a result of this, maintenance staff can directly inquire and control the maintenance work, which contribute maintenance work efficiency up and to reduce maintenance staff load.

3. Development of step-by-step construction technology for large-scale transport operation control system by ADS.
   (i) The minimum autonomous unit of the ADS is a station system, which is possible to operate even by the station system alone. Station systems are constructed step by step then, after completion of all stations, connected with the line section management system. Furthermore, line section system is expanded to whole railway network. This is called step-by-step system construction technology.
   (ii) In order to assure the stable system operation in the system environment where different kinds of operation modes of online system and test system coexist on the network, management of operation mode, data-filtering technology to protect online system from test system, and step-by-step system construction technology were developed.

The ATOS has been completed and applied to the railway network in the Tokyo metropolitan area, which is a high traffic density line section, and the intended purpose has been sufficiently reached. This system has been applied to the 19-line area as originally planned, further expansion of the line section, and replacement of the system of the line section originally introduced is in progress. Here, step-by-step system construction and partial replacement of the system, which is a feature of the ADS, work effectively.

References

[1] K. Mori, S. Miyamoto, and H. Ihara, "Proposal of Autonomous Decentralized Systems Concept", The Journal C of The Institute of Electrical Engineers of Japan, vol. 104-C, no. 12, pp. 303–310, 1984.

[2] K. Kera, K. Bekki, K. Fujiwara, and F.Kitahara, "Assurance Technology in Large Transport Control Systems", Technical Report of IEICE. FTS99-29 (1999).

[3] I.-L. Yen, R. Paul, and K. Mori, "Toward Integrated Methods for High-Assurance Systems", IEEE Computer, vol. 31, no. 4, pp. 32–34, 1998.

[4] F. Kitahara, K. Kera, and K. Bekki, "Autonomous Decentralized Traffic Management System", Proceedings 2000 International Workshop on Autonomous Decentralized System (Cat. No.00EX449), pp. 87–91, 2000.

[5] F. Kitahara, H. Katano, T. Ono, Y. Kakumoto, K. Kikuchi, M. Shinomoto, "Distributed Management for Software Maintenance in a Wide-Area Railway System", Proc. of ISADS'97, Berlin, Germany, 1997.

[6] K. Kera, *et al.* "Autonomous Decentralized System Technology in Large Transport Control Systems", The 40th Japan Joint Automatic Control Conference, 1997.

[7] K. Kera, K. Bekki, K. Fujiwara, F. Kitahara, and K. Kamijo, "Assurance System Technologies Based on Autonomous Decentralized System for

Large Scale Transport Operation Control System", IEICE Transactions on Communications, Vol. E83-B, No. 5, pp. 1085–1093, 2000.

[8] K. Kera, K. Bekki, K. Fujiwara, K. Kamijo, and F. Kitahara, "Assurance Technology for Growing System and Its Application to Tokyo Metropolitan Railway Network", IEICE, Vol. E84-D, No. 10, pp. 1341–1349, 2001.

[9] M. Orimo, K. Mori, and H. Ihara, "Evaluation of System Division Based on Functional Reliability", Proceedings of the Automatic Control Conference, vol. 28, No. 2, 1992.

[10] K. Kera, K. Bekki, K. Kamijyo, and K. Mori, "Adaptive Step-by-step Construction Technique Achieving High Assurance for Tokyo Metropolitan Railway Network", 6th IEEE International Symposium on High Assurance Systems Engineering, HASE 2001, pp. 228–233, 2001.

[11] K. Kera, K. Bekki, K. Mori, I. Masumoto, "High Assurance Step-by-step Autonomous Construction Technique for Large Real Time System", 7th IEEE International Symposium on High Assurance Systems Engineering, HASE 2002, pp. 79–86, 2002.

Chapter 5

ADS fault tolerant property in air-traffic control systems

Tadashi Koga[1]

Abstract

In this chapter, an automatic dependent surveillance (ADS) application with fault tolerant property will be discussed. In narrow sense, fault-tolerant technologies guarantee continuous or uninterrupted operations of systems, units or devices under faulty conditions. In broader sense, fault-tolerant technologies guarantee to accomplish continuous or uninterrupted tasks under restricted conditions. An ADS technology for air-traffic radar systems accomplishes continuous or uninterrupted aircraft surveillance by sharing resources among radars with network. It guarantees safe and effective air-traffic operation in congested airspaces.

5.1 Introduction/Backgrounds

The demand for air traffic in the world is growing year by year. The number of air-traffic passengers in 2017 is increasing 6.8% compared to passengers in 2015. To accommodate this demand, the number of air flights is increasing. The congestion in airspace becomes heavier.

As the number of aircraft is increasing, air-traffic control (ATC) radar systems are required to obtain updated and detailed information from aircraft in order to provide safe and efficient ATC services in congested airspaces [1]. Secondary surveillance radar (SSR) is one kind of ATC radars that provide aircraft position and identification to air-traffic controllers. The growth of air traffic brings two problems to SSR. The first problem is radio-frequency (RF) congestion and system-load growth due to increasing the number of signals between aircraft and SSR ground stations (GSs). The heavy congestion interrupts radar surveillance service. The second problem is the interrogator identifier code (II code) shortage problem. Both problems hinder continuous aircraft surveillance in SSR. Therefore, some countermeasures are required since ATC systems are not allowed to interrupt its services.

[1]Electronic Navigation Research Institute, Japan

To solve these problems, we have been proposed autonomous decentralized surveillance system and the autonomous continuous target tracking technology (ACTTT) [2]. In the proposed technologies, SSRs are connected to a network and autonomously coordinate for continuous tracking by sharing target information.

The proposed technology reduces data traffic in air. Then, the technology enables radars to provide continuous services to aircraft under aircraft-congested environments.

In Sections 5.2–5.4, the evaluation results and practical experiments of proposed system architecture and technology are shown. First, to evaluate the performance of the technologies, we conducted simulation under two to four site environments. The results show that RF congestion and system load can be reduced by proposed technologies. Second, to confirm the II code coordination capability, we conducted practical experiments by using real SSR systems. The results show that proposed technologies achieve continuous aircraft surveillance in the real application.

The remains of the section are organized as follows. In Sections 5.2 and 5.3, SSR and emerging problems are described. In Sections 5.4 and 5.5, the autonomous decentralized surveillance system architecture for radar system and ACTTT are presented. In Section 5.6, the evaluation results and practical experiments are shown in Section 5.7. In Section 5.8, we conclude the chapter.

5.2 Air-traffic control radar system

5.2.1 SSR Mode S

SSR includes two elements: an interrogative GS and a transponder on board of the aircraft. The transponder answers to the GS interrogations giving its range and its azimuth. SSR is different from the primary surveillance radar.

The development of the SSR occurs with the use of Mode A/C and then Mode S for the civil aviation. Mode A/C transponders give the identification (Mode A code) and the altitude (Mode C code). Consequently, the GS knows the three-dimension position and the identity of the targets. Mode S is an improvement of the Mode A/C as it contains all its functions and allows a selective interrogation of the targets thanks to the use of a unique address coded on 24 bits as well as a bidirectional data link which allows the exchange of information between air and ground. Figure 5.1 shows aircraft surveillance by Mode S. Several countries had started deploying SSR Mode S in the 1990s. As of 2016, more than a hundred SSR Mode S are in operation in Europe [3]. In Japan, more than 16 SSR Mode S is in operation. Figure 5.2 shows the radar coverage in Japan.

5.2.2 Mode S surveillance protocol

Currently, most of the Mode S radars are using the multisite surveillance protocol [4,5]. With this protocol, the GS takes two steps to achieve aircraft surveillance.

First, the GS is searching and acquiring targets with all-call interrogation. The GS uses all-call interrogation to acquire aircraft newly appearing in the coverage. The GS transmits all-call interrogation periodically. The all-call interrogation does

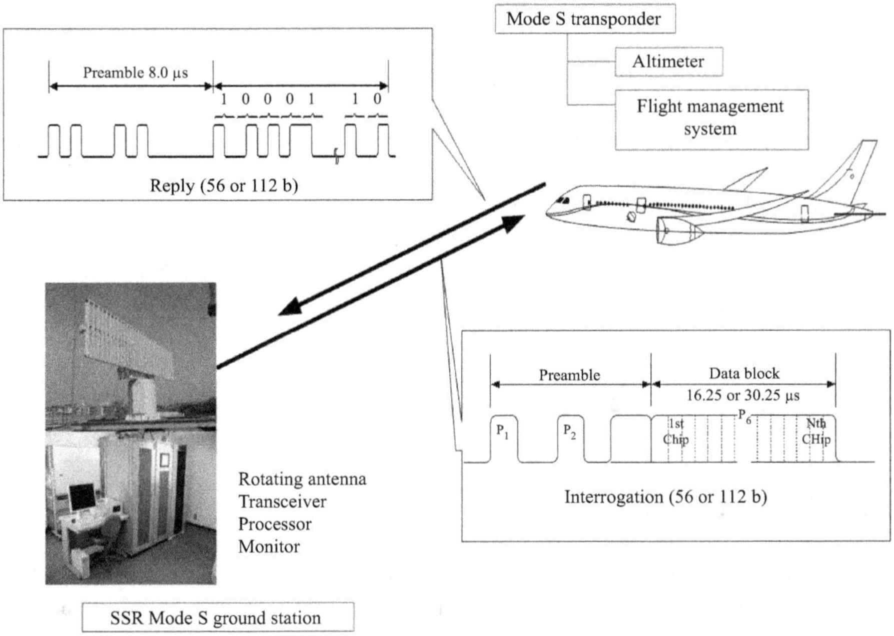

Figure 5.1 Aircraft surveillance by SSR Mode S systems

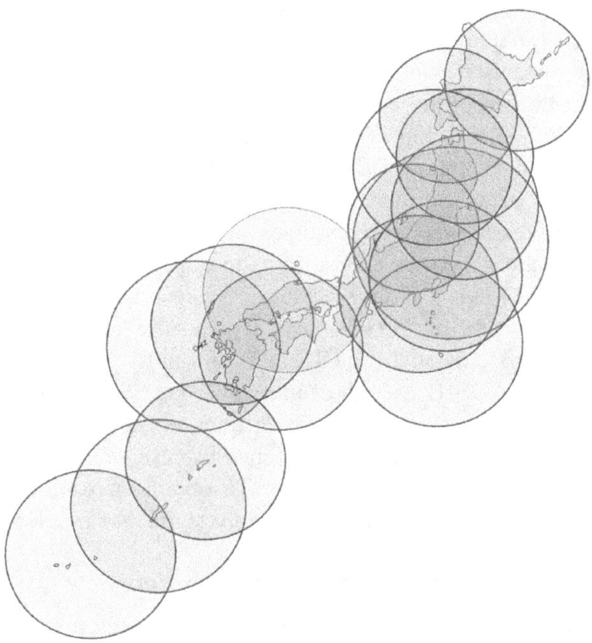

Figure 5.2 The coverage of air-traffic control radars in Japan

not contain destination address. All targets send a reply when receiving an all-call interrogation. By receiving all-call replies from targets, the GS acquires rough target positions and identification numbers.

Second, after acquisition, the GS start tracking targets with aircraft information. The GS is scheduling interrogation time and expected reply arrival time. The GS transmits roll-call interrogation at the time that radar antenna is facing the target. The roll-call interrogations contain the destination address. Several targets in the antenna beam receive interrogations but only one target with the designated address sends back a reply to the GS. Roll-call interrogations contain lockout command, which orders aircraft not to send back all-call replies for 18 s. By lockout, the GS suppresses all-call reply from acquired aircraft.

The multisite protocol is designed for independent GS operation. Therefore, in the overlapping coverage, each GS independently performs all-call and roll-call interrogations.

5.3 Emerging problems

Because most of the SSRs are operated independently with separated networks, two emerging problems in SSR Mode S should be considered with the rapid increase in air traffic and dynamically changed situations.

5.3.1 RF congestion problem

There are two changes in SSR Mode S operational environments. The first is downlink aircraft parameters (DAPs). There is a growing demand for SSR Mode S data link. In the 1990s, the European states launched DAPs project which use Mode S data link [4]. DAPs provide air-traffic controllers and systems with additional aircraft information such as selected altitude, roll-angle, and magnetic heading. To downlink a parameter, the GS needs to send an interrogation. As the number of DAPs equipped aircraft increase, the number of replies increases. If multiple GS independently downlink the same parameters, excessive replies deteriorate the RF signal environment [6].

The second is the appearance of automatic dependent surveillance-broadcast (ADS-B) system. ADS-B is a promising surveillance system that will play a major role in ATC in the future. In ADS-B, aircraft measures own precise positions by satellite navigation systems such as Global Positioning System, and periodically broadcasts messages. ADS-B uses the same signal as SSR Mode S reply signals. ADS-B uses random access protocol, which does not control message transmissions. Therefore, it is susceptible to interference by SSR replies. ADS-B should be operated in radio-signal silent environment. To obtain an ADS-B friendly environment, SSR Mode S is required to reduce reply signals as much as possible.

5.3.2 Interrogator identifier shortage problem

II code is set in interrogation and reply signals. It makes possible transponders identify the source site of interrogation and GS distinguish the destination site of

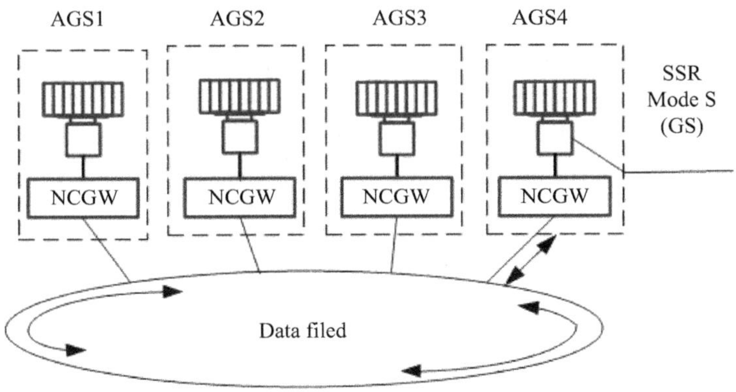

Figure 5.3 Autonomous decentralized Mode S network

reply. In International Civil Aviation Organization standards, 4 b space is prepared in interrogation and reply signals for II code [1]. Fifteen II codes are available for GS. II code should be assigned without conflicts between neighbouring GS that have overlapping coverage.

As the number of SSR Mode S GS increases, SSR operator is not able to assign II codes without conflicts between neighbour GS. If the same II code is assigned to neighbouring Mode S GS without any coordination, the GS is not able to achieve continuous aircraft surveillance in overlapping area. For instance, GS is not able to acquire inbound aircraft in overlapping area, what is worth that both GSs are not survey aircraft at the boundary of coverage. To keep continuous surveillance in overlapping coverage by using the same II code, the GS is required to have II code coordination function.

5.4 Autonomous decentralized surveillance system

Safe and effective operation of the ATC under congested air traffic relies on accurate and timely airspace situational awareness supported by surveillance systems. Most of the current SSR surveillance operations are separated by different networks. Therefore, it is necessary to integrate the separated SSR systems to meet the requirements of ATC applications. To construct a tracking information sharing infrastructure that enables seamless sharing of target report messages and other relevant information through network can satisfy the surveillance system needs and improve the tracking performance. Some Mode S network approaches are proposed in [5]. They adopt central processor architectures which are difficult to satisfy online expansion, online maintenance and fault tolerance requirements. To assure the continuous tracking with reduction of data traffic in air, an autonomous decentralized surveillance system architecture is proposed, as shown in Figure 5.3.

The system is composed of multiple autonomous ground sites (AGSs) and a data field (DF). There is no central processor in the system.

5.4.1 Autonomous ground site

The AGS is composed of two subsystems, Mode S GSs and network control gateways (NCGW). GS which has its own surveillance area obtains aircraft position, identification number, time and other information by sending interrogations and receiving replies. NCGW sends the target report message to the DF. At the same time, NCGW stores the information of all detected targets and the target report messages from its neighbour AGSs which have the overlapping coverage areas to control the operation of GS.

5.4.2 Data field

The AGSs in the system are connected through DF that serves as a communication medium of coordination between the AGSs. It can be a local area network (LAN) or LANs connected by wide area network. All necessary data is broadcast into the DF, where the data logically circulates in the DF. The data moves around the application modules in the node. This results in excluding the mutual dependency among AGSs.

5.5 Autonomous continuous target tracking technology

The technology runs in AGS and reduces replies in the overlapping coverage. Figure 5.4 shows the target relies at the overlapping coverage. In the multisite protocol, multiple GSs are monitoring a target in the overlapping coverage. This multiple surveillance produces excessive interrogations and replies. And in the multisite protocol, in order to start roll-call interrogation, GS needs the rough position and identification number. On the contrary, ACTTT obtains them from neighbouring GS, then it starts roll-call interrogation without all-call interrogation. Therefore, ACTTT can eliminate all-call replies.

Moreover, ACTTT coordinates interrogation in overlapping coverage with neighbouring GS. GS stops acquiring a target if the other GS is acquiring the target. Therefore, ACTTT can reduce roll-call replies in the overlapping coverage.

5.5.1 Autonomous data sharing

AGSs share target information by the target report messages. The AGS sends to the network a target report message for each target per scan while AGS is monitoring targets. The AGS receives the target report messages from the neighbour AGSs through the network. Then, the AGS stores those messages in the target tables. The AGS also stores the target report messages of own site in the target table. The AGS updates these tables when receiving the target report message or maintain the contents of the tables periodically.

5.5.2 Autonomous judgement

The autonomous judgement determines targets and timings with balancing the surveillance load of each AGS.

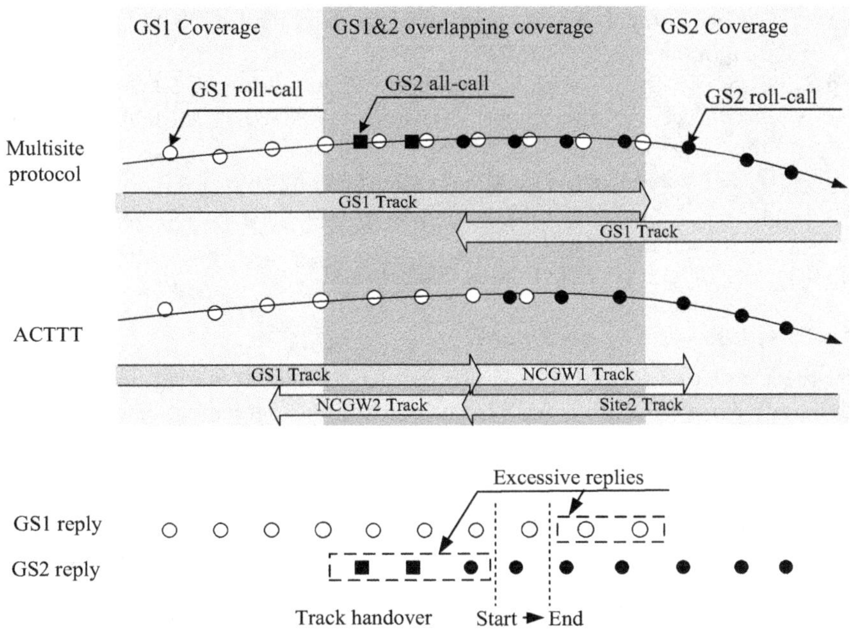

Figure 5.4 Surveillance in the overlapping coverage

The autonomous judgement is composed of following two sub-processes:

1. Surveillance load assessment
   The AGS computes the surveillance load $S(j)$ of own and neighbour AGS in every Tas seconds. First, AGS computes the surveillance load $T(i)$ of target #i as follows:

$$T(i) = \frac{2 \cdot R(i)}{C} + T_{delay}$$

   where $R(i)$ is the range between the target #i to GS, C is the speed of light, and T_{delay} is the processing delay in the aircraft transponder or AGS.
   Then, the surveillance load at AGS #j $S(j)$ is computed as follows:

$$S(j) = \sum_{i=0}^{N} T(i)$$

   where N is the number of targets monitored by AGS #j for the last one rotation period.
2. Acquisition target decision
   The acquisition target decision determines targets that AGS is going to acquire at the following scan. To simplify explanations, we consider a network with two AGSs: AGS1 and AGS2.

[Step 1] AGS1 compares the surveillance load of own AGS S(1) and that of neighbouring AGS2 S(2).

[Step 2] If S(1) < S(2), AGS1 try to acquire targets from AGS2. AGS1 chooses the target with the biggest T(i) from the AGS2 target table. Lets name this target as Tgt1.

[Step 3] AGS1 computes S(1) and S(2) after handing over Tgt1.

[Step 4] If S(1) > S(2), AGS determines to acquire Tgt1 at the next scan. Then, AGS1 returns Step1.

[Step 4′] If S(2) ≧ S(1), AGS1 stops this process.

5.5.3 Autonomous agreement

After the acquisition target decision process, AGS1 quickly starts acquiring targets without any permission from AGS2. The autonomous agreement determines AGS that stops monitoring targets.

1. Consistent with judgement
   When AGS1 successfully acquires targets, AGS1 adds those targets to the AGS1 target table. At this time, AGS1 increments the target acquisition number (TAN) and registers it to the table. AGS checks the surveillance status (SS) of the target. Then, AGS sends target report messages with TAN and SS to network.
2. Consistent with judgement
   By receiving the target reports with the greater TAN than that of own AGS2 target table, AGS2 notices that the AGS1 starts roll-call interrogation. AGS2 updates the target tables. Then, AGS2 stops monitoring targets if the target satisfies the following conditions:
   (i) The SS is core.
   (ii) The TAN of AGS1 is bigger than that of AGS2.
   (iii) The surveillance load of own AGS2 is larger than that of AGS1.
   AGS2 stops acquiring the target and drops the target from the own target table. However, AGS2 obtains target information from AGS1.
3. Inconsistent with judgement
   The decision timing of the TAN and the surveillance load is different in each AGS. Those values are inconsistent in two AGSs. If two AGSs decisions are inconsistent, both AGS continue monitoring the target. But, if the TAN of the AGS1 is equal to that of AGS2, both AGS obeys the following rules. AGS1 checks the Site IDentification number (SID) where SSR operator assigns a unique number to each AGS. The AGS with the greater SID takes the responsibility of monitoring the target. The AGS with the smaller SID stops monitoring the target at the next scan.

5.5.4 Autonomous boundary target handover

If targets are not handed over until, they are close to the boundary of the coverage, the AGS force to hand over those targets to neighbour AGSs regardless of the surveillance load before the AGS loses sight of targets. Otherwise, the AGS lose targets at the boundary for a while. It interrupts continuous target tracking.

The autonomous boundary target handover achieves continuous target handover at the boundary of the coverage.

1. Boundary detection
   The AGS detects that targets are close to the boundary by referring to the surveillance map. The surveillance coverage map is composed of three-dimensional cells. A cell contains the SS at the cell area. There are four statuses in the surveillance map. Therefore, a coverage is classified into four areas: surveillance core area, surveillance boundary inside area, surveillance boundary outside area, and out-of-coverage.
2. Boundary target status notice
   The AGS checks the SS when the AGS obtains target information by roll-call interrogation. If a target is in the boundary inside, the AGS notice the fact by setting the SS to "boundary-inside" in the target report messages. Neighbour NCGWs know that the targets are in the boundary inside area by the target report message.
3. Boundary target acquisition
   Neighbour AGS has to acquire targets at the following scan regardless of the surveillance load of own site.
4. Double boundary target
   After the boundary target acquisition if the AGS detects that the acquired target is in the boundary inside, the AGS notice the fact by setting the SS to "boundary-inside" in the target report message. The neighbour AGS does not stop but continue acquiring.

5.6 Simulation

To evaluate the RF performance of the proposed technology, we conduct computer simulations.

5.6.1 Model

In this simulation, we accomplish simulations by up to four GSs with 200 NM coverage and 4 s scan period. Figure 5.5 shows the locations of GSs.

We randomly deploy 500–2,000 aircraft in a simulation area of 800 NM by 800 NM. Each aircraft is moving at the speed of 720–880 km, which is typical cursing speed of the commercial aircraft. An aircraft changes its direction randomly every 2,000 s. One flight of aircraft is 16,000 s (4,000 scans).

5.6.2 Simulation results

5.6.2.1 RF load

Figure 5.6 shows the variation of the number of reply signals in one simulation. The black line is multisite and grey line is the proposed technology. The number of reply signals with the proposed technology is always less than that of multisite surveillance. The proposed technology reduces reply signals through the whole period.

Figure 5.7 shows the relation between the number of aircraft and the total number of reply signals in the simulation. The black line shows the number of replies with

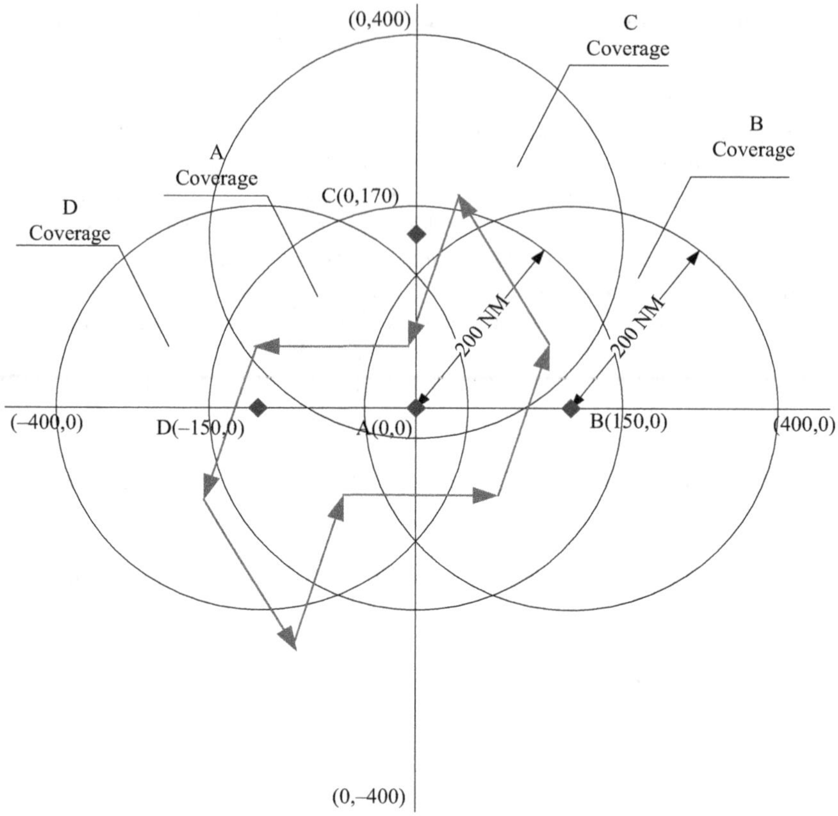

Figure 5.5 The ground station locations in the simulation

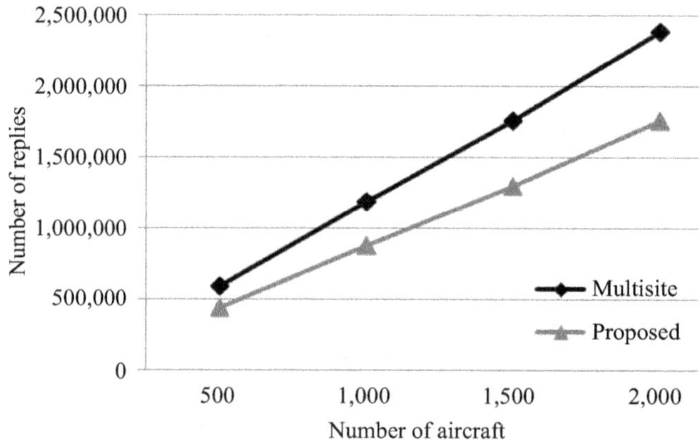

Figure 5.6 The relation between number of aircraft and replies

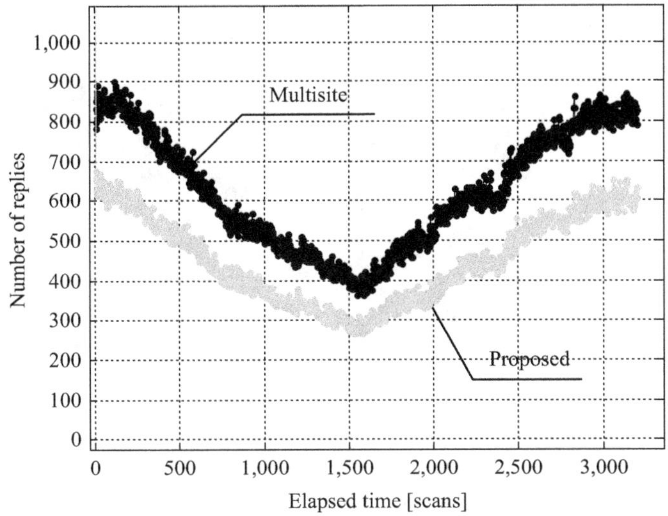

Figure 5.7 The variation of the number of replies

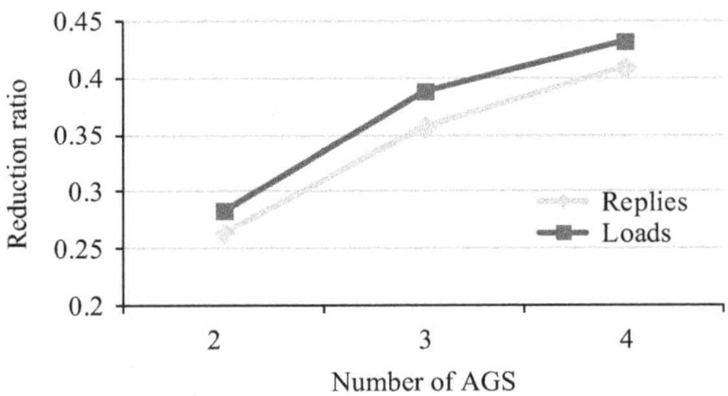

Figure 5.8 The variation of the number of replies

multisite surveillance protocol. The grey line shows with the proposed technology. Under 2,000 aircraft and two AGS environment, the number of reply signals in the multisite (N_m) is 2.38 million. On the other hand, that of proposed technology (N_p) is 1.75 million. Here we define the reduction ratio $R_r = (N_m - N_r)/N_m$. The R_r with two sites and 2,000 aircraft is 26%. Then, we achieve three and four site simulations and compute the reduction ratio under environments. The reduction ratio is 36% in three sites and 41% in four sites, as shown in Figure 5.8. The reduction ratio is increasing as the number of sites is increasing.

5.6.2.2 System load

Figure 5.9 shows the system load under two site and 2,000 aircraft simulation during a whole simulation period. The *X*-axis shows the simulation elapse time, and the *Y*-axis

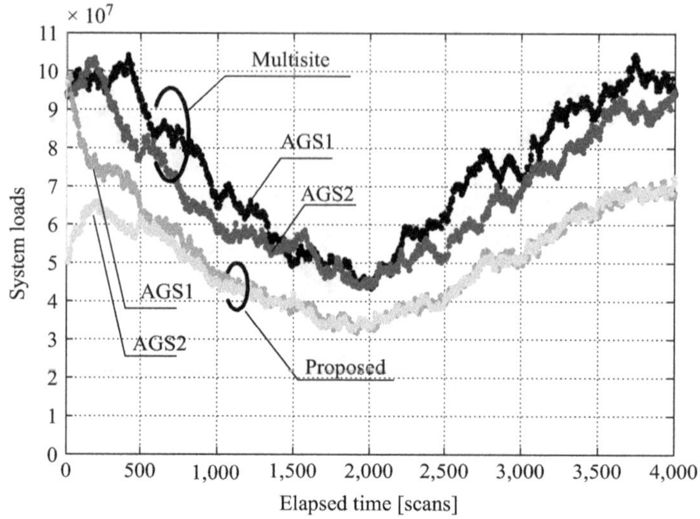

Figure 5.9 System load

shows the system load. The dark lines show the system load by multisite. The system loads of two AGS have a difference and vary time to time. Sometimes, AGS1 system load is heavier than that of AGS2. On the other hand, the system loads of two AGS with the proposed technology are almost equal. In addition, both system loads are lighter than that of multisite. Here we define the system load reduction ratio, $R_s = (L_m - L_p)/L_m$ is the total sum of system loads in AGSs by multisite. L_p is that of the proposed technology. The R_s with two sites and 2,000 aircraft is 28%. Then, we achieve three and four site simulations and compute the reduction ratio under environments. The reduction ratio is 39% in three sites and 43% in four sites. The system load reduction ratio is increasing as the number of sites is increasing.

5.7 Practical experiments

5.7.1 Network structure

To evaluate by monitoring real aircraft in the air, Electronic Navigation Research Institute developed an experimental network system with ACTTT. The network is composed of two GSs. One GS is located in the ENRI headquarter in Chofu, Tokyo. The other GS is located in the ENRI branch office in Iwanuma, Miyagi. Currently, several functions have already been implemented.

The coverage of two GSs at 40,000 ft is shown in Figure 5.10. The plus sign (+) indicates the locations of Mode S GSs. A black solid line shows the edge of the coverage of the Chofu GS, and the grey line shows that of the Iwanuma GS. The distance between two GSs is approximately 160 NM.

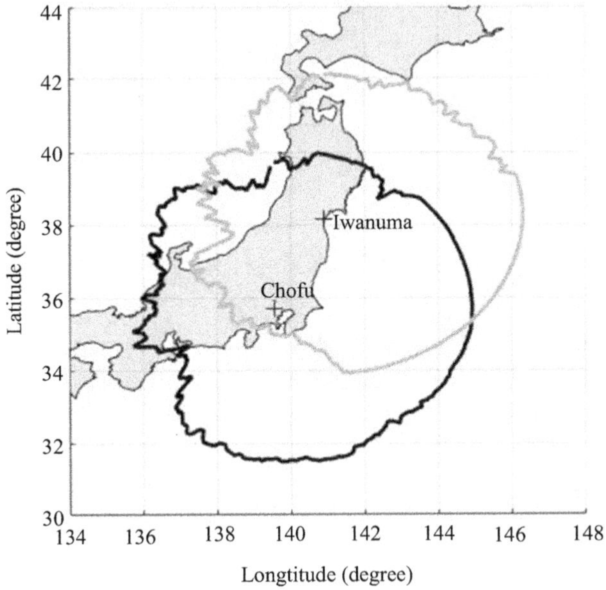

Figure 5.10 The coverage of ADCT experimental network

Two GSs are connected through ground network. Two NCGWs exchange target information with each other. In this experiment, Iwanuma and Chofu GS are assigned with the same II code.

5.7.2 Experiment results

Figure 5.11 shows the tracks by ACTTT. The grey and black solid lines show a track of aircraft flying from north to south. The grey line means that Iwanuma GS is acquiring the target and the black line means that Chofu GS is acquiring the target. Iwanuma AGS is acquiring the target. Then target enters into overlapping area. Two AGSs calculate the system load by using aircraft information from both AGSs. Chofu AGS determines to acquire the target and starts surveillance. Then Iwanuma AGS determines to hand over the target to Chofu AGS and releases the target. Finally, Chofu AGS acquires the target. There is no interruption during the tracking.

To show the validity of ACTTT, the aircraft surveillance by multisite (conventional) protocol is adopted. Figure 5.12 shows the tracking of aircraft flying from north to south. From beginning, Chofu GS acquires the target. However, when the target exits in the border of Chofu coverage area, both GSs are not able to monitor the target. After a few tens of seconds, Iwanuma AGS acquires the target. The practical experiments show that proposed system and technology assure continuous target tracking even when adjacent GSs keep the same II code.

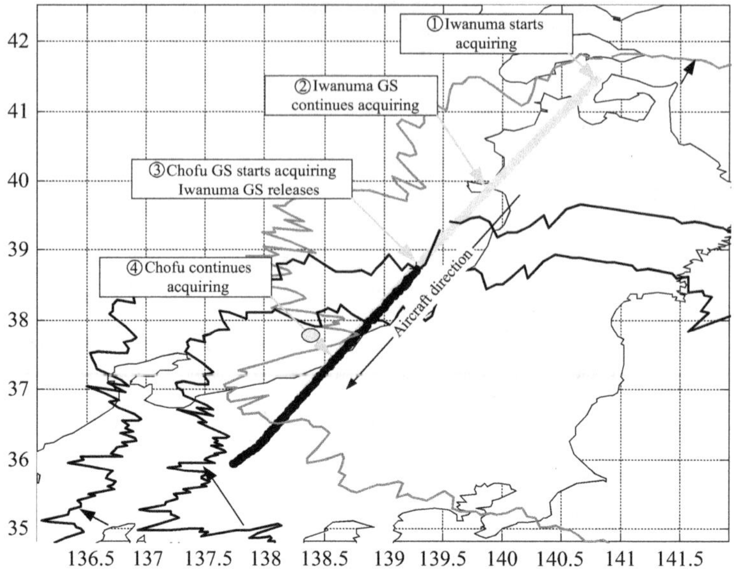

Figure 5.11 Target track by ACTTT

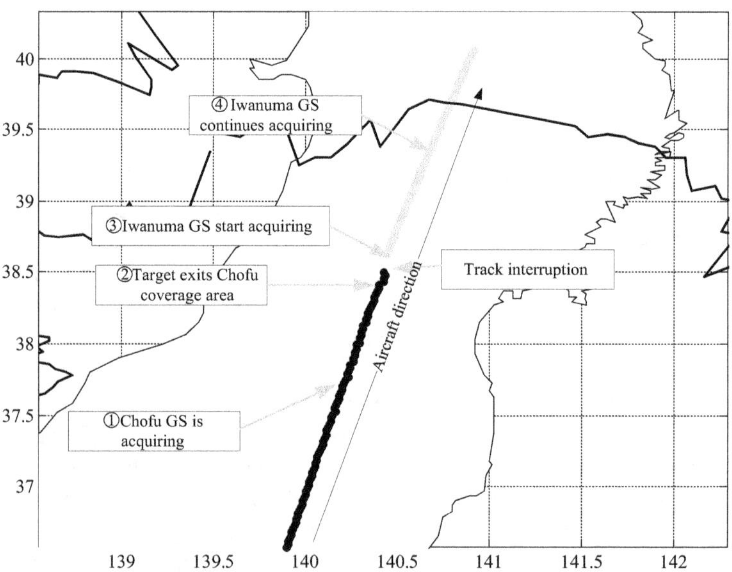

Figure 5.12 Target track by multisite protocol

5.8 Conclusion

In this chapter, an ADS technology with fault-tolerant property is shown. The proposed autonomous decentralized surveillance system architecture and ACTTT are evaluated by computer simulations and practical experiments.

The technologies reduce data traffic in air–ground communication with keeping continuous target tracking. The simulation results show the proposed technology reduces approximately 40% data traffic compared to the multisite protocol under four site environments. The simulation results show that as the number of GS increases, the more RF congestion and system load are reduced.

Moreover, the practical experiments of real SSR systems show that proposed technologies enable GS to hand over targets and achieve continuous tracking even in the situation that two adjacent GSs keep the same II code. The proposed technologies also guarantee to accomplish continuous or uninterrupted tasks under restricted conditions.

References

[1] Y. Bar Shalom, "Multitarget-Multisensor Tracking: Applications and Advances," Artech House, ISBN 978-0890065174, Norwood, 1992.

[2] T. Koga, X.D. Lu, K. Mori, "Autonomous Continuous Target Tracking Technology for Safety in Air Traffic Radar Systems Network," IEEE Proc. of 6th International Symposium on Service-Oriented System Engineering (SOSE2011), pp.235–240, USA, 2011.

[3] EUROCONTROL, "Concept of operations Mode S in Europe 2.0," EUROCONTROL, 1996.

[4] ICAO, "Aeronautical Telecommunications Annex 10 vol. IV," Fourth edition, ICAO, 2007.

[5] ICAO, "Manual of SSR Systems," Doc 9684, Second edition, ICAO, 1998.

[6] M. Lazarus, "The Great Radio Spectrum Famine," IEEE Spectrum, Vol. 48, Issue 10, pp.26–31, 2010.

Chapter 6

An agile manufacturing model based on autonomous agents

Leo van Moergestel[1]

Abstract

This chapter describes the implementation of an agile autonomous agent-based manufacturing system based on a set of cheap production machines placed in a grid. This grid contains production machines, represented by agents, capable to perform certain production steps and also an infrastructure for transporting the products to be made. Products to be made are also represented by agents. Many different products can be made in parallel, each product having its own sequence of production steps and its own path along the production machines. The whole manufacturing is based on interaction of autonomous agents living in a distributed environment. This chapter discusses the basic design considerations and includes a description of a simulation model of the transport system used during production. The chapter ends with a discussion of using agent technology during the whole life cycle of a product, where a life-cycle agent becomes the basis of Internet of Things technology.

6.1 Introduction

The industry has had several revolutions. The first revolution was the use of steam power to facilitate production. The second revolution was the introduction of production lines based on the use of electrical energy and resulting in mass production. The rise of computer technology resulted in the third revolution. Many production tasks were automated, programmable logic controllers, distributed control systems and robots were introduced on the production floor. The latest revolution is the integration of information technology in the production process as a whole. This has been described by the term industry 4.0 [1] or cyber physical systems [2].

In industry 4.0, modern production systems are influenced by new demands such as time to market and customer-specific small quantity production. In other

[1]HU Utrecht University of Applied Sciences, Institute for ICT, The Netherlands

words, the transition time from product development to production should be minimal and small quantity production must be cheap. To fulfil these requirements, we need to develop new production methods. Such a new approach means new production hardware as well as co-designed software.

At the Utrecht University of Applied Science, we have developed special production platforms that are cheap, agile and easily configurable [3]. These platforms can operate in parallel. We call these platforms equiplets and a collection of these equiplets is called a production grid. The idea behind the concept is that we need a production system that is capable of producing a lot of different products in parallel. This is what we call multiparallel manufacturing.

The software infrastructure for such a production grid is highly responsible for the agile and diverse way of production. In this chapter, we present a possible realization of the software infrastructure. This model is based on agent technology. Though we based our model on our own designed production hardware, the agent-based approach can be useful in other production environments as well.

The concept of using a collection of cheap machines is comparable to the research done around 1980, where the focus was on using cheap microprocessor-based computer systems to cooperate in a multiprocessor or multicomputer environment but because the focus is now on manufacturing real physical products, there are many differences and also many specific problems to be solved to make the concept work.

6.2 Manufacturing concepts and technologies

This section is dedicated to manufacturing technologies. The main reason to have this section is to give a background and reference models for the concepts introduced in the later part of this chapter. The section contains the following subsections: production concepts, push-driven versus pull-driven manufacturing, lean manufacturing and agile manufacturing.

6.2.1 Production concepts

In this chapter, the words production, manufacturing and assembling are used intermixed. There are some subtle differences in the meaning of these words.

- Production is used as the most generic term. It can be used for material products as well as things like software, ideas, theories etc.
- The word manufacturing has its origins in Latin: 'manus' meaning hand and 'facere' meaning to make. Though it literally means make by hands, it is now also applied to situations where material things are made using machines.
- Assembling is used for the type of manufacturing where components or subparts are put together using several possible techniques to make the final product.

When we take a closer look at making things, we may distinguish three approaches for production.

1. Making a single product. This is normally the case in situations where a specific one time product is needed. Examples are ships, some buildings and special items, where a single unit is needed.
2. Continuous production is a type of production, where there is a continuous stream of output. This type of production is encountered in the chemical industry.
3. Batch production. This type of production is mostly used to make a number of similar products. Normally, the number is rather big as we shall see shortly.

When we look at the production infrastructure, especially the infrastructure for batch processing, again three different approaches can be seen.

1. Dedicated production line. This is a concept that is widely used and fits the need for cheap mass production. The machinery that is used is dedicated to do one specific task in the production.
2. Flexible manufacturing system (FMS). An FMS has, as its name suggests, some flexibility to react to changes. The most well-known change is changing to new product types. This change can be incorporated in the production machines. This concept was introduced in the time that cost-effective computer numerical control machines became available. Another type of flexibility is routing flexibility. This concept is based on the fact that there might be several machines to perform the same operation on a product. This introduces choices for the product in following the production line.
3. Reconfigurable manufacturing system (RMS). An RMS is a manufacturing system that is designed for fast changes, both in hardware as well as software components, in order to quickly adjust production capacity and functionality in response to sudden changes in market or in changes in requirements.

The main goal of RMS is to achieve cost-effective responsiveness. An RMS can adapt easier and faster to changes than an FMS. The drawback is that RMS is more complicated.

To achieve a high level of reconfigurability, the system should meet the following characteristics in its design [4]:

- Modularity. The basic components are modules that can easily be exchanged or replaced.
- Scalability. To adapt to changing demands, scalability is an important characteristic.
- Integrability. Modules can be easily integrated in the system.
- Convertibility. Changes to the production system are easy to achieve.
- Customization. Adaptation of the system to specific needs is possible.
- Being diagnosable. To prevent that searching for failing modules takes a long time, modules and the system itself should be diagnosable.

The cost effectiveness of RMS is achieved by designing a system with an adjustable structure, and around a part family. An adjustable structure enables system scalability in response to market demands and system/machine adaptability to new products. The structure may be adjusted at system level by adding new machines and at machine level by adding/removing machine software [5].

6.2.2 Push-driven versus pull-driven manufacturing

Standard mass production is mostly push driven. It means that the expected purchase of a product is anticipated for and products are pushed into the market.

Pull-driven waits for demands for a product and at the moment the production is started, it is sure that the market will accept it.

6.2.3 Lean manufacturing

Lean manufacturing has its origins in Japan. The production at Toyota has been the model for lean manufacturing [6]. The concept is based on five steps.

1. What is the value of the product from the customers' perspective?
2. Discover where in the production process this value is added.
3. Determine the waste in the process, remove it and shorten the duration of the lead time.
4. Apply pull-driven production instead of push-driven production.
5. Keep the waste away and try to optimize the process.

The challenge is to discover what really adds value to the product. For example, keeping a lot of products or half products in stock is considered waste. From the client perspective, it does not matter how big the stock is. The time the client has to wait for his product is a very important issue. So, the production stream should be optimal with a minimum of internal delay. The aforementioned steps result in a set of best practices.

6.2.4 Agile manufacturing

In response to the customers demand, manufacturing companies have to focus on low cost, high quality and rapid responsiveness [4]. A new paradigm called Agile Manufacturing was invented. It focuses on agility, i.e. the quick and accurate response to changes in the market and technology while controlling production cost. A slightly adapted definition from Goldman [7] is as follows:

Definition [Agile manufacturing] *An agile manufacturing system is a system that is capable of operating profitably in a competitive environment of continually and unpredictably changing customer requirements.*

In the next section where standard production automation is discussed, we will discover that this type of production is not agile by itself.

6.3 Standard production automation

Standard production software is mostly designed for batch production or continuous production. Continuous production can be considered as an endless batch.

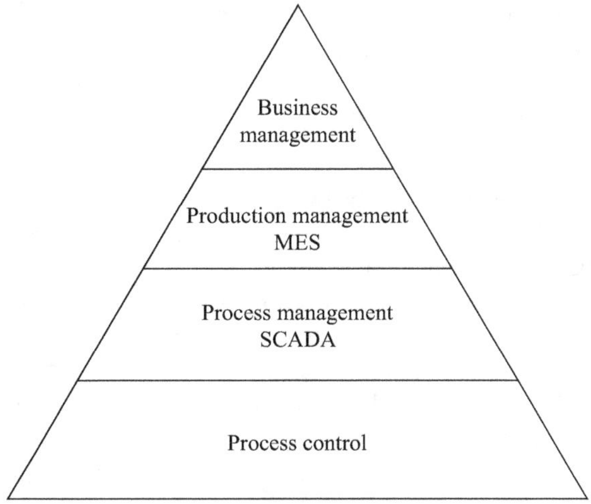

Figure 6.1 Automation pyramid

These production approaches are characterized by the fact that it is bulk processing. Many items or a large quantity of the same products are produced. We will discuss standard automation software, the way it is used, its properties and its shortcomings.

6.3.1 Standard automation software

The software for standard production systems is based on a layered model. This model is mostly referenced as the automation pyramid [8] (Figure 6.1).

In the following, we will give a short explanation of the layers in this pyramid:

1. At the top, we find the business management software. This is the software level where the orders for production come in and where the connection with clients, suppliers and other important things in the outside world is handled.
2. The production management layer software is mostly covered by software systems called MES. MES is an abbreviation of Manufacturing Execution System. This software enables high-level control over production facilities in a broad sense.
3. Process management layer is the software that supervises the process control devices. It will also collect production data. The software in this layer is mostly referred to as SCADA, which is an abbreviation of supervisory control and data acquisition.
4. The process control layer is responsible for the actual production process itself. In an automated environment, the software controls motors, heating devices, robot arms, etc.

6.3.2 Properties of standard automation

The standard production automation model is designed for producing large quantities of the same product. Normally, these products are produced in batches.

Definition [Batch process] [9] *A process is considered to be a batch process if the process consists of a sequence of one or more steps or phases that must be performed in a defined order. The completion of this sequence of steps creates a finite quantity of finished products. If more products are to be created, the sequence must be repeated.*

To make a product (that is a real hardware thing, not software or a service), one needs raw material and actions to work on this raw material. In many cases, the raw materials are actually components that are used as parts of the final product. Normally, these components are also produced by a production process.

As stated earlier, the production automation model is designed for producing large quantities of the same product. Two approaches of batch production exist. To understand what the two approaches are, consider a product that is made by a sequence of actions performed by machinery or craftsmen. These actions are also called production steps. In Section 6.4.2, a precise definition of a production step will be given.

Here we mention two different approaches.

1. Stepwise approach: The production starts and the first action of the sequence is performed. This leaves a set of incomplete products. Then, the production environment is changed. This could be an adjustment of the machinery to perform a new type of action or the use of other tools by craftsmen to perform an action. This is repeated for all necessary steps, and in the last step, the final product is created. The approach is also appropriate for production of one single product. Because we need storage for the intermediate sets of incomplete products after every step, the approach is used for small-scale production. A second property of this type of production is that there is a long delay between start of production and final completion of the products. The investment on machinery however is lower than in the next approach, because we adapt (as far as possible) the machines and/or craftsmen after each step to the next step.
2. Pipeline approach: In this approach, all machineries (or craftsmen with specific tools) are available, and the product to be made is handed over to the machine that is capable to perform the next action for the production.

For both the cases, there is an optimum for the size of a batch. Though this size depends on several factors, the easiest approach is to consider the cost of the overhead of batch switching and the cost of storage and inventory.

The optimum batch size is given by the point where the storage and inventory cost plus the batch cost for a given size are at a minimum. In practice, this is more complicated as the price of raw material may fluctuate and other parameters such as market demand influence the optimum [10].

Between these two types, there is also a hybrid approach where the pipeline approach is used to make half-products that are collected to form a set and then handed over to another pipeline that will build the actual product. This situation occurs when production platforms produce components that are used as 'raw material' by other production platforms (that could be owned by a different company).

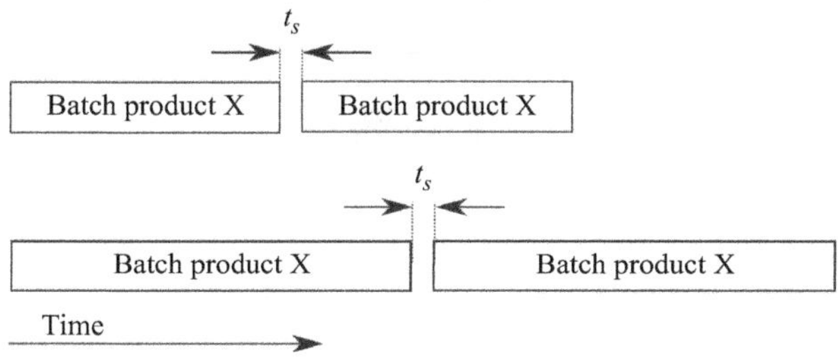

Figure 6.2 Batch switch overhead when switching to a new batch of the same product

6.3.3 Batch switches and new products

A batch is a quantity of products produced without interruption. The size of a batch should be big enough to be cost effective but should be limited because of maintenance and because of production supply fill-up. For every different batch, a software configuration of normally the lower two layers of the automation pyramid should be available.

The transition from one batch to the next one is called a batch switch. There are two types of batch switches.

1. A switch between batches of the same product and
2. A switch to a batch of a different product.

The overhead of a switch of the first type is not so large, but still some time is required. This time is normally used for preventive maintenance of the production equipment, for filling up component trays, etc. What also should be done is allocate resources for the next batch and adjust the software configuration of the lowest two layers to these newly allocated resources. In Figure 6.2, we use *ts* to denote the time for this switch. As can be seen in the figure, a larger batch introduces relative smaller overhead because is mostly independent of the batch size.

A switch of type two takes a longer time because in addition to *ts* we also need time *tp* to reconfigure the software on the lower two layers and perhaps the hardware of the lowest layer (see Figure 6.3). The production equipment will get different software to operate, and also the SCADA system needs to be reconfigured to match the new situation. Sometimes, it means that a production platform or parts of it are unavailable for some time because of the reconfiguration that should also be tested of course. As a consequence, batch switching of this type introduces a lot of overhead and production (and the final product) is cheaper if we produce batches of the same product and not a sequence of batches for different products. There are situations where switching between different batches is incorporated in the production of a manufacturing plant. Consider for example the food industry. To produce a batch of peanut butter, a production line could be set up. There are however variants to standard peanut butter (with honey, pieces of peanuts, chocolate, etc.). Normally, the market for these variants is not the same, meaning that the

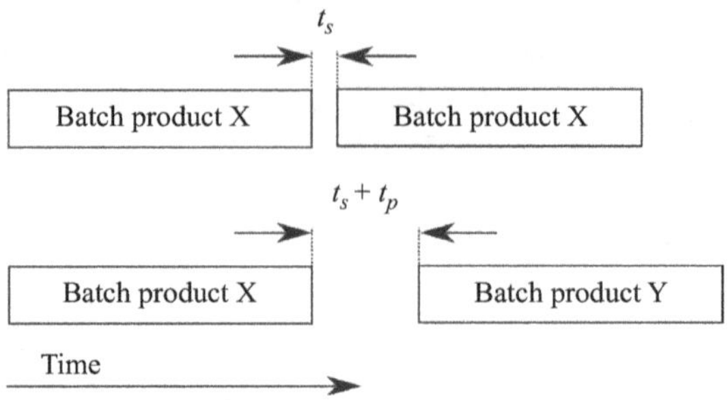

Figure 6.3 Batch switch overhead when switching to a batch of a different product

amount of production of these variants could be lower. It is in that case not a good idea to set up separate production lines for these variants. It is more cost effective to produce a batch of variant A, next a batch of variant B, and several batches of the most popular standard product. The same production line should be used in that situation, so switching between different batches becomes compulsory. All batches are also big by themselves to be cost effective, because even switching to a new but similar batch introduces overhead.

The introduction of a new product means development and testing of new configurations for the three lower software layers. Sometimes, we need new software for our production system or even new hardware. The transition from product development to producing can be time consuming, because in the development stage, we did not use the production equipment that is used during the real production. So, after development of a new product, an extra step is needed to switch to production on the production equipment. To test this transition, we also need to use the final equipment so it will not be available for normal production for some time.

6.3.4 Summary

Our investigation in properties of standard batch production automation can be summarized in four properties.

1. Huge batches for cost-effective production,
2. Small overhead introduced by batch switching of the same product,
3. Large overhead introduced by switching to another product and
4. Hard transition from product development to product production.

After identifying the four properties of standard batch production, we mention six weak points or shortcomings of standard production.

1. Standard manufacturing is suitable for mass production, but small quantities or even single unique products according to user requirements are not advisable.
2. Standard manufacturing uses costly dedicated production machinery, which should be used at a high load to make it cost effective.

3. Pipelined batch production is vulnerable for failures of manufacturing equipment.
4. Most standard manufacturing is push driven. This can result in overproduction and waste of money, materials, resources and time.
5. The transitions from concept to product to mass production take several steps and might take too long to be competitive in the market.
6. Most SCADA and MES implementations are not well suited for decentralization and could introduce a single point of failure.

In the next section, equiplet-based manufacturing will be introduced. The goal of this type of manufacturing is not to be a replacement of standard production. The goal is to offer a cost-effective solution for the situations where standard production is inadequate.

6.4 Equiplet-based production

The basic production platform for the new agile production system is the equiplet. The concept of an equiplet has been introduced by Puik [3].

Definition [Equiplet] *An equiplet is a reconfigurable manufacturing device that consists of a standard base system upon which one or more frontends with certain production capabilities can be attached.*

The frontends give the equiplet the possibility of production. It means that at the moment the frontend is attached to the equiplet, and certain production steps can be accomplished.

Every frontend has its specific set of production step capabilities.

A picture of an equiplet with a delta-robot frontend is shown in Figure 6.4. A delta-robot is a special type of robot, which can perform fast pick and place actions. With this frontend, the equiplet is capable of pick and place actions. A computer vision system is part of the frontend. Using this vision system, the equiplet can localize parts and check the final position they are put in.

The first field of application of the concept was building micro-devices with a three-dimensional structure (in contrast with the two-dimensional approach used in placing electronic components on printed circuits). In this case, one could think of steps to pick up a component and place it at a certain position (pick-and-place). Applying adhesive could be an option for this pick-and-place step. A local computer system is available on the equiplet for running control software depending on the applied frontends. Software configuration and management of an equiplet should be simple and easy.

As already mentioned, we call a collection of equiplets a production grid or in short a grid. The equiplets in a grid do not necessarily have the same frontend. Some frontends are unique, other frontends are available on several equiplets. The production process as a whole will dictate which frontends are needed. In the upcoming subsections, the properties of this agile production grid will be discussed, and the enablers for this type of production will be mentioned.

Figure 6.4 An equiplet with a delta-robot frontend

6.4.1 Properties of equiplet-based production

The most important properties of the production approach are solutions to the shortcomings of standard production.

6.4.1.1 Small-scale production

To produce small-scale batches or even unique single products, standard batch production automation is inadequate because of the shortcomings mentioned and summarized in a previous section. To make a single product, we should guide a product along the set of equiplets that offer the required steps for the product to be made. At the same time other products, requiring different sets of steps can also be made, assuming that access to the equiplets is adequately scheduled. When we use the concept of equiplets, one should think of multiple production systems capable of producing many different products in parallel. We call this multiparallel production.

At any moment, we can start the production of a new or different product.

This type of production does not introduce the overhead of batch switching and is capable of starting the production of a different product during the time another product is produced.

6.4.1.2 Time to market

As mentioned in section about the shortcomings of standard production, the transitions from concept to mass production might take too long. It means that the time to market might be too long. The time to market is the time that it will take for a newly developed product to go into mass production. From an economic point of view, this time should be minimal. Normally, new products are developed at the Research-and-Development department. Then, the production automation team will search for ways to make the transition to mass production. This phase is sometimes referred to as upscaling. To test this upscaling, we also need to use the production floor equipment for test batches.

To make the transition from product development at the Research-and-Development department much easier, equiplets are used in the product development as well as in the final production process. So development, production automation and testing will be combined. The extra step of upscaling or adapting to the real production system is absent. The production is done by the same equipment and software as in the product development phase. This alleviates the aforementioned time to market problem [11].

6.4.1.3 Reliability

The grid production system is less vulnerable for failing production machinery, because equiplets can offer redundancy and the software architecture that will be described shortly is decentralized. System faults will not block other still operating parts of the grid.

6.4.2 Enablers for the equiplet-based production

Developments of the last few decades support the realization of this equiplet-based production. We mention five of them.

1. Internet and fast computer networks: Our model is a distributed system where communication between the components should be fast and reliable.
2. Interactive web technology: This technology helps to involve the user of the product in the design and requirements phase.
3. Powerful micro-systems: All systems should have computing power to support a multitasking environment. This is not a problem anymore in modern processor designs.
4. 3D printing: This technology enables making possibly unique parts for a product at low cost and low quantity.
5. Availability of cheap mobile robots: In the past, attempts have been made to implement agile manufacturing, but the flexible transport infrastructure turned out to be a big problem. Nowadays, cheap mobile robot platforms are available that can be used to implement flexible transport.

Every equiplet is capable to perform one or more production steps. A definition of a production step is as follows [12]:

Definition [production step] *A production step is an action or group of coordinated or coherent actions on a product, to bring the product a step further to its final realisation. The states of the product before and after the step are stable, meaning that the time it takes to do the next step is irrelevant for the production as a process (not for the production time) and that the product can be transported or temporarily stored between two steps.*

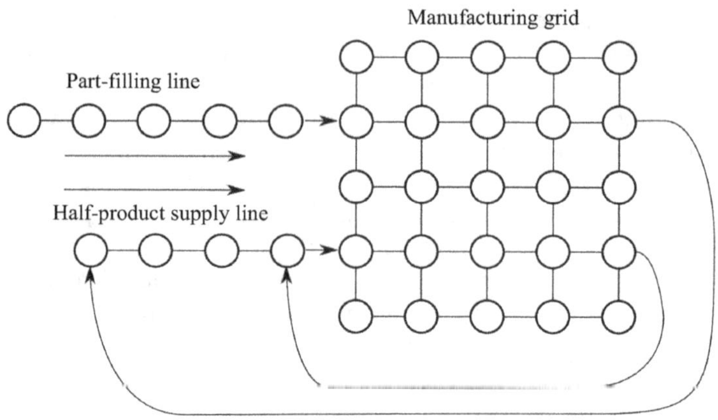

Figure 6.5 Grid production setup

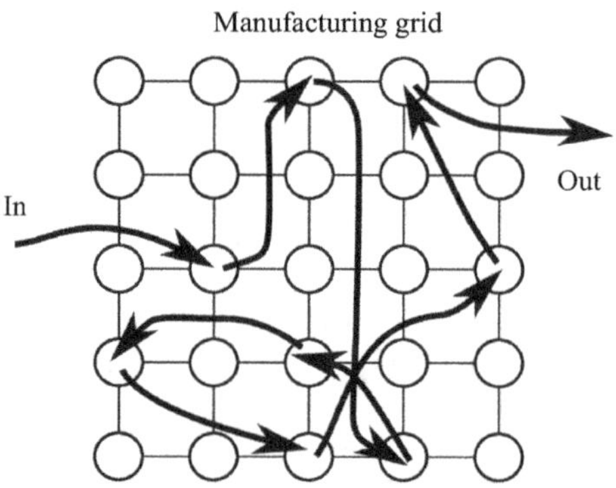

Figure 6.6 Path for a certain product

To create a product, a set of production steps should be performed. To accomplish this, a set of equiplets should be used in a certain order. This is done by moving platforms that can transport components as well as the product itself from equiplet to equiplet. In Figure 6.5, this setup is shown. The equiplets are placed in a grid. The transport platform is first loaded with components and will enter the grid where the components are handled by the equiplets to create a product (or product part or half product). When this is done, the product will be finished or in case of product parts or half products, the grid can be re-entered to handle different product parts to make the final product.

Different products need different production steps in their own specific order. The transport between the equiplets for a certain product will look like the path depicted in Figure 6.6. This particular path is actually a production line for that specific product [12]. The strength and versatility of the system is that every product can have its own path in the grid, resulting in a unique product. Complex products consisting of a set of half products can be built using the same principle. In that case, multiple paths should be followed and the grid should be re-entered. The approached used in that case is that the half products are created first by their own product agent. The half product is transported to the grid again in combination with other half products, needed to construct the final product. This is also shown in Figure 6.5.

The software entities that control the production are software agents [13]. An equiplet is represented by an equiplet agent and every product to be made is represented by its own product agent. The robot platforms used for transport are also special cases of equiplet agents. The production step they perform is transport between equiplets. In the next section, we will discuss the usage of agent technology as the basis of the software infrastructure.

6.5 Software infrastructure of the manufacturing system

The motivation to use agent technology for the software infrastructure of the manufacturing grid can be found in [12]. A short introduction to agents is presented in the following sections:

6.5.1 Agents

There are many definitions of what an agent is. We use here a commonly accepted definition by Wooldridge [13].

Definition [Agent] *An agent is an encapsulated computer system that is situated in some environment and that is capable of flexible, autonomous action in that environment in order to meet its design objectives.*

The realization of the software is actually a system with more agents with different roles.

6.5.1.1 Multiagent systems

A multiagent system (MAS) consists of two or more interacting autonomous agents. Such a system is designed to achieve some global goal. The agents in an MAS should cooperate, coordinate and negotiate to achieve their objectives. When we consider the use of an MAS, we should specify abstract concepts such as:

- Role: What is the role of a certain agent in an MAS? Perhaps an agent has more than one role.
- Permission: What are the constraints the agent is tied to?
- Responsibility: This means the responsibility an agent has in achieving the global goal. A global goal consists in most situations of a set of sub-goals. An agent can be responsible for achieving one or more sub-goals.
- Interaction: Agents interact with each other and the environment.

6.5.2 Multiagent production system

To realize our system, we define two agent roles. These roles are the main roles in the system, and we should investigate if we need some extra minor roles. Every

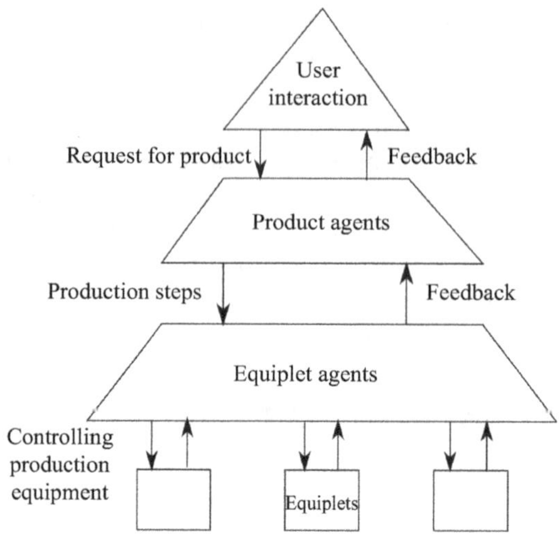

Figure 6.7 Multiagent production system

product is represented by a product agent, and every equiplet with a certain frontend is represented by an equiplet agent. The product agent has the intention of the product being produced. The intention of the equiplet agent is to accomplish production steps. Every product is made according to a certain set of production steps, while every equiplet is capable to perform a certain set of production steps.

6.5.2.1 The agent-based automation pyramid

The automation pyramid for this design is given in Figure 6.7.

The product agent operates at what was in standard production called the MES layer, while the equiplet agent is more closely tied to the hardware. This approach results in a purely pull-driven manufacturing system.

To create a product, a product agent is generated. This agent knows what production steps should be taken and the components to be used. The product agents allocate a transport system and collect the components. These components might be made by 3D printers. Next, it will visit the equiplets required to perform the production steps. The product agent can discover the equiplets needed by looking at a blackboard system where the equiplets have published their production steps. Before an equiplet is chosen, the product agent will first investigate if the equiplet is capable to perform the production steps, given the parameters involved. To do so, the equiplet will run a precise simulation of the step with the parameters given to discover the possibility and the time needed to bring the production step to an end. It will then inform the product agent about success or failure. In case of success, the product agent adds the equiplet to its list of equiplets to be visited during production.

When the actual real production step is performed, the product agent will also be informed about success or failure, but also the production parameters that have been used. This might be the exact temperature, or the amount of adhesive used, etc. Finally, the product agent has a complete production log of the product it represents.

In our model, the creation of a product agent can be done by using a web-interface where the end user can specify his/her product to be made [12]. By using this approach, the implementation of manufacturing as a service can be accomplished. A user selects properties of a product or makes a design for a product that will result in the creation of the product agent.

6.5.2.2 Summary of the agile production system

- Cheap and reconfigurable systems offer production steps.
- A product agent needs steps to create a product.
- A product agent selects equiplets for the production.
- This selection results in a path along the equiplets, placed in a grid.
- A transport system is used to follow the production path.
- Several different product agents can be active in the grid at the same time.
- After production, the product agents have a complete production log of the product.

6.5.3 Human interaction

Until now, the equiplet agent has not been completely characterized. This could be a piece of software, but also a human-operated equiplet fits in this concept. In the latter case, a piece of interaction software is needed to instruct and interact with the human operator and to participate in the multiagent-based production system. One could ask if it is now easily possible to let a software equiplet agent takeover the action of the human operator based on the production steps model we presented in this paper. A gradual takeover could be accomplished by learning the software agent step by step – the production steps.

6.6 The transport system

One of the problems to be solved is the system that will transport the product to be made along the equiplets. Research and simulation has shown that a system of moving platforms might be a good solution [14]. This model fits nicely in the agent-based concept. The moving platform can be considered as an equiplet that should be reserved for the whole path a product has to follow during production. By this is meant the path followed after entering the grid until the product or sub-product leaves the grid. Though moving platforms offer a flexible transport mechanism, the solution in not simple. The agents controlling the platform are informed by the product agent which equiplet to visit next. The platform should be moved to that equiplet, while other platforms are also moving around in the grid. To get a grip of this situation, a simulator has been developed to study the behaviour of the moving platforms and to discover a good distributed agent-based system for controlling the whole set of platforms. In [14], the system is described in more detail. The main features are discussed here.

The grid is considered as a graph, where equiplets are the nodes and the transport routes the edges.

The simulation model presented in this section opens the possibility to explore the behaviour of the production system as a whole, taking into account, the transport as well as the time to perform a production step. The model is based on the production model described in the previous section. It means that at random times an autonomous entity enters the grid with a list of steps to perform, resulting in a list of equiplets to be visited. This is comparable to a group of people shopping in a shopping centre, where they need to buy items available in different shops. Everybody is doing this autonomously and according to their own specific list.

To make the simulation versatile, a decision was made to use a graph approach for the description of the grid. The advantage is that all kind of interconnected nodes can be simulated including a grid so this approach is more powerful and can also be used in a grid where some of the interconnections are obstructed or impossible to use.

The simulation is started with three different information files. These files are XML files so human- and computer-readable.

- The file maps.xml describes the structure of the grid, actually the structure of the graph.
- The steps needed for a certain product.
- The products to be made.

A map consists of nodes and equiplets, where an equiplet is actually a node offering production steps. Both nodes and equiplets have a unique id, an x-coordinate and y coordinate. A node can also be an entry point and/or exit point of the grid. Equiplets have a set of at least one production step. This way, all kind of production infrastructures fitting in our production model can be expressed.

The simulation is controlled by a central clock. The simulation is not a real-time simulation, but by using this clock as the central heartbeat, a lot of concurrency problems could be prevented.

A path finding solution is in case of this particular simulation one of the challenges. The production system is based on autonomous entities, actually the product agents, that share the production grid, each having a specific goal, and each making the product it represents. The way this goal should be accomplished should fit in the common goal of the system, a versatile agile production system.

6.6.1 Implementation

The simulation has been implemented as two components. First, there is the core system that actually performs the simulation. The second component is a graphical user interface (GUI) that will show in detail the working of the production system. It is possible to use the core system without the GUI if a lot of simulation runs can be made to generate data that can be studied afterwards.

Java has been used as the language for implementation. It is not considered to be the fastest language, but it fits well in modern software engineering concepts. The fact that many multiagent platform implementations are also based on Java was the second reason to use this language, because this simulation can also become part of the production software that is actually a MAS based on Jade. Jade is a Java-based multiagent programming environment.

6.6.1.1 Core system

This section describes the functional requirements as well as some aspects of the technical implementation.

The following list of requirements shows the most important aspects of the core system functionality:

- De-simulation is driven by a central clock, used by all software components of the system.
- Paths between nodes can be unidirectional or bidirectional.
- There can be only one transport robot on a path between nodes.
- The core simulation program should run with or without the GUI.
- A node can have an equiplet attached, but this is not required.
- The grid description is supplied by an XML file.
- A grid should have at least one entry node and also at least one exit node.
- The production requirements and other data are supplied by XML files.
- The data generated by the simulation should be stored in a file that can be analysed afterwards.

The central clock is implemented using standard patterns of object oriented programming, the singleton pattern combined with an adapted version of the observer pattern. The singleton pattern prevents the existence of multiple clock objects and the observer pattern makes it possible to inform other objects about a new clock tick.

Several solutions for path finding were investigated to find a solution that was usable in our system. The original choice was an implementation of flood fill (based on the Dijkstra algorithm). The reason for this choice is that it would be easy to implement a decentralized as well as a centralized solution. In our solution, this approach turned out to be not usable in bigger grids with a lot of transport robots. It had to do with the way the programming language Java is managing memory making the simulation memory-intensive and slow as well as a problem with preventing deadlocks. So, a less memory intensive solution was sought for. This solution was completely decentralized. This decentralization fits well in the agent-based concept. A transport robot looks for free nodes in its direct neighbourhood and then checks if for all the nodes attached to these nodes if an equiplet is available offering the production step needed. If a node with the right equiplet is available, the robot will move in that direction, otherwise a random available direction will be used where the same approach is applied avoiding nodes that were already visited. It turned out to work; however, the solution was far from optimal as one might expect. The memory usage was low and no deadlocks occurred under a load of 80%, but some paths turned out to be very inefficient. Finally, based on this experience, a solution has been chosen. This solution uses a special map that was generated telling for every node how far it was away from a certain production step. By using this information, the path finding turned out to work well.

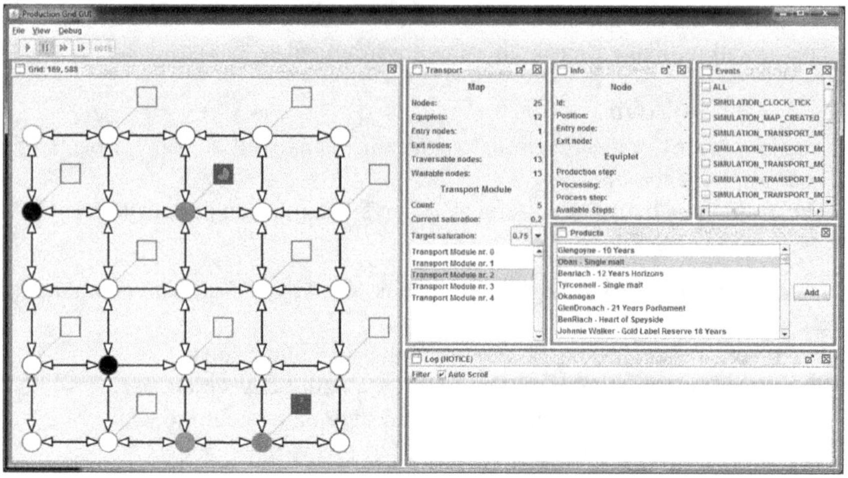

Figure 6.8 The GUI of the simulation

6.6.1.2 Graphical user interface

The GUI should offer possibilities to visualize and even control the simulation while it is running by showing a map of the grid. The simulation can be paused and the objects of the simulation (like equiplets, nodes, transport robots) can be inspected by clicking on it. The progress of the execution of a production step by an equiplet is visible during the simulation. Several windows showing additional information should be available. In Figure 6.8, the GUI is shown.

The GUI has been implemented using the Java-swing library. Swing is widely used for GUI-based Java applications and turned out to fit our requirements.

6.7 Benefits beyond production, the life-cycle agent

After completing the product, a product agent is available that collected information about the design as well as every single production step. An important aspect of our research is the investigation on what roles the product agent can play in other parts of the life cycle of a product. It should be a waste of information by just throwing the information collected by the product agents away. The concept presented in this section is a so-called life-cycle agent that will be the representative of the product in cyberspace. These agents could reside within the product itself or living in the cloud and being connected with the product every now and then or continuously. The product agent thus evolves to what we will call the life-cycle agent.

6.7.1 Design and production

As stated earlier, the design of a product will be greatly influenced by the individual end-user requirements. Cost-effective small-scale manufacturing will become more and more important.

The manufacturing system based on a grid of cheap and versatile production units called equiplets is already described in the previous sections. Important is the fact that the product agent is responsible for the manufacturing of the product as well as for collecting relevant production information of this product.

This concept is the basis for the roles of the product agent in later phases of the life cycle. The product agent carries the product design as well as the production data and can be viewed as the software entity that represents the product possibly in cyber space. From now on, the product agent will be the life-cycle agent in the other phases of the life cycle of a product.

6.7.2 Distribution

Life-cycle agents can negotiate with logistic systems to reach their final destination. Logistic applications based on MASs already exist [15]. Information of product handling and external conditions, like temperature, shocks etcetera can be measured by cheap wireless sensors and collected by the life-cycle agent in its role of guidance agent during the transport or after arrival at the destination. The handling and external conditions during transport can be important during product use, especially for product quality, maintenance and repair.

6.7.3 Use

The role of the life-cycle agent during the use of the product could focus on several topics. The first question one should ask is: who will benefit from these agents, that is who are the stakeholders. In a win–win situation, both the end user as well as the manufacturer could benefit from the information. If a product is a potential hazard (in case of misuse) for the environment, the environment could also be a winner if the agent is capable of minimizing the effects of misuse or even prevent it.

In the next subsections, several topics of usage of the life-cycle agent are proposed.

6.7.3.1 Collecting information

A life-cycle agent can log information about the use of the product as well as the use of the subsystems of the product. Testing the health of the product and its subsystems can also be done by the agent. These actions should be transparent for the end user. If a product needs resources like fuel or electric power, the agent can advise about this. An agent can suggest a product to wait for operation until the cost of electric power is low, i.e. during the night. It depends of course on the type of device if this should be implemented.

6.7.3.2 Maintenance and repair

Based on the logging information about the product use and the use of the subsystems, an agent can suggest maintenance and repair or replacement of parts. Repairing a product is easier if information about its construction is available. Also the use of a product or the information about transport circumstances during distribution can give a clue for repair. An agent can also identify a broken or malfunctioning part or subsystem. This could be achieved by continuous monitoring, monitoring at certain intervals or a power-on self-test.

An important aspect of complex modern products is the issue of updates or callbacks in case of a lately discovered manufacturing problem or flaw. In the worst situation, a product should be revised at a service centre or the manufacturing site.

Information about updates or callbacks can be sent to the life-cycle agent that can alert the end user in case it discovers that it fits the callback or update criteria. This is a better solution for a callback than globally advertising the problem and alerts all users of a certain product when only a subgroup is involved.

6.7.3.3 Miscellaneous

Use of life-cycle agents could result in transparency of the status of a product after maintenance by a third party. The agent can report to the end user what happened during repair so there is a possibility to check claimed repairs. Of course, the agent should be isolated from the system during repair to prevent tampering with it.

Recovery, tracking and tracing in case of theft or loss are also possible by using this technique.

When the end user wants to replace a certain device by a new one, the life-cycle agent can give advice about the properties the replacing device should have, based on what the life-cycle agent has learned during the use phase.

6.7.3.4 Internet of Things

All possibilities mentioned for the use phase fit nicely in the concept of the Internet of Things (IoT). The life-cycle agent is actually the software that lets the device communicate and act in the environment.

6.7.4 Recycling

Complex products will have a lot of working subsystems at the moment the end user decides it has come to the end of its life cycle. This is normally the case when a certain part or subsystem is broken. The other remaining parts or subsystems of the product are still functional, because in a lot of complex products, the mean times between failure of the subsystems are quite different. The life-cycle agent is aware of these subsystems or components and depending on the economic value and the remaining expected lifetime these components can be reused. This could be an important aspect of 'green manufacturing'. An important issue here is that designers should also take in account the phase of destruction or recycling. Disassembly and reuse of subsystems should be a feature of a product for this approach to be successful.

The life-cycle agent can reveal where rare or expensive material is situated in the product, so this material can be recovered and recycled. This way the product agent can contribute to the concept of zero waste. *Zero waste is just what it sounds like – producing, consuming and recycling products without throwing anything away.*

Another advance of having a life-cycle agent at hand in case of recycling is the fact that the life-cycle agent has the information how a product is constructed (the agents was at that moment the product agent). This is helpful when a product must be taken apart. For certain steps, a kind of undo-steps should be carried out to dismantle a product.

6.8 Summary

In this chapter, a global overview of the agile agent-based manufacturing system has been presented. The differences with standard production systems and special features have been discussed. As stated earlier, the proposed system is not meant as a replacement for large-scale production, but as a new paradigm for small-scale agile production fitting in the concepts of industry 4.0. The concept can also be useful for situation beyond manufacturing where IoT turns the world into a domain of distributed autonomous systems.

So far we have working equiplets, a webinterface for designing simple products and a working agent platform to control the grid, but the transport system is still under development.

References

[1] Brettel, M., Friederichsen, N., Keller, M. and Rosenberg, M., *How Virtualization, Decentralization and Network Building Change the Manufacturing Landscape: An Industry 4.0 Perspective*, International Journal of Mechanical, Aerospace, Industrial, Mechatronic and Manufacturing Engineering, volume 8, number 1, 2014, pp. 37–44.

[2] Rajkumar, R., Lee, I., Sha, L. and Stankovic, J., *Cyber-physical Systems: The Next Computing Revolution*, Proceedings of the 47th Design Automation Conference (DAC), Anaheim, California, 2010, pp. 731–736.

[3] Puik, E. and van Moergestel, L.J.M., *Agile Multi-Parallel Micro Manufacturing Using a Grid of Equiplets*, Proceedings of the International Precision Assembly Seminar (IPAS 2010), 2010, pp. 271–282.

[4] Koren, Y. and Ulsoy, G., *Vision, Principles and Impact of Reconfigurable Manufacturing Systems*, Powertrain International, volume 5, number 3, 2002, pp. 14–21.

[5] Koren, Y., Heisel, U., Jovane, F., *et al.*, *Reconfigurable Manufacturing Systems*, Annals of the CIRP, volume 48, number 2, 1999, pp. 527–540.

[6] Shingo, S., *A Study of the Toyota Production System*, Productivity Press, New York, 1989.

[7] Goldman, S.L., Nagel, R.N. and Preiss, K., *Agile Competitors and Virtual Organizations – Measuring Agility and Infrastructure for Agility*, Van Nostrand Reinhold, New York, 1995.

[8] Vogel-Heuser, B., Kegel, G., Bender, K. and Wucherer, K., *Global Information Architecture for Industrial Automation, ATP*, Automatisieringstechnische Praxis, volume 1, number 2, 2009, pp. 108–115.

[9] Shaw, W.T., *Computer Control of BATCH Processes*, EMC Controls, Houston, TX, 1982.

[10] Sarker, R.A. and Khan, R., *An Optimal Batch Size for a Production System Operating Under Periodic Delivery Policy*, Computers & Industrial Engineering, volume 37, number 4, 2013, pp. 711–730.

[11] Puik, E., van Moergestel, L.J.M. and Telgen, D.H., *Cost Modelling for Micro Manufacturing Logistics when using a Grid of Equiplets*, International Symposium on Assembly and Manufacturing (ISAM 2011), 2011, pp. 1–4.

[12] van Moergestel, L.J.M., *Agent Technology in Agile Multiparallel Manufacturing and Product Support*, thesis Utrecht University, 2014.

[13] Wooldridge, M., *An Introduction to Multiagent Systems, Second Edition*, Wiley, Sussex, UK, 2009.

[14] van Moergestel, L.J.M., Puik, E., Meyer, J.-J., *et al. A Versatile Simulation Model of Agile Production by Autonomous Systems*, Proceedings ISADS 2017, 2017.

[15] Burmeister, B. and Haddadi, A. and Matylis, G., *Application of Multi-Agent Systems in Traffic and Transportation*, IEEE Proceedings on Software Engineering, volume 144, number 1, 1997, pp. 51–60.

Conclusion of part 2

In this part, the applications of autonomous decentralized systems to control systems of important social infrastructures which have an extremely high mission criticality are shown. In Chapters 2, 3 and 4, the applications for a railway control system and a variety of assurance technologies are described. In Chapter 5, the application for air-traffic surveillance and its fault-tolerant technology is described. Chapter 6 has its focus on the role of autonomous systems in manufacturing.

In Chapter 2, online test technology for an autonomous decentralized railway signal control system is shown. Here a heterogeneous real-time autonomously integrating architecture, its system configuration technology and ensuring safety technology are described. With this technology, huge and complex control systems can be set up step by step and partially replace the system.

In Chapter 3, the technologies that realize safe and stable operation under any circumstance are described. For example, it is in improvement of systems or coexistence with existing systems, flexible correspondence is possible. How this need was solved was described, giving an actual example.

In Chapter 4, technologies which integrate control and information utilizing autonomous decentralized system are shown. The characteristics of autonomous decentralized system such as a data-driven architecture and online property make easy-to-develop large-scale and complex information-based control system step-by-step possible. As an example of its application, ATOS (Autonomous Decentralized Transport Operation Control System), which is a Tokyo area transport operation control system, has been mentioned.

In Chapter 5, an ADS technology with a fault-tolerant property is discussed. The proposed technologies reduce data traffic in air–ground communication with keeping continuous target tracking in an air-traffic surveillance system.

In Chapter 6, an agile manufacturing model based on autonomous decentralized systems is described. This model offers interesting possibilities for modern manufacturing where the role of the end-user is increasing resulting in the need for affordable production of small batches or even single unique products. Extending the concept may also result in reuse of product parts and reducing waste.

In this way, the autonomous decentralized systems can flexibly reconfigure the system while securing the safety and stability of the system with high mission criticality. It is expected that autonomous decentralized systems are useful to build social infrastructure systems that can quickly adapt to changes in the social environment and contribute to the improvement of people's quality of life.

Part 3

Developing ADS technologies and applications leading to innovation in lifestyle

Overview

We are living today in a society controlled and managed by various systems. Our lifestyles are fully determined by the advancement of such systems. A large system, either a social or an engineering system, is a collection of interacting elements or component, in which the behaviour of any element depends on its direct or indirect interactions with the other elements. A system can be considered from three major viewpoints: the architecture, the interface, and the behaviour. Systems can be designed using different approaches, such as centralised, decentralised, top-down, and bottom-up approaches. ADS (autonomous decentralised system) is a set of concepts and technologies developed to design and manage large engineering systems. The same concepts and technologies are now being used in the large social and economic systems.

ADS concepts and technologies were initially developed in the 1970s and were applied in high-speed trains and transportation systems. International symposium on ADS was established in 1993 to disseminate and publicise ADS and related research and practice. Over the past two decades, ADS has not only evolved and advanced conceptually and technologically but also penetrated and found applications in many areas of social, economic, business, and healthcare systems. ADS also blends with many current technologies, such as service-oriented computing, where participating parties are loosely coupled and autonomously cooperative; cloud computing, where distributed computing units are transparent to the users and can elastically scale out and scale in; message-based integration systems, where messages are sent by content code instead of by destination address; and distributed intelligence, such as the Internet of Intelligent Things (IoIT), where computing, communication, and data resources can be distributed and collaboratively complete coherent missions. These advances lead to innovations in our daily activities of our lifestyles.

In this part of the book, we present a number of new studies that represent the latest development and advancement of ADS concepts and technologies in different application areas.

Chapter 7 in this part is on railway ticketing services, which represents a key application area of ADS concepts and technologies. Ticketing services are directly used by ordinary people in their daily life, and the innovation of making the system more friendly and usable changes the way the people use the transportation system and thus their lifestyle. In this study, a new ticketing system is developed to address problems in the conventional ticketing systems, such as the difficulties of collaboration among different systems, fulfilling diversity needs of users, and upgrading the system. The new system is more convenient for the passengers and is more cost effective. ADS concepts and technologies are perfect for this system, as it involves a heterogeneous system of autonomous devices, collaboratively operated through messaging and transactions. The system consists of units, such as integrated circuit card readers, terminals, automatic fare-collection gates, ticket-vending machines, fare-adjusting machines, station servers, and a central server. Except for the central server, all the other devices are autonomous and can be seamlessly added and removed without impacting the operations of the other devices, because ADS supports loose coupling and content-based messaging. The study further extended the content-based messaging technology into the autonomous cooperative processing technology (ACPT) and autonomous decentralised data consistency technology (ADDCT). ACPT makes autonomous decisions based on locality information. This ensures the high-speed operation of each subsystem by dividing a process and combining the results from the divided subprocesses according to the characteristic of each transaction. Even if the decisions are made autonomously and using locality information, ADDCT guarantees data consistency under the heterogeneous real-time architecture by cooperation among the system's units and by determining the residence time of data in each autonomous unit according to the data distribution. The evaluation results show that the new ticketing system meet the diverse requirements and can easily be scaled out to more subsystems.

Chapter 8 in this part presents the extension of ADS concepts to service-oriented computing and cloud computing. The overall system studied in this chapter is considered to have three layers: architecture, interface, and behaviour. The system architecture is based on the service-oriented architecture and follows ADS concepts. The system interface defines the interconnection among the units through standard service interfaces. The units' behaviours and functionalities are autonomous and decentralised. Each unit is considered to be a robot and a service connected to a cloud-computing environment and is called Robot as a Service (RaaS). RaaS concepts and related technologies bridge the ADS to the current design and integration patterns of heterogeneous systems. RaaS emphasises standard interfaces, platform independence, loose coupling, serverless local computing surrounding a cloud-computing environment, and elastic scalability with scaling in, scaling out. The behaviours of the RaaS units are defined independently of each other, and they cooperate in loosely coupled manner.

RaaS is not only a set of concepts. Related technologies have been developed to implement RaaS. A VIPLE (Visual IoT/Robotics Programming Language Environment) has been developed based on RaaS concepts to support the easy implementation and deployment of RaaS through visual programming. VIPLE uses a JavaScript Object Notation – JSON-based standard interface to connect different hardware platforms. Physical robots and IoT devices running on Intel and ARM (Advanced RISC Machines) architectures have been implemented. Virtual devices represented in the Unity game engine and in HTML 5 Web are implemented to simulate the execution of RaaS. VIPLE can read any types of sensors on the devices, process the sensory data to generate the control data, and send the control data to the devices. It is also possible to perform local computing on the devices, if they have adequate computing capacity. VIPLE is based on π-calculus and has the full capacity of a complex programming language, such as C++, C# and Java, plus the process control and event-driven programming capacity. It can connect to multiple RaaS units and process sensory data from different units jointly to make coordinated decision and control. Different case studies have been developed, including encoding/decoding, service orchestration, traffic simulation, and maze navigation.

RaaS and VIPLE have been used in education for teaching computer science concepts, programming, control, communication, and embedded systems in various schools and universities.

Chapter 9 in this part presents a smart phone app that helps JR East (East Japan Railway Company) passengers to obtain necessary information easily. The app utilises various kinds of location technologies and devices, such as GPS, Wi-Fi, and ultrasonic beacons, to provide real-time data unique to railway business operators, such as operational status, delay of trains, location information, and occupancy of trains, as well as well as entertainment contents consisting of e-books, games, etc., offered by the shops in its station buildings and around its stations.

The entire system consists of a number of major components. The client is the smart phone. It connects to the ground system through Wi-Fi and Internet. The ground system offers various content data, stores the user profile, and accesses logs. The ground system connects to the trains to obtain information such as the occupancy and temperature in the train. It also connects to external systems to obtain other information, including each train's operational status, the locations of trains, their departure and arrival times, as well as public service information such as weather.

The app was first made available in March 2014, and user accesses have been traced and analysed. Several millions of users in different age groups have been using the app, and many of them are using it actively. Access logs show that the content mainly requested is real-time information related to the use of railways. In addition, the questionnaire survey revealed that around 80% of the respondents are satisfied with this information service and around 90% of respondents are willing to continue to use this app.

Chapter 10 in this part studies large social systems and how ADS concepts can contribute to the development and management of such large systems, just like ADS was initially applied to manage large engineering systems in a decentralised manner.

The technological innovations, particularly in information and communications technology, have made major impacts on personal life and on enterprise systems. However, their impact on social control systems has been marginal so far. Although there are many similarities between large engineering systems and social systems at the high levels, the differences are significant, as social systems have more non-deterministic factors and are less predictable than the engineering systems. ADS concepts are based on autonomous and decentralised approaches, which can better deal with the nondeterministic factors and unpredictable issues in social systems.

This chapter also develops the concepts of social control systems to deal with the centralised regulations and decentralised, local, and cultural enforcement. Based on an in-depth autonomy of social systems, this new approach is proposed to apply ADS concepts to form the dynamic social systems, including the preparation, implementation, local expansion, and global-expansion processes. Several case studies, including large systems in engineering and local and global governments, are presented to illustrate and validate the proposed approach.

Simulation is a critical step in the system design process. It is used in almost every sector of industry, business, economy, and society, and it affects nearly every aspect of our life. With the support from the latest computing and communication technologies, simulation becomes even more powerful and more useful. However, today's systems are often very large, decentralised, and interactive. The system behaviours are always complex and nondeterministic. For such systems, distributed simulations are required, and the distributed simulations themselves are very difficult to build individually.

Chapter 11 in this part proposed and applied an emerging technology, called Internet of Simulation (IoS), to assist in understanding the design and implementation processes of complex ADSs. IoS extends the concepts of Internet of Things (IoT) with virtual things and then bridges the virtual and physical worlds. The core characteristics of IoS cover three areas of research: the relationship to IoT as a network of virtual things, the iterative and hierarchical concepts of Simulation as a Service and Workflow as a Service for massive-scale co-simulation, and the interconnectivity between IoS and devices within the physical world of IoT. Based on IoS template, engineering and design challenges can be significantly improved in the design process. The study also shows that IoS can be useful for large-scale simulation. It can be used for improving autonomy by providing ready access to detailed simulation services. Such a scheme facilitates unsupervised learning for decentralised systems with increasingly autonomous systems.

Chapter 7
Railway ticketing services (Suica)

Akio Shiibashi[1]

Abstract

Conventional automatic fare collection systems (AFCSs) present difficulties in collaborating with different systems, fulfilling various needs, and upgrading. Each terminal was designed to provide a single function such as ticketing, checking, or collecting in a stand-alone configuration, equipped with an embedded system. Upon designing the new AFCS, three points were expected: solutions to those problems, improved convenience for the passengers, and reduced maintenance costs. To satisfy such requirements, the contactless integrated circuit cards (IC cards) were proposed as the alternatives of the paper tickets, and the IC card ticket system named Suica has been developed and utilised by East Japan Railway Company, based on the concept of autonomous decentralised systems. Gate control, transaction process, and value-added services have been integrated in this system. Nowadays, it is becoming more and more important to integrate control, information, and service into a system. This integration enables not only the exchange of messages among heterogeneous systems but also the creation of adaptive integrated systems that satisfy a wide variety of user requirements. This chapter presents the development of Suica system as a case study.

7.1 Introduction

With the contactless integrated circuit cards (IC cards), the IC card ticket system (ICCTS) introduced in this chapter is a typical example of embedded service by autonomous decentralised system [1–3]. The conventional automatic fare collection systems (AFCSs) with the magnetic paper tickets present difficulties in:

1. collaborating with different systems because each terminal is in a stand-alone configuration;
2. fulfilling various needs because each terminal is designed to provide a single function such as ticketing, checking, or collecting; and
3. upgrading because each terminal is equipped with an embedded system.

[1]JR East Mechatronics Co Ltd., Japan

Therefore, the customers have to go to the exact terminals every time they want something. Sometimes, they have to be in queues several times if their purposes refer to multiple functions which are not defined by them but the terminals. For maintenance, the servicepersons have to visit every terminal to upgrade its embedded software – sometimes by replacing the built-in memories.

The ‘new’ AFCS was first expected to solve those problems. Networking the multifunctional terminals was the first solution, but network was neither fast nor reliable enough for transactions. The gates, for example, should hold the customers as long as the automatic teller machines if they should confirm reliability in the same way. Hence, the concept of autonomous decentralised systems (ADS) is introduced to the new AFCS, where IC cards, with more capacity and better security than the magnetic paper tickets, are adopted as the media in the data fields (DFs); this new AFCS is named the ICCTS.

In the ICCTS, the ticket vending machines which used to sell magnetic paper tickets only are replaced to provide IC card tickets, top up the balance and the ticket values onto them, and print the transaction records. The gates which used to check the paper tickets at entrances and collect them at exits are replaced to calculate and deduct the fares and even top up the balance if necessary. The terminals are connected to each other in a network and enable such collaborated tasks of fare calculation and transactions. The network is also used for upgrading the ‘opener’ software in each terminal.

The ICCTS also improves convenience for the passengers. The passengers no longer have to purchase a single paper ticket and insert it to the gates for every journey; they just top up the balance onto the IC cards and tap-and-go through the gate where the exact fare is deducted. Replacement of paper tickets by IC cards reduces the mechanical parts in terminals, which results in reduction of maintenance costs as well. The system named **Suica** was originally developed only as a fare collection system for East Japan Railway Company (JR East) and has gradually expanded to support private railways and e-commerce as well. Suica owes such expansions as well as the originally expected features for the specific system architecture: autonomous decentralised system (ADS) (Table 7.1).

To enable such features, there exist heterogeneous requirements. Both quick response and high reliability are mandatory in fare collection and e-commerce. In the ICCTS, the users tap their cards onto the readers. In other words, it is the users who govern the time in which the cards and the readers are communicable. Then the quicker is the better because the terminals cannot control the way of tapping. Meanwhile, the values spent at the terminals are very equivalent to cash, and the system can never miscount the value, that is a certain length of time is necessary to process the transactions. To resolve such time constraints in different service process requirements, the DFs are differentiated by different time ranges, and the heterogeneous-timed DFs are constructed in ICCTS.

7.2 System structure

As described in Section 7.1, the concept of ADS is introduced to ICCTS to achieve both high performance and high reliability. ICCTS consists of IC cards, terminals

Table 7.1 Comparison of conventional AFCS with ICCTS

Viewpoint	**Conventional AFCS**	**ICCTS**
Collaboration with different systems	Unavailable in a stand-alone configuration	Available through a network • Terminal to terminal, e.g. fare calculation • Card lifecycle management system • Seat reservation system • Train controlling system
Fulfilment of various needs	Poor with single functionality	Better with multiple functionality
Upgrades	On the spot because of a stand-alone configuration	Available through a network
Improved convenience for the passengers	Poor with single functionality	Rich with multiple functionality, e.g. e-commerce
Reduction of maintenance costs	Costly with enormous mechanical parts	Less costly with less mechanical parts

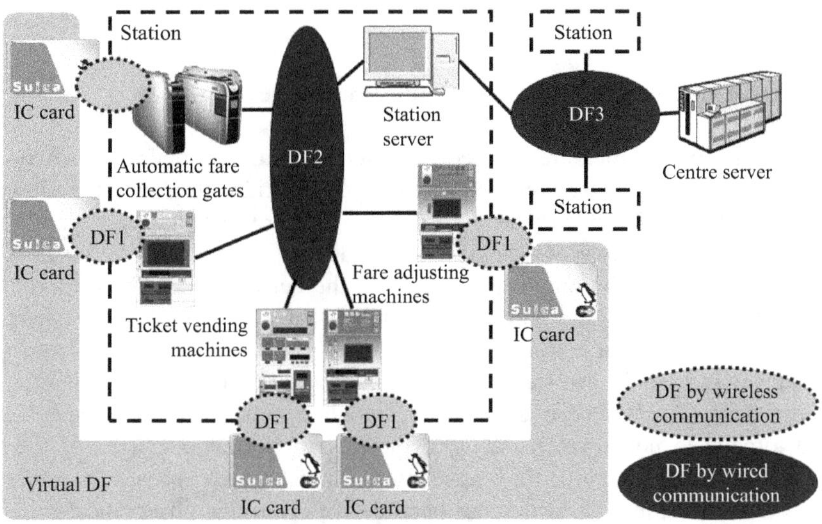

Figure 7.1 IC card ticket system. Copyright © 2009 IEICE [4]

(e.g. automatic fare collection gates, ticket vending machines, and fare adjusting machines), station servers, and a central server. Each station has several terminals and a station server, and they are connected to each other via local area network within the station. Station servers are connected to the central server via a wide area network.

There are three DFs distributed among the subsystems (Figure 7.1):

1. DF1 between an IC card and a terminal,
2. DF2 among terminals and a station server within a station, and
3. DF3 among station servers and the centre server throughout the stations.

These three DFs have different transmission methods with the respective time ranges determined by the needs of the subsystems. The subsystems attached to DF1 have functions for payment, fare collection, and gate control. Here real-time processing is required in order to avoid congestion. Therefore, the total processing time in DF1 is designed to be under 0.2 s – this is a requirement for what are called control systems [5]. The station servers and the centre server do not serve the hurrying passengers directly but deal with their accounts. Therefore, the data in DF2 and DF3 are transmitted through the wired network hourly and daily. The wired network and the time range for storing this data have been adopted to assure reliability of the data as required for information systems.

In addition, a 'virtual DF' is applied to the system. This 'virtual DF' is the field where IC cards are actually carried by the passengers. Here an IC card is regarded as a medium of data transmission, and the data are accessed when needed at the terminals.

These DFs have different characteristics and are named 'Heterogeneous DFs', and the total structure is named 'heterogeneous autonomous decentralised system structure'. This structure is one of the key properties of the ICCTS. The heterogeneous DFs are adopted to assure (1) high performance (real-time operation) and (2) fault tolerance. This chapter introduces two technologies which are necessary for the properties of real-time processing and transactions.

7.3 Autonomous cooperative processing technology

The fare systems in Japan are very complicated; each transit operator provides an independent fare table based on the exact distance in kilometres, and the total fare is the summation of each fare with some exceptions. What makes more complicated is 'commuter passes' which allow free rides within the registered area in the designated period. In the conventional AFCS, the customers looked up the destination in the fare tables and purchased the tickets; it was up to them to take commuter passes into consideration. The customers who travelled further from their commuter pass areas had to step into the terminals in front of the exit gates to pay the extra fares. The ICCTS is expected to calculate such fares at the gates so that the customers do not have to step into the ticket vending machines or the fare adjusting machines. To fulfil this expectation, the gates are expected to process an IC card ticket within 0.2 s, which has been derived from the observance of human behaviours, while there are trillions of fare combinations. As the entrance gates cannot predict where the customers will get off, it is considered to be the exit gates that calculate the fare. However, it is far above the specifications. To solve this problem, fare calculation is divided to two steps at an entrance and an exit, using autonomous cooperative processing technology (ACPT).

In the heterogeneous-timed-DF architecture, each subsystem cooperates with other subsystems and distributes processes autonomously based on the local information. ACPT was proposed to guarantee high-speed processing in each subsystem by dividing a process and combining the results from the divided sub-processes according to the characteristic of each transaction.

7.3.1 Technology

The processing time at the terminals in DF1 is required to be under 0.2 s. In this short time, they need to detect, authenticate, read, judge, write, and verify IC cards [6] in addition to fare calculation, which is the most time-consuming of all these processes.

There are two problems with fare calculation. The first problem is the highly complicated fare system in Japan where fares are based on distance in kilometres, while they are fixed at flat rates or based on simple zones in other countries. In Japan, there are fare combinations equal to the number of combinations of stations. Moreover, some passengers hold 'commuter passes' which allow unlimited rides within the passenger's predetermined commuting zone. The fare calculations must take such passes into consideration if it is less expensive.

The second problem is that the passengers' destinations are unpredictable. In a conventional magnetised ticket system, passengers have to check the fare table, buy their own tickets, and pay additional fare upon exiting if they did not pay enough with the original tickets. In ICCTS, the passengers no longer need to buy tickets in advance – it is very convenient for them, but the terminals at the entrances have no way to know their destination. It takes a long time if the fare calculation is done entirely at the exit; the terminals must scan the complete fare list including the fares from the entrance station to the exit station.

ACPT was proposed to resolve these two problems using a virtual DF [7]. In this technology, the fare calculations are divided into two steps (upon entrance and exit), and the necessary information is transmitted by cards carried by passengers who move within the virtual DF. The procedure is shown in Figure 7.2. In this case,

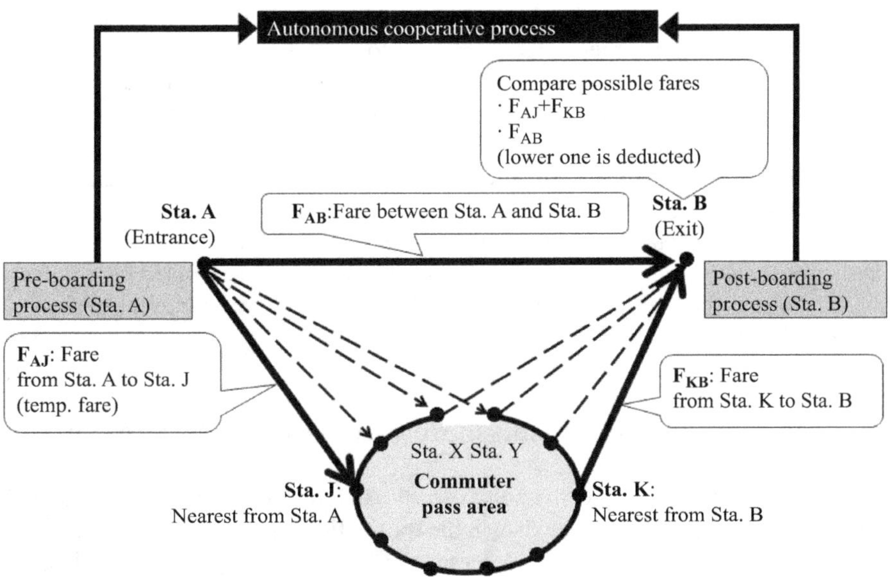

Figure 7.2 Autonomous cooperative processing technology. Copyright © 2009 IEICE [4]

a passenger has a commuter pass, which is valid from Station J to Station K and travels from Station A to Station B, both of which are outside of the valid commuter pass area. Within the commuter pass area, Station J is the nearest station to Station A, and Station K is the nearest to Station B. There are two possible fares: (1) the direct fare from Station A to Station B or (2) the sum of the fares from Station A to Station J and from Station K to Station B.

When the passenger enters Station A, the terminal determines that it is out of the valid commuter pass zone and selects Station J, the nearest within the valid commuter pass area. Then, it writes on the IC card that the cardholder gets on at Station A, and that the temporary fare to the nearest Station J is F_{AJ} (pre-boarding process).

When the cardholder exits from Station B, the terminal judges that it is also out of the valid commuter pass zone and selects Station K, the nearest within the valid commuter pass area (post-boarding process). Then it calculates the fare between Station K and Station B (F_{KB}) and compares two possible fares: (1) the direct fare between Station A and Station B (F_{AB}) and (2) the sum of F_{AJ} and F_{KB}. The less expensive fare is deducted from the IC card (autonomous cooperative process).

7.3.2 Evaluation

The effectiveness of ACPT is shown in Figure 7.3. These are the results from simulations comparing the process time with and without ACPT. With ACPT, the fare calculations are divided into two stages: upon entrance at Station A and upon exit at Station B. Without ACPT, all the fare calculations are done upon exit from Station B. The ‘calculating time’ is the simple calculating time, and ‘process time’ is the total time for a passenger to go through a gate, including both ‘calculating time’ and waiting time at gates. Each station is supposed to have ten gates.

Figure 7.3(a) shows the times in relation to the number of stations. ACPT is superior (spending less time in calculating and processing) when the number of stations exceeds 43. Figure 7.3(b) shows the times as the number of the transactions. ACPT is superior when there are over 42,700 transactions per day.

According to these results, the greater the numbers of stations and transactions, the more effective ACPT is.

The number of stations in the Suica system was 647, and the number of the transactions was approaching 20 million per day when it was implemented. Thus, ACPT is most effective in the situation that the numbers of stations and transactions are huge [7].

7.4 Autonomous decentralised data-consistency technology

High performance at the gates is achieved by ACPT as described in Section 7.3. However, another problem still exists in the transactions; no matter how quickly the terminals process, there are the customers who behaves much faster than them. In such cases, the IC cards are not processed correctly. Some customers are blocked at the next gates because ACPT cannot work without the previous record. Other customers might be able to escape from payment, which results in inconsistency of

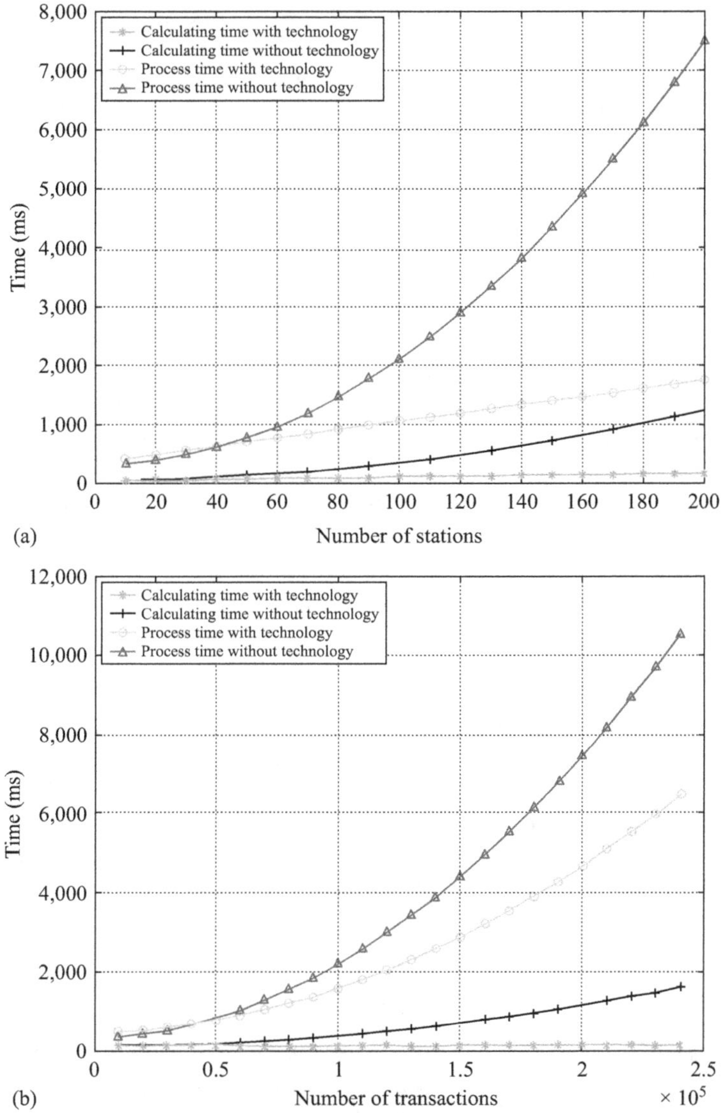

Figure 7.3 Comparison of calculating time and process time with and without ACPT: (a) by number of stations and (b) by number of transactions. Copyright © 2009 IEICE [4]

clearing data between the customers and the stores. To solve this problem, the transaction data are stored both in IC cards and the centre server, compared to each other, and used for complementation if necessary, using autonomous decentralised data consistency technology (ADDCT).

The main objective of Suica is to assure the fluidity of passengers. To guarantee real-time processing at high-transaction rates, incomplete data are passed and coexists with complete data in DFs. ADDCT was proposed to achieve data consistency under the heterogeneously timed DFs architecture by cooperating with each other and determining the residence time of data in each DF autonomously according to the data distribution.

7.4.1 Technology

ICCTS uses wireless communications at DF1. It is very convenient for passengers, but it is prone to instability in communication due to improper card use. This section introduces ADDCT, which recovers missing data caused by such unstable communication [6]. This technology has two derivatives: single-layered data consistency technology [8] and multi-layered data consistency technology [9,10].

In ICCTS, an IC card can communicate with a reader at the gate while it is within the communication area. The time of staying within the communication area depends on the holder's behaviour. According to the statistics, the minimum required time is 0.2 s [6,11,12], the time period which ACPT targets. However, some passengers do not handle the cards properly, and the card thus cannot be processed.

The way ICCTS detects the end of process is shown in Figure 7.4. Each reader unit updates its data when it receives a 'data-process completed' signal from the IC card. This last signal is transmitted near the border of communications area, so the process is not always completed successfully when the passengers handle their IC cards improperly. In this case, although the data in the IC card is updated, the reader has not received any signal indicating that the update is complete. This is a problem called 'data missing'. A conventional magnetised ticket system would shut the gate and thereby completely control the contact signals; thus, 'data missing' was practically unheard of. However, if ICCTS had to shut the gate, serious congestion would occur, which could lead to accidents. Opening gates even in the case of failures have been a problem with ICCTS.

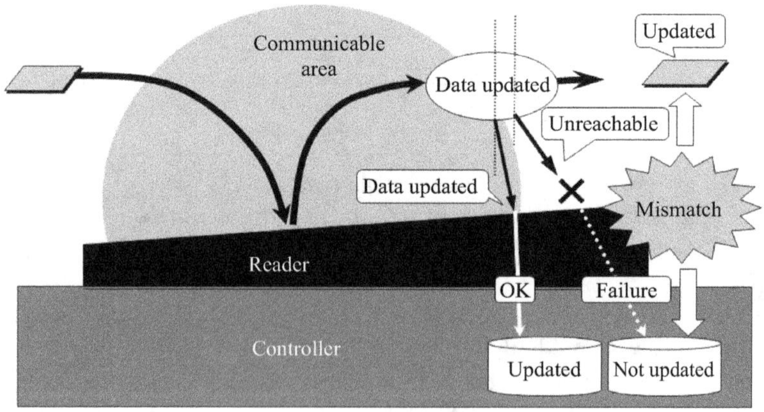

Figure 7.4 Data processing at the reader. Copyright © 2009 IEICE [4]

ADDCT is a technology that recovers the data from a failure, considering data consistency. Here, the data in the readers are treated as 'temporary data' even when the readers cannot catch the 'data-process completed' signal. If the next process is completed normally, ADDCT checks the consistency of the data and revises the 'temporary data' as 'definite data'.

Samples of ADDCT are shown in Figure 7.5. Subsystems are classified into several layers. For example, ICCTS has three layers: (1) the gate at Layer 1, (2) the station server at Layer 2, and (3) the centre server at Layer 3. Since the ADDCT application runs at each subsystem, 'missing data' has only to meet the partner data to be recovered. If not, data will be broadcast to the lower layer, where ADDCT application runs again. This combination of ADDCT at multiple layers is called multi-layered data consistency technology.

Figure 7.5 can be modelled as Figure 7.6. Here, three data (Data 1, Data 2, and Data 3) are created first. Then, they are stored at Gate11 in Layer 1 for time t_1, and then ADDCT application runs for time t_{1p}. While being stored, Data 4 to match Data 1 catches up at Gate11 at Layer 1. Hence, ADDCT application adjusts the data, and the rest (Data 2 and Data 3) are broadcast to DF2. Data 5 from Gate12 in

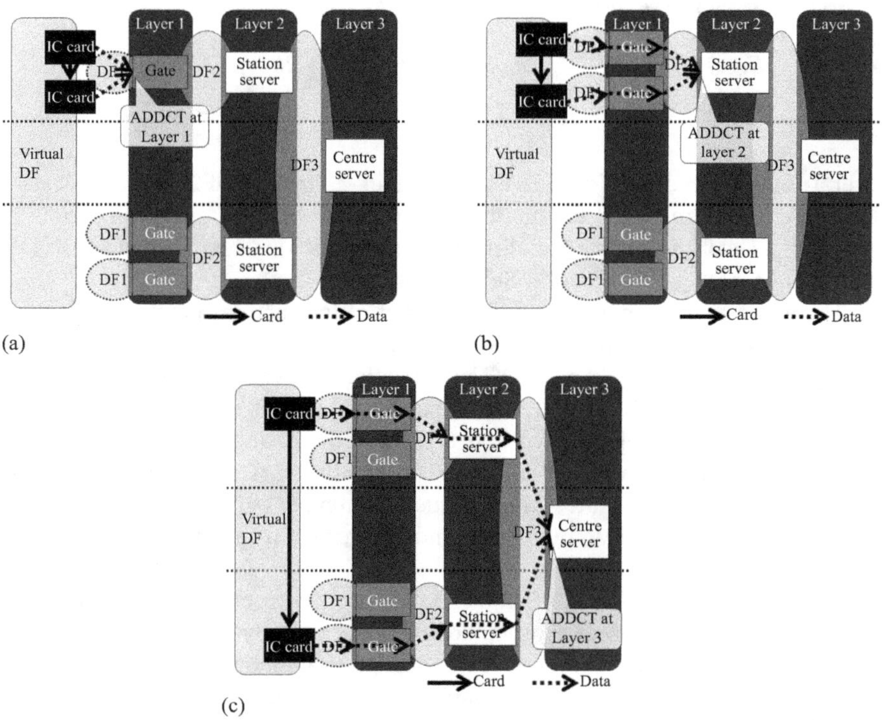

Figure 7.5 ADDCT application running at Layer 1/2/3: (a) ADDCT application running at Layer 1, (b) ADDCT application running at Layer 2, and (c) ADDCT application running at Layer 3. Copyright © 2009 IEICE [4]

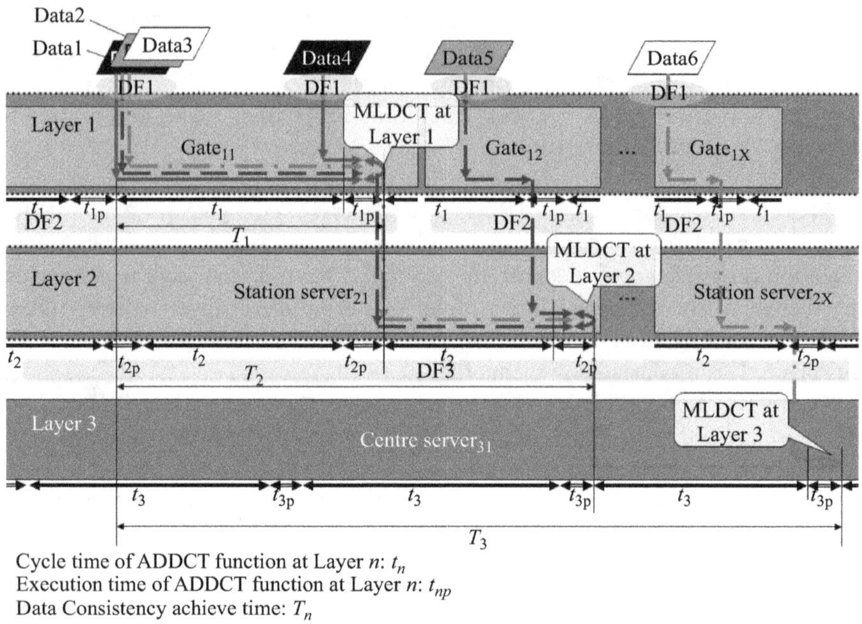

Figure 7.6 ADDCT model. Copyright © 2009 IEICE [4]

Layer 1 joins them at Station Server21 at Layer 2 through DF2 to match Data 2. Those three data are stored for the time t_2, and then ADDCT application runs for the time t_{2p}. Data 2 is adjusted, and the last one (Data 3) is again broadcast to DF3 where Data 6 comes from Station Server2x at Layer 2.

The number of layers results in the number of transactions per node; the fewer layers result in the more transactions per node.

Figure 7.7 shows how the ADDCT application recovers ‘missing data’. This is an example of the ADDCT application running at Layer 3. A passenger has an IC card with the value of 1,000 yen and travels from Station A to Station B. The possible fare of 130 yen is written at a gate in Station A. The gate ‘broadcasts’ the data to the DFs as ‘definite’ with a sequential number (#14 in this sample) when the process is successfully completed, and the reader receives the ‘data-process completed’ signal (Figure 7.7(a)).

Suppose that a reader at Station B fails to receive the signal indicating the completion of data processing, although it has actually been processed in the passenger’s IC card itself, in this case, the gate autonomously broadcasts the unconfirmed data to the DFs as ‘temporary’ (sequential number 15) (Figure 7.7(b)).

If the passenger uses the same card and completes the processes upon entrance at Station C, the data numbered 16 is ‘definite’. The centre server checks those sequences and changes its status from ‘temporary’ to ‘definite’ if the ‘temporary’ record is surrounded by ‘definite’ data without inconsistency (Figure 7.7(c)).

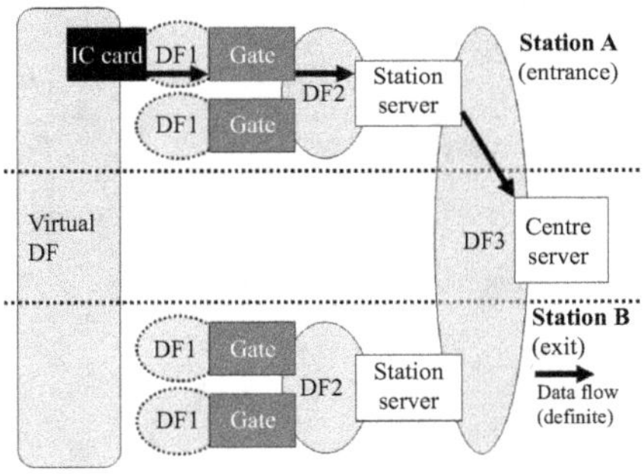

Seq.#	Status	Date and time	Used	Left
				1,000
14	**Definite**	**2008/04/01 10:00**	**–130**	**870**

(a)

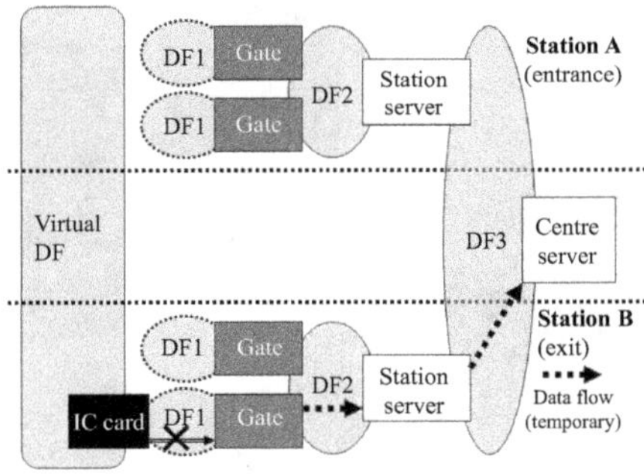

Seq.#	Status	Date and time	Used	Left
				1,000
14	Definite	2008/04/01 10.00	–130	870
15	**Temporary**	**2008/04/01 10.30**	**-50**	**820**

(b)

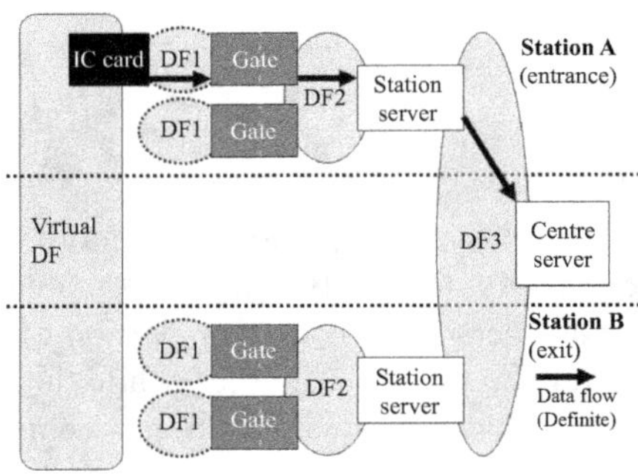

Seq.#	Status	Date and time	Used	Left
				1,000
14	Definite	2008/04/01 10.00	–130	870
15	Temporary >>> **Definite**	2008/04/01 10.30	-50	820
16	Definite	2008/04/02 10.00	–130	690

(c)

Figure 7.7 Data recovery by ADDCT: (a) definite data #14 is processed; (b) temporary data #15 is created; and (c) definite data #16 is created, and ADDCT recovers #15. Copyright © 2009 IEICE [4]

With ADDCT, IC cards can escape from being voided even if passengers have caused 'data missing'. The data are recovered before being blacklisted, that is this technology assures reliability of data through the integration of autonomous processes at the terminals and at the centre server.

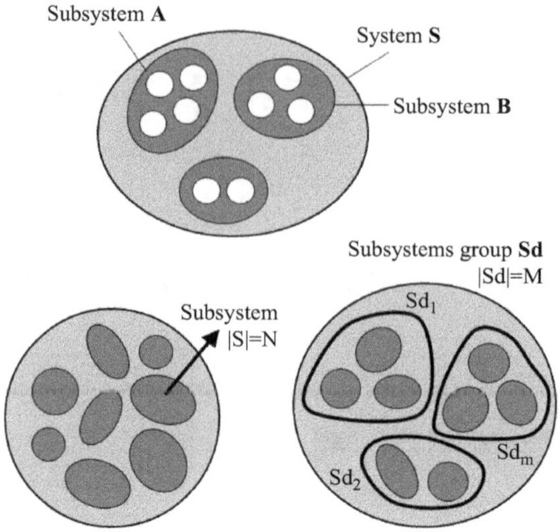

Figure 7.8 Function reliability evaluation model. Copyright © 2009 IEICE [4]

7.4.2 Evaluation

7.4.2.1 Function reliability

The autonomous decentralised system guarantees smooth operation even with partial failure. Therefore, function reliability is more a suitable criterion to evaluate the architecture, since it considers the functioning portion of subsystems with the ability to cooperate and integrate with one another, while the conventional method of reliability evaluation judges whether total function is achieved or not [13–15].

A system model for evaluating function reliability is shown in Figure 7.8. It shows a system consisting of several subsystems (or subsystem groups), and each subsystem has certain functions. Each unit is evaluated by amount of function.

This section takes 'function reliability' into consideration for evaluation of ICCTS. The basic functions of the ICCTS are to check for invalid cards and calculate fares. Since these functions are related to data found in the cards, each subsystem has 'consistent data'. The achievement of functionality in terms of ICCTS is defined as the 'consistency of data', which means how much data the servers and the cards have in common (Figure 7.9).

Accordingly, ICCTS is modelled as seen in Figure 7.10. Each pellet shape seen within the subsystem represents one piece of consistent data. In reality, ICCTS servers save the data simply as backup and use them to facilitate greater reliability. This simulation focuses on modelling cards and terminals since each terminal checks invalid cards and calculates fares.

7.4.2.2 Evaluation results

The effectiveness of ADDCT in enhancing data reliability is shown in Figure 7.11, according to simulations based on actual transactions. The amount of data which a

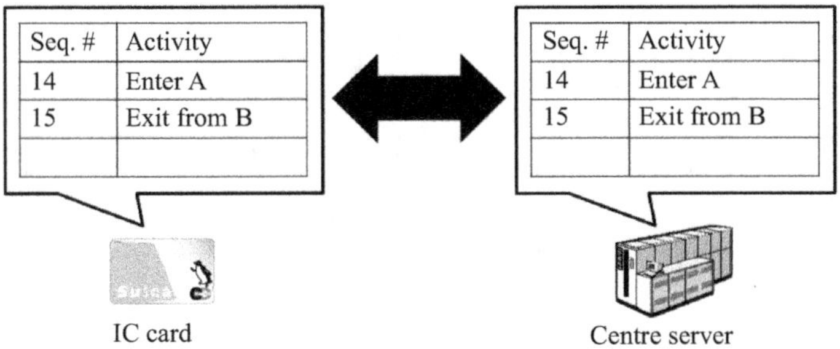

Figure 7.9 Consistency of data. Copyright © 2009 IEICE [4]

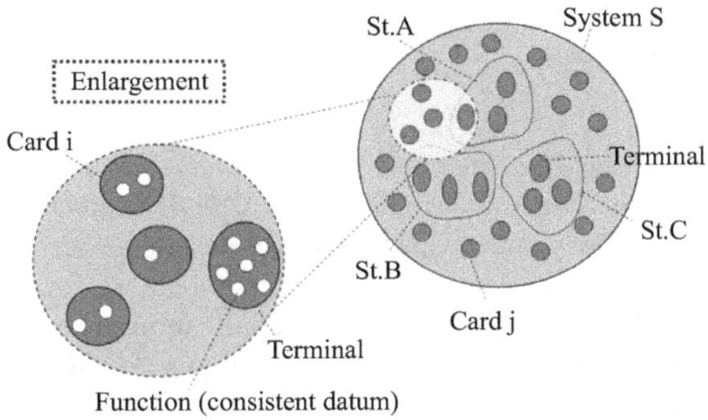

Figure 7.10 Model for evaluating function reliability. Copyright © 2009 IEICE [4]

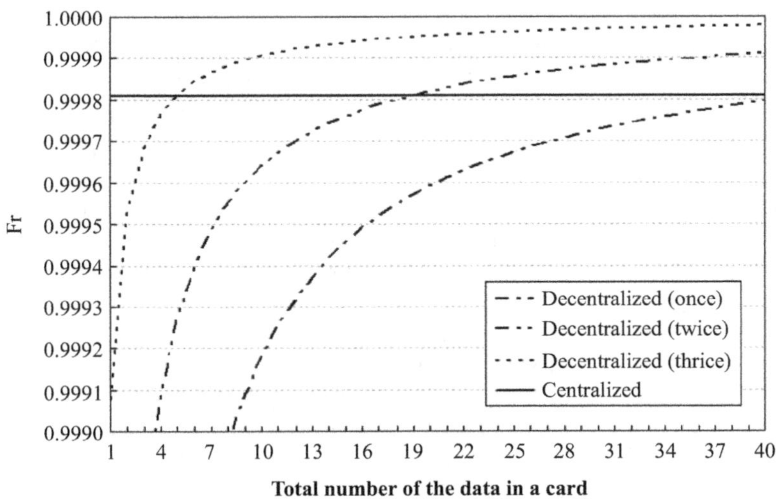

Figure 7.11 Function reliability with/without ADDCT. Copyright © 2009 IEICE [4]

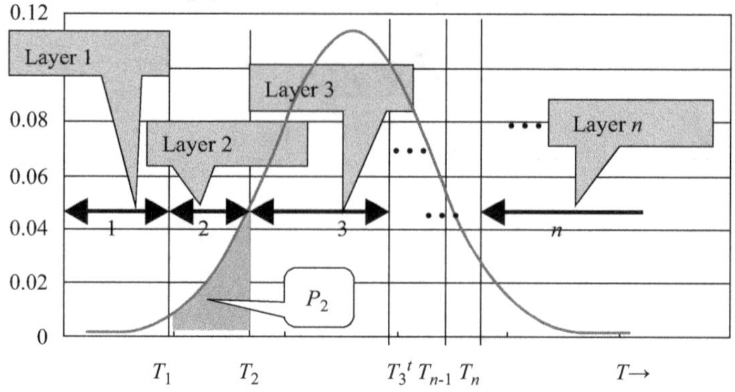

Figure 7.12 Distribution rate of passengers' travelling times. Copyright © 2009 IEICE [4]

card is able to store is shown along the X axis, and the average function reliability is plotted along the Y axis. Function reliability in ICCTS is evaluated in terms of data consistency, which defines stable operation of a system in terms of how well it guarantees the provision of reliable data. In the simulation, four cases are examined: one system each which runs the ADDCT application once a day, twice a day, and three times a day; and a centralised system without ADDCT.

ADDCT can make up for as much missing data as there is available recording space in the card. The more records a card is able to store, the more 'missing data' are accepted. The results from the simulations prove that if there is sufficient data, autonomous decentralised system structure is more reliable with ADDCT. However, it was found that there is no apparent merit in increasing the number of records in one card to more than 20. Hence, it is very important to determine the appropriate memory size. In the Suica system, each IC card is capable of keeping 20 records at a time, and the ADDCT application runs three times a day. According to Figure 7.11, under these conditions, the system scores the highest function reliability. This means that the effectiveness of ADDCT is proven both practically and theoretically [8].

Then, what if ADDCT application runs in each subsystem at each layer? To evaluate this, a sample distribution of the passengers' travelling times is prepared as seen in Figure 7.12. The data in such distribution is divided into groups based on T ($0<T<T_1$, $T_1<T<T_2$, and so on), and each group is assigned to a layer where it is recovered. The simulations are intended to find the most effective configuration to deal with the transactions: the number of layers and the time each layer holds the data.

Its effectiveness is shown in Figures 7.13 and 7.14. Figure 7.13 shows the result from a 2-layered model where T_1 ranges from 1 to 59 min, while T_2 is fixed at 60 min. T_{exp}, the expectation of recovery time, indicates the minimum value of 33.82 when $T_1 = 31$. This means, in the most effective configuration of 2-layered structure, the ADDCT application at Layer 1 should be set to run 31 min after the flow starts.

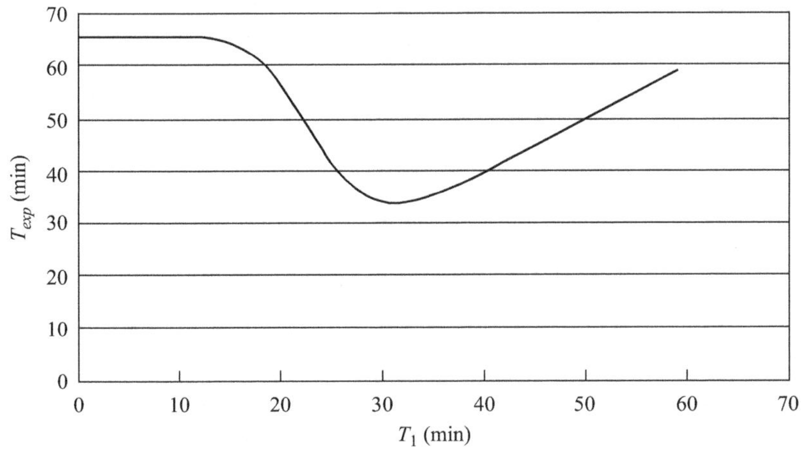

Figure 7.13 T_1 vs. T_{exp} in 2-layered models. Copyright © 2009 IEICE [4]

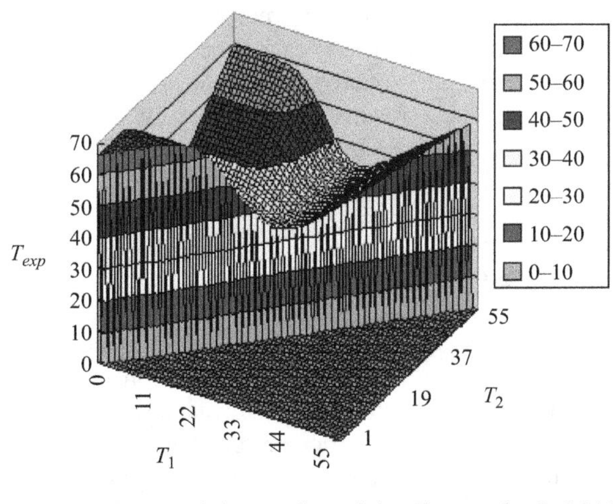

Figure 7.14 T_1 vs. T_{exp} in 3-layered models. Copyright © 2009 IEICE [4]

Figure 7.14 shows the result from the 3-layered model, where T_1 and T_2 ranges from 1 to 59 min, while T_3 is fixed at 60 min. T_{exp} indicates a minimum value of 29.72 when $T_1 = 26$ and $T_2 = 34$. According to these results, in the most effective configuration of 3-layered structure, the ADDCT application at Layer 1 should be set to run 26 min after the flow starts at Layer 1, and the ADDCT application at Layer 2 should be set to run 34 min after. In addition, T_{exp} is less in a 3-layered model than in a 2-layered model, that is the 3-layered model is more efficient in this input flow.

T_{exp} depends on the number of layers (indicated as n) and the timing of the application (indicated as T_n). The way to design the most efficient system is to find n and T_n to minimise T_{exp}. To compare the results from the simulations more easily,

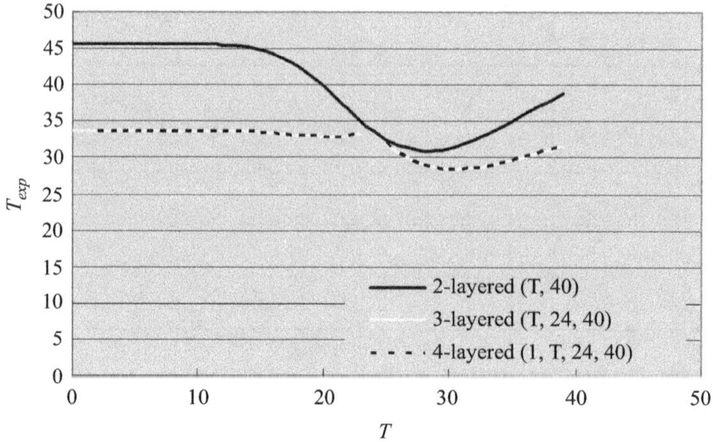

Figure 7.15 T_{exp} *in 2-, 3-, and 4-layered simulations. Copyright © 2009 IEICE [4]*

Figure 7.15 shows the most appropriate parameters in each layered simulation. The minimum values of T_{exp} are 30.96 in the 2-layered model, 28.43 in the 3-layered, and 28.44 in the 4-layered. The 3-layered model performs better than the 2-layered or the 4-layered models. These results reveal two facts: more layers work more efficiently, and too many layers become useless because the nodes cannot gather enough diffused data to match up [9,10].

7.5 Best designing of the system

In the previous sections, ACPT and ADDCT are introduced as implementation samples of the ADS concept to the ICCTS. This section introduces how specifications are determined to implement high performance and high reliability to the actual designing of the systems, in particular focusing on the size of the communicable area between an IC card and a reader. The size is very important because it governs the processing time.

7.5.1 System modelling

High performance and high reliability are the very important requirements in ICCTS, which has been repeatedly mentioned in this chapter. Once these requirements are satisfied, service requirements such as fluidity and service continuity are assured. One of the most important factors is the distance between an IC card and a reader. Readers can process IC cards only while the IC cards are within the communicable area of the readers; the readers can no longer process the IC cards which go out of the area. If an IC card goes out of the area in a transaction, it leads to the missing data described in Section 7.4. Infinitively powerful readers and processors could capture the IC cards and avoid missing data, but it is not realistic, especially in the cases depending on

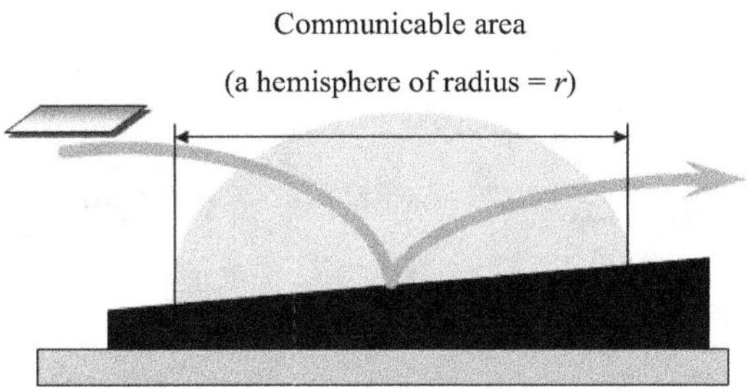

Figure 7.16 Distance between an IC card and a reader

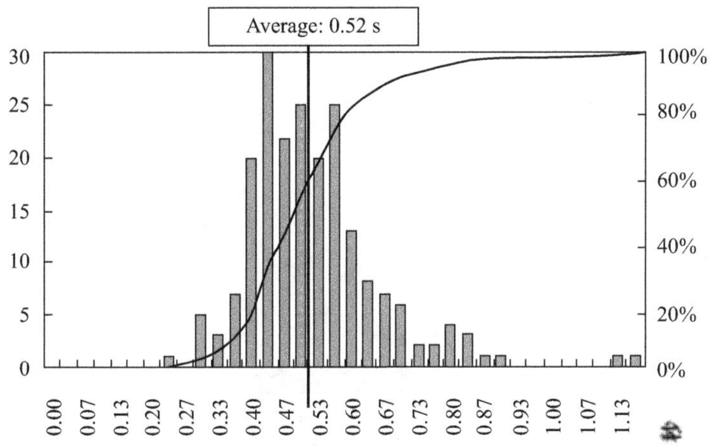

Figure 7.17 The variety of the time in which humans hold IC cards within the communicable area

human behaviours. Figure 7.16 is a model of a communicable area between an IC card and a reader. It is regarded as a hemisphere of radius $= r$.

When a person goes through a gate with an IC card, the reader on the gate has to process the IC card before the IC card goes out of the communicable area. The actual distribution of the time in which humans hold IC cards within the area is observed in Figure 7.17. This can be approximated as normal distribution.

7.5.2 Evaluation

To evaluate the model, 'service continuity' (reliability) and 'fluidity' (performance) are proposed.

7.5.2.1 Service continuity

Service continuity is defined as the complement of error rate. The error rate is a function of process time, and the process time is a function of radius of communicable area, that is service continuity is a function of radius of communicable area.

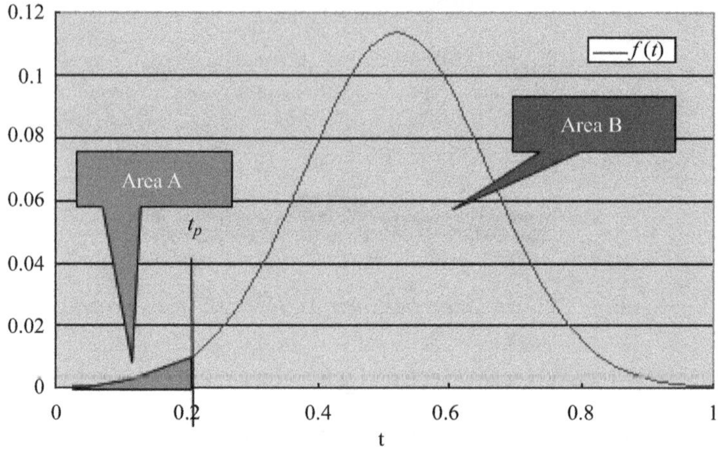

Figure 7.18 Distribution and probability of the time within the communicable area (r = 0.1)

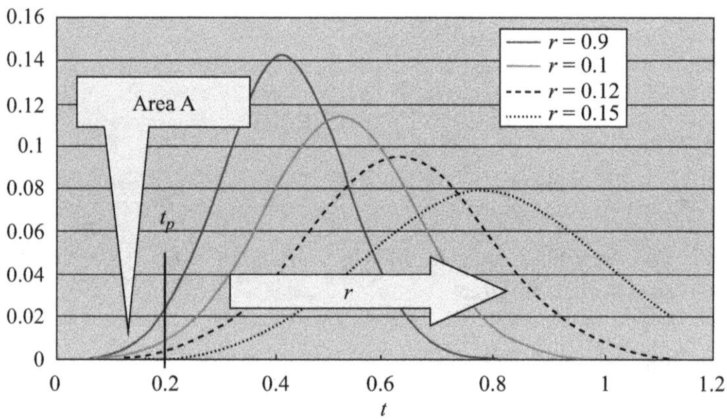

Figure 7.19 Distribution and probability of the time within the communicable area by time/radius

Figure 7.18 is the normal distribution which approximates Figure 7.17. Suppose that t_p is the processing time of a reader. When a person holds the IC card longer than t_p, the transaction is successful. When a person takes the IC card away from the area before t_p, the transactions fails. Then successful rate of transactions is depicted as Area A while failure rate is depicted as Area B.

In Figure 7.18, the radius is fixed at $r = 0.1$. Figure 7.19 shows the distributions by various radius r. The longer r leads to the bigger Area B.

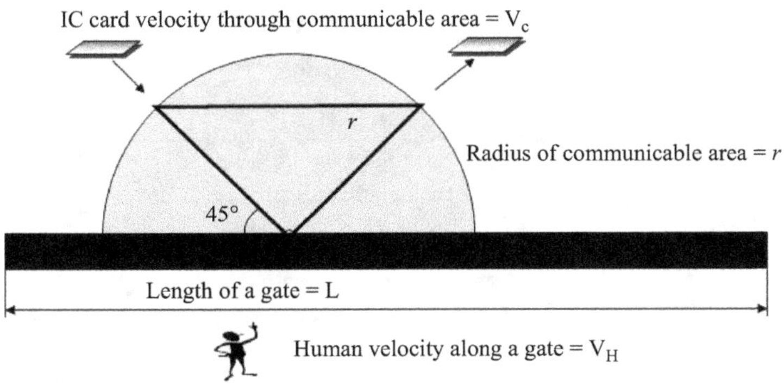

Figure 7.20 Behaviours of IC card in a communicable area and person along a gate

7.5.2.2 Fluidity

Fluidity is defined as the number of processed IC cards per passing person in unit time. Ideally high performance increases the numbers of both processed IC cards and passing persons. However, in reality, more IC cards are left uncompleted in the process when persons go through more quickly. In other words, reliability gets lower when more persons pass through a gate. Figure 7.20 is a model of a person and an IC card at a gate. Suppose that the velocity of an IC card is a function of the holding person's velocity. Then the times in which an IC card is within the communicable area of a reader and in which a person is along the gate can be calculated. Fluidity of IC cards is defined as reciprocal of the time in which an IC card is within the communicable area of a reader, while fluidity of person is defined as reciprocal of the time in which a person is along the gate. The entire fluidity is defined as fluidity of IC cards divided by fluidity of person.

7.5.2.3 Evaluation results

Figure 7.21 shows service continuity and fluidity along the radius of communicable area upon a certain condition. As the radius gets bigger, service continuity increases, while fluidity decreases. Service continuity and fluidity cross at $r = 0.072$; this is the best size of communicable area derived from the condition. The actual size of communicable area is 0.09 – that is the actual readers are designed to have more importance on service continuity or reliability at the cost of fluidity.

7.6 Conclusion

In this paper, the advancement of ADS technologies in fare-payment systems was surveyed in accordance with their changing and growing requirements and their corresponding system structures in the last 30 years. The Suica system operated by JR East is one of the most advanced implementations of ADS, and it has been

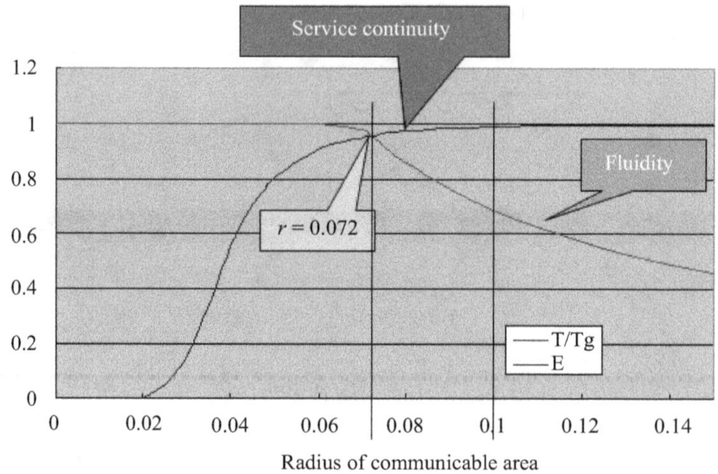

Figure 7.21 Service continuity and fluidity along radius of communicable area

utilised not only for information transaction processing in fare calculation and e-commerce but also for controlling the gates under 0.2 s.

Suica started as an ICCTS within Tokyo metropolitan area in 2001. It was just a new AFCS at that time. However e-commerce use is added in 2004. Meanwhile, similar ICCTS gradually started in major cities in Japan. They started as independent AFCS within the cities but have eventually expanded to interoperation – one local IC card can be used in other cities and vice versa. Such expandability owes to the ADS concept – one ICCTS consisting of many internal subsystems becomes one subsystem of nationwide interoperation.

The future enhancement of ICCTS is considered to be in 'cloud'. Thanks to networking multifunctional terminals, AFCS became more convenient and cost-effective. However, the functions are still installed in each terminal locally; servicepersons no longer visit every terminal to upgrade but download new software one by one when necessary. The terminals need expensive secured hardware to store confidential data. When ICCTS is proposed, communications and processor performances are too poor to host online transactions at the gates. Recent development of communications and processor performances indicates a possibility to enable such online transactions, on the standpoint of performance. The next step is to establish a standard to evaluate reliability of cloud – regarding as a big system consisting of numerous tiny subsystems. If all the functions and data are stored in cloud which all the terminals access to, upgrading become easier and less expensive, and then more convenient service can be proposed.

These research activities of autonomous decentralised system extend to computer, communications, and control technologies, and their integration. It is expected that autonomous decentralised system created in Japan will play an active role in collaboration of academia, industry, and government around the globe.

References

[1] Mori, K., Miyamoto, S., and Ihara, H.: 'Proposition of Autonomous Decentralized System Concept', Trans. IEE Japan, vol. 104C, no. 12, pp. 303–340, 1984.

[2] Mori, K.: 'Autonomous Decentralized Software Structure and Its Application', IEEE Proc. of FJCC86, pp. 1056–1063, 1986.

[3] Mori, K.: 'Introduction of Autonomous Decentralized System', Morikita Pub. Co., Tokyo, Japan, 2006.

[4] Mori, K., Shiibashi, A.: 'Trend of Autonomous Decentralized System Technologies and Their Application in IC Card Ticket System', IEICE Trans. Commun., vol. E92-B, no. 2, pp. 445–460, 2009.

[5] Shirakawa, Y., Shiibashi, A.: 'JR East Contact-less IC Card Automatic Fare Collection System "Suica"', IEICE Trans., vol. E86-D, no. 10, pp. 2070–2076, 2003.

[6] Shiibashi, A.: 'Autonomous Decentralized High-speed Processing Technology and the Application in an Integrated IC Card Fixed-line and Wireless System', IEICE Trans., vol. E88-D, no. 12, pp. 2699–2707, 2005.

[7] Shiibashi, A., Mizoguchi, N., Mori, K.: 'High-Speed Processing in Wired-and-Wireless Integrated Autonomous Decentralized System and Its Application to IC Card Ticket System', Innov. Syst. Softw. Eng., vol. 3, no. 1, pp. 53–60, 2007.

[8] Shiibashi, A., Kuroda, T., Yamana, M., Mori, K.: 'Research of Reliability Technology in Heterogeneous Autonomous Decentralized Assurance Systems', ISADS, Sedona, US, pp. 207–214, 2007.

[9] Shiibashi, A., Yamana, M., Mori, K.: 'Multi-layered Data Consistency Technology in IC Card Ticket System', Assurance Symposium, Yokohama, Japan, pp. 1–8, 2007.

[10] Shiibashi, A., Maruyama, Y., Yamana, M., Mori, K.: 'Multi-layered Data Consistency Technology in IC Card Ticket System', ADSN, Toronto, Canada, pp. 58, 2007.

[11] Naka, Y.: 'Study on complicated passenger flow in a railway station', Railway Technical research report, no. 1079, 1978.

[12] Imai, A.: 'Examination of Parameter of Size of Automatic Fare Collection Gate', Omron Technics, vol. 12, no. 1, pp. 25–40, 1972.

[13] Mori, K., Miyamoto, S., Ihara, H.: 'On Evaluation of Function and Reliability of Distributed Control System', Soc. Instrum. Control Eng. Trans., vol. 20, no. 4, pp. 314–321, 1984.

[14] Kera, K., Bekki, K., Mori, K.: 'Step-by-Step System Construction Technique with Assurance Technology – Evaluation Measure for Step-by-Step System Construction', 22nd International Conference on Distributed Computing Systems Workshops, p. 101, Vienna, Austria, 2002.

[15] Matsumoto, M., Mori, K.: 'Assurance Evaluation Technology for an Autonomous Decentralized ATC System', IEICE, vol. J86-D-I, no. 1, pp. 14–22, 2003.

Chapter 8

Robot as a Service and its visual programming environment

Yinong Chen[1] and Gennaro De Luca[1]

Abstract

Robot as a Service (RaaS) is a cloud-computing unit that facilitates the seamless integration of robots and embedded devices into Web and cloud-computing environments. The RaaS concepts can be applied to different types of IoT applications, including cyber-physical systems, autonomous decentralized systems, serverless computing systems, and Internet of Intelligent Things. This section presents the design and implementations of a number of RaaS units, as well as a Visual IoT/Robotics Programming Environment that can visually program RaaS units through a drag-and-drop style. Multiple physical robots and simulated robots are implemented. The platform independence was ensured through a standard interface defined in a JavaScript Object Notation object. The development and testing of several sample applications is shown.

8.1 Introduction

The Internet of Things (IoT) refers to uniquely identifiable objects (things) and their virtual representations in an Internet structure [1]. The concept was initially applied in the radio-frequency identification tag to mark the electronic product code (Auto-ID Lab). The concept of the IoT is extended to refer to the world where physical objects are seamlessly integrated into the information network, and where the physical objects can become active participants in business processes [2]. The Internet of Intelligent Things (IoIT) deals with intelligent devices that have adequate computing capacity. Distributed intelligence is a part of the IoIT [3]. According to Intel's report, there are 15 billion devices that are connected to the Internet, in which 4 billion devices include 32-bit processing power, and 1 billion devices are intelligent systems [4].

In addition to IoT, a number of related concepts and systems have been proposed to take advantage of Internet and cloud computing. A cyber-physical system (CPS) is a combination of a large computational and communication core and

[1]Computer Science and Engineering Department, Arizona State University, USA

physical elements that can interact with the physical world [5–7]. CPSs can be considered the extended and decentralized version of embedded systems. In CPSs, the computational and communication core and the physical elements are tightly coupled and coordinated to fulfill a coherent mission. The US National Science Foundation issued a program solicitation in 2008 on CPSs, envisioning that the CPSs of tomorrow would far exceed those of today in terms of adaptability, autonomy, efficiency, functionality, reliability, safety, and usability. Research advances in CPSs promise to transform our world with systems that respond more quickly (e.g., autonomous collision avoidance), are more precise (e.g., robotic surgery and nano-tolerance manufacturing), work in dangerous or inaccessible environments (e.g., autonomous systems for search and rescue, firefighting, and exploration), provide large-scale, distributed coordination (e.g., automated traffic control), are highly efficient (e.g., zero-net energy buildings), augment human capabilities, and enhance societal well-being (e.g., assistive technologies and ubiquitous healthcare monitoring and delivery) [8].

An autonomous decentralized system (ADS) is a distributed system composed of modules or components that are designed to operate independently but are capable of interacting with each other to meet the overall goal of the system. ADS components are designed to operate in a loosely coupled manner, and data is shared through a content-oriented protocol. This design paradigm enables the system to continue to function in the event of component failures. It also enables maintenance, and repair is to be carried out while the system remains operational. Autonomous decentralized systems have a number of applications including industrial production lines, railway signaling, and robotics [9,10].

Robot as a Service (RaaS) is a cloud-computing unit that facilitates the seamless integration of robots and embedded devices into Web and cloud-computing environments [3,11,12]. In terms of the service-oriented architecture, a RaaS unit includes services for performing functionality, a service directory for discovery, and service clients for the user's direct access. The current RaaS implementation facilitates SOAP and RESTful communications between RaaS units and the other cloud-computing units. Hardware support and standards are available to support RaaS implementation. For example, the Devices Profile for Web Services defines implementation constraints to enable secure Web service messaging, discovery, description, and eventing on resource-constrained devices between Web services and devices. The recent Intel IoT-enabled architectures, such as Galileo and Edison, make it easy to program these devices as Web services. From different perspectives, an RaaS unit can be considered a unit of the IoT, IoIT (which have adequate computing capacity to perform complex computations [3]), a CPS (which is a combination of a large computational and communication core and physical elements that can interact with the physical world [4]), and an ADS. Scheduling tasks and allocating resources in such environments require new strategies and techniques. The gang scheduling and real-time scheduling techniques in ad hoc distributed systems and on the elastic cloud environment ensure that related tasks are scheduled in groups and are running simultaneously [13–15], which can be used to coordinate with the intelligent devices.

RaaS is an important extension to cloud computing by adding distributed computing capacity into the centralized computing model in cloud computing. This model is similar to serverless computing [16], in which the resources used are not in the cloud, but outside the cloud. This addition is critical for excessive computing requirements, such as big data processing. Figure 8.1 illustrates the spiral model of computing system development.

We started our computing era with centralized computing using mainframe computers in the 1950s. We moved to distributed computing using personal computers and workstations in the 1980s. Cloud computing hides the detail and the complexity of distributed computing and presents the combined computing resources as a centralized environment. Even though cloud computing can provide any computing capacity that a client wants, it is beneficial to have a part of the computing distributed to the front end computer. In the case of IoIT and RaaS, using the computing capacity of the devices can reduce the computing requirement of cloud computing, the communication delay, and the response time for clients.

The development of IoT and RaaS is not only a hardware issue. It involves device level coding, which is difficult and time-consuming. The recent developments in executable business logic, visual workflow, and software integration have greatly simplified software development [17]. The same techniques are being applied to IoT and RaaS development. A number of visual programming environments have been developed. MIT App Inventor [18] uses drag-and-drop style puzzles to construct phone applications involving not only computation but also reading of sensory values. University of Virginia and Carnegie Mellon's Alice is a 3D game and movie-development environment [19]. It uses a drop-down list for users to select the available functions in a stepwise manner. App Inventor and Alice allow novice programmers to develop complex applications using visual composition at the workflow level. First released in 2016, Intel IoT Service Orchestration Layer is the latest visual programming tool that allows the developers to combine

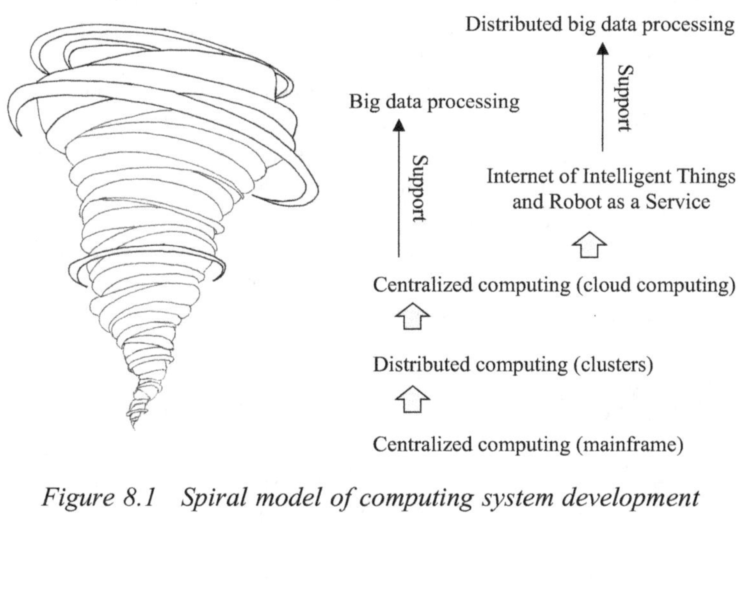

Figure 8.1 Spiral model of computing system development

the functionalities of IoT devices through visual workflow [20]. It can be conveniently used for programming IoT devices based on Intel Edison, Galileo, and other platforms. Microsoft Robotics Developer Studio (MRDS) and visual programming language (VPL) are specifically developed for robotics applications [21], which is a milestone in software engineering, robotics, and computer science education from many aspects. Microsoft VPL is service-oriented, it is visual and workflow-based, it is event-driven, it supports parallel computing, and it is a great educational tool that is simple to learn and yet powerful and expressive. Sponsored by two Innovation Excellence Awards from Microsoft Research in 2003 and 2005, Dr. Yinong Chen participated in the early discussion of a service-oriented robotics developer environment at Microsoft. MRDS and VPL were immediately adopted at Arizona State University (ASU) in developing the freshman computer science and engineering course CSE101 in 2006. The course grew from 70 students in 2006 to over 350 students in 2011. The course was extended to all students in the Ira A. Fulton Schools of Engineering (FSE) at ASU and was renamed FSE100, which is offered to thousands of freshman engineering students now.

Unfortunately, Microsoft stopped its development and support for MRDS and VPL in 2014 [22], which leads to our FSE100 course, and many other schools' courses using VPL, without further support. Particularly, the current version of VPL does not support LEGO's third generation of EV3 robot, while the second generation NXT is out of the market.

To keep our course running and also help the other schools, we take the challenge and responsibility to develop our own visual programming environment at ASU. We name this environment Visual IoT/Robotics Programming Environment (VIPLE), standing for Visual IoT/Robotics Programming Language Environment.

ASU VIPLE is based on our previous e-Robotics development environments [23]. It is designed to support as many features and functionalities of MRDS and VPL as possible, in order to better serve the MRDS and VPL communities in education and research. To serve this purpose, VIPLE also keeps a similar user interface, so that the MRDS and VPL development communities can use VIPLE with little additional learning. VIPLE does not replace MRDS or VPL. Instead, it extends MRDS and VPL in their capacities in multiple aspects. It can connect to different physical robots, including EV3 and any robots based on off-the-shelf processors.

The rest of the paper is organized as follows. Section 8.2 presents an overview of the RaaS design and the VIPLE environment. Section 8.3 outlines the VIPLE programming environment, its foundation, and building blocks. Section 8.4 presents and discusses the platforms implemented and the RaaS design both in hardware and in simulation. Section 8.5 illustrates the applications of VIPLE in different applications. Section 8.6 concludes the chapter.

8.2 System overview

To facilitate the development of RaaS and applications, we designed and implemented an environment that consists of VIPLE on the backend computer, JSON

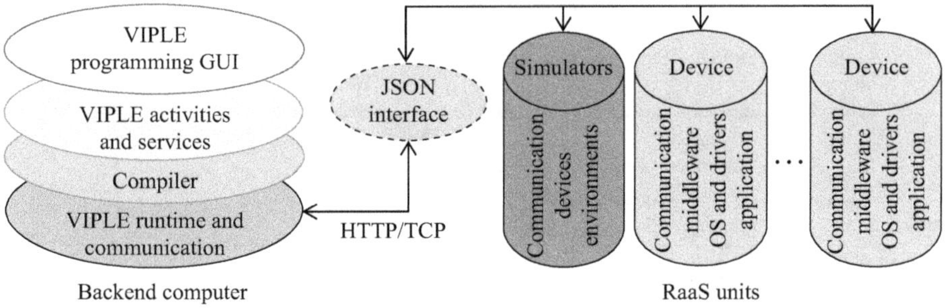

Figure 8.2 RaaS development and application environment

(JavaScript Object Notation) interface, middleware, and operating system and drivers on the devices, as shown in Figure 8.2.

VIPLE consists of four major components: (1) a visual programming interface for the developers to draw the applications, (2) a repository of built-in activities and services to simplify the development jobs, (3) a compiler that translates the visual code/flowchart into the executable, and (4) the runtime that executes the code and communicates with the RaaS units through HTTP (Hypertext Transfer Protocol) and TCP (Transmission Control Protocol) with the simulators and the physical devices. The data interfaces between VIPLE and the simulators and the devices are defined in a JSON object, which plays a vital role in the platform independence of the environment. Any simulator and physical devices can be added into the system, as long as they can provide a TCP socket or a Web socket for the communication and encode and decode the JSON object to understand the semantics. Figure 8.3 shows an example of the JSON object.

In the following section, we will explain the key components of the system, VIPLE, RaaS design, including physical devices and simulator.

8.3 VIPLE: Visual IoT/Robotics Programming Environment

The basic building blocks of an ASU VIPLE program (diagram) is listed and explained in Figure 8.4. They offer a complete set of activities described in process control theory, such as π-calculus [24,25].

Table 8.1 lists the basic constructs in π-calculus, their graphic representations [24], and the VIPLE activities that implement the constructs.

The usability of a language largely depends on the availability of library functions or called services. Figure 8.5 shows the three sets of ASU VIPLE services.

The first set contains the general services, including input/output services (simple dialog, print line, text to speech, and random), event services (key press event, key release event, custom event, and timer), and RESTful services.

The second set is the generic robotic services. VIPLE offers a set of standard communication interfaces, including Wi-Fi, TCP, Bluetooth, USB, localhost, and WebSocket interfaces. The data format between VIPLE and the IoT/Robotic devices is defined as a standard JSON object. Any robot that can be programmed to

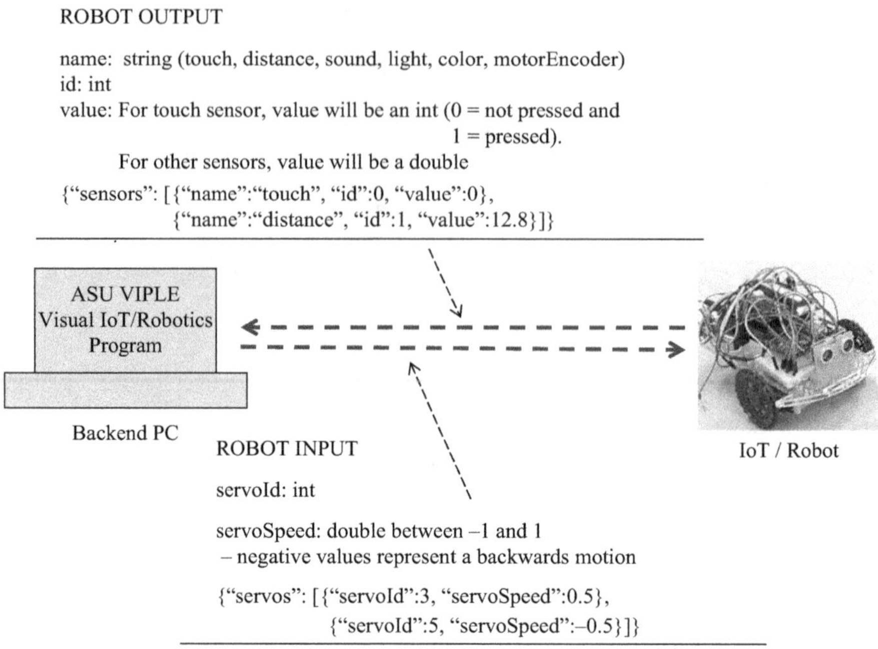

Figure 8.3 JSON object example containing sensor sending and moto control data

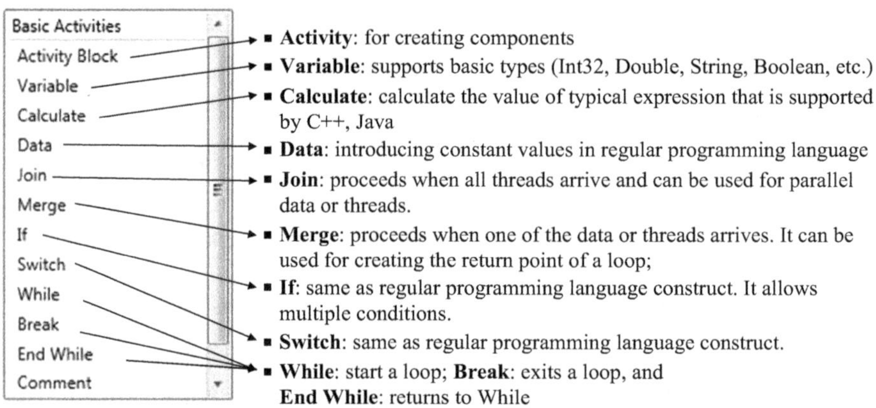

Figure 8.4 Activities: ASU VIPLE versus Microsoft VPL

support one of the communication types and can process the JSON object can communicate with VIPLE and be programmed in VIPLE. As shown in the second part of Figure 8.5, all VIPLE services that start with "Robot" are generic robotic services. We can use these services to program our simulated robots and custom-built physical robots.

Table 8.1 Basic constructs in π-calculus, their graphic representations constructs, and VIPLE activities

π-Calculus construct	π-Calculus construct diagram	VIPLE activities
Sequential process $S ::= new\ a(A\|B).\mathbf{0}$ $A ::= \tau_A.\overline{a}\langle x\rangle.\mathbf{0}$ $B ::= a(x).\tau_B.\overline{b}\langle y\rangle.\mathbf{0}$	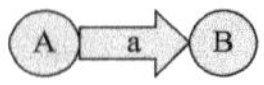	Activity block Variable Calculate Data
Split (parallel processes) $n \geq 2$ $S ::= new\ a1a2\ldots an(A\|B1\|B2\|\ldots\|Bn)$ $A ::= \tau_A.\overline{a1}\langle x\rangle.\overline{a2}\langle x\rangle.\ldots.\overline{an}\langle x\rangle$ $Bn ::= an(x).\tau_{\mathrm{Bn}}.\overline{bn}\langle y\rangle$	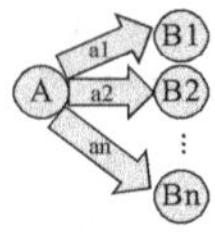	Activity block Variable Calculate Data
Join $n \geq 1$ $S ::= new\ a1a2\ldots an(A1\|A2\|\ldots\|An\|B)$ $An ::= \tau_{\mathrm{An}}.\overline{an}\langle x\rangle$ $B ::= a1(x1).a2(x2).\ldots.an(xn).\overline{b}\langle y\rangle$	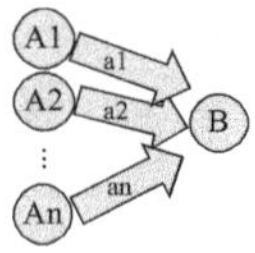	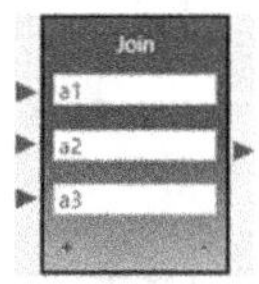
Merge $n \geq 1$ $S ::= (A1\|A2\|\ldots\|An\|B)$ $An ::= \tau_{\mathrm{An}}.\overline{a}\langle x\rangle$ $B ::= !a(x).\overline{b}\langle y\rangle$	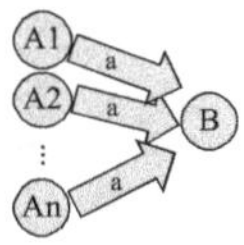	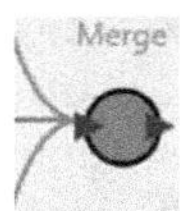
Choice (if) $n \geq 2$ $S ::= new\ a1a2\ldots an(A\|B1\|B2\|\ldots\|Bn)$ $A ::= \textbf{\textit{if}}\ bexpr1\ \textbf{\textit{then}}\ \overline{a1}\langle x\rangle\ \textbf{\textit{else}}$ $\textbf{\textit{if}}\ bexpr2\ \textbf{\textit{then}}\ \overline{a2}\langle x\rangle$ $\textbf{\textit{else if}}\ldots \textbf{\textit{else if}}\ bexpr(-1)\ \textbf{\textit{then}}\ \overline{a(n-1)}\langle x\rangle$ $\textbf{\textit{else}}\ \overline{an}\langle x\rangle\ Bn ::= an(x).\tau_{\mathrm{Bn}}.\overline{bn}\langle y\rangle$	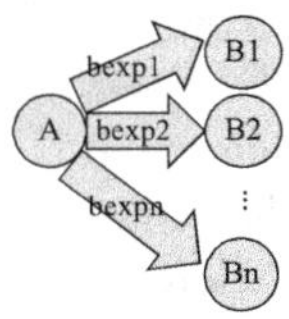	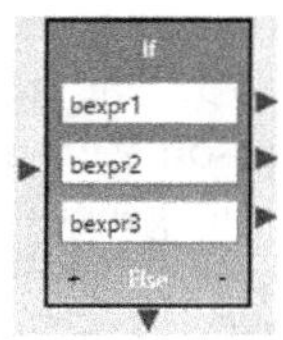
Choice (switch) $n \geq 2$ $S ::= new\ ca1a2\ldots an(C\|A\|B1\|B2\|\ldots\|Bn)$ $C ::= \overline{c}\langle z\rangle$ $A ::= c(z).\ \textbf{\textit{if}}\ exp1 = z\ \textbf{\textit{then}}\ \overline{a1}\langle x\rangle$ $\textbf{\textit{else if}}\ exp2$ $= z\ \textbf{\textit{then}}\ \overline{a2}\langle x\rangle\ \textbf{\textit{else if}}\ldots \textbf{\textit{else if}}\ exp(n-1)$ $= z\ \textbf{\textit{then}}\ \overline{a(n-1)}\langle x\rangle\ \textbf{\textit{else}}\ \overline{an}\langle x\rangle$ $Bn ::= an(x).\tau_{\mathrm{Bn}}.\overline{bn}\langle y\rangle$	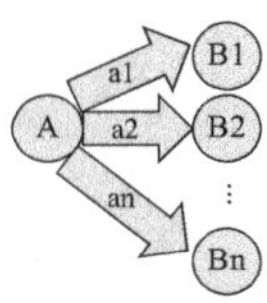	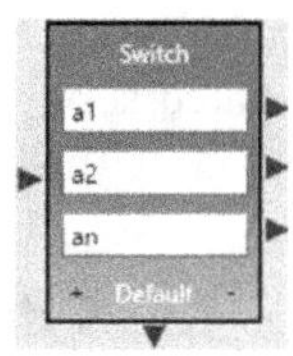
Loop (while … end while) $S ::= A\|B\|C$ $A ::= d(y).\ \textbf{\textit{if}}\ bexpr\ \textbf{\textit{then}}\ \overline{a}\langle x\rangle\ \textbf{\textit{else}}\ \overline{c}\langle y\rangle$ $B ::= a(x).\tau_B.\overline{b}\langle y\rangle$ $C ::= b(y).\overline{d}\langle y\rangle$	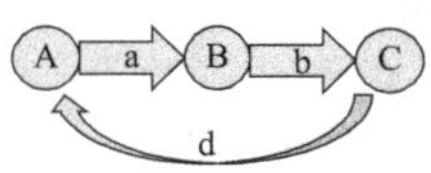	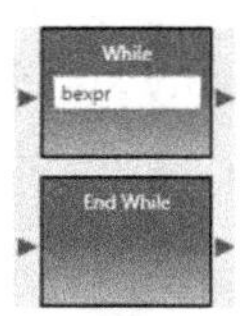

General-purpose and event services	Generic robotic services	Vendor-specific robotic services
Services	Robot	
Code Activity	Robot Color Sensor	Lego EV3 Brick
Custom Event	Robot Distance Sensor	Lego EV3 Color
Key Press Event	Robot Drive	Lego EV3 Drive
Key Release Event	Robot Holonomic Drive	Lego EV3 Drive for Time
Print Line	Robot Light Sensor	Lego EV3 Gyro
Random	Robot Motor	Lego EV3 Motor
RESTful Service	Robot Motor Encoder	Lego EV3 Motor by Degrees
Simple Dialog	Robot Sound Sensor	Lego EV3 Motor for Time
Text to Speech	Robot Touch Sensor	Lego EV3 Touch Pressed
Timer	Robot+ Move at Power	Lego EV3 Touch Released
	Robot+ Turn by Degrees	Lego EV3 Ultrasonic

Figure 8.5 ASU VIPLE services

The third set is the vendor-specific services. Some robots, such as LEGO robots and iRobots, do not offer an open communication and programming interfaces. In this case, we can offer built-in services in VIPLE to access these robots without requiring any programming efforts on the device side. Currently, the services for accessing LEGO EV3 robots are implemented, so that VIPLE can read all EV3 sensors and control EV3 drive-motors and arm-motors, as shown in the third part in Figure 8.5. For those who do not want to build their own robots, they can simply use VIPLE and an EV3. The addition of EV3 services to VIPLE is significant, as it allows the Microsoft VPL developers who used NXT robots now to use the new EV3 robots.

The generic robot services allow the developers to use VIPLE to connect to an open architecture robot. In Microsoft VPL, DSS services developed specifically for MRDS can be added into the VPL service list. In ASU VIPLE, RESTful services can be accessed in VIPLE diagrams. As RESTful services are widely used in today's Web application development, the access to RESTful services extends the capacity of VIPLE to a wide range of resources. ASU VIPLE does not have simulated services at this time, although the generic robot services can be used to interface with a custom-simulation environment.

8.4 RaaS design and implementation in different platforms

A number of RaaS units have been implemented based on different hardware platforms, as shown in Figure 8.6. Three simulators have been developed to work with VIPLE to visualize the execution of robots in the maze. They can be used in the testing phase of RaaS design or in the case where physical robots are not available. The Unity simulator shows a simulated robot in a 3D maze, while the Web simulators are developed with HTML 5 and can run on any device and communicate with

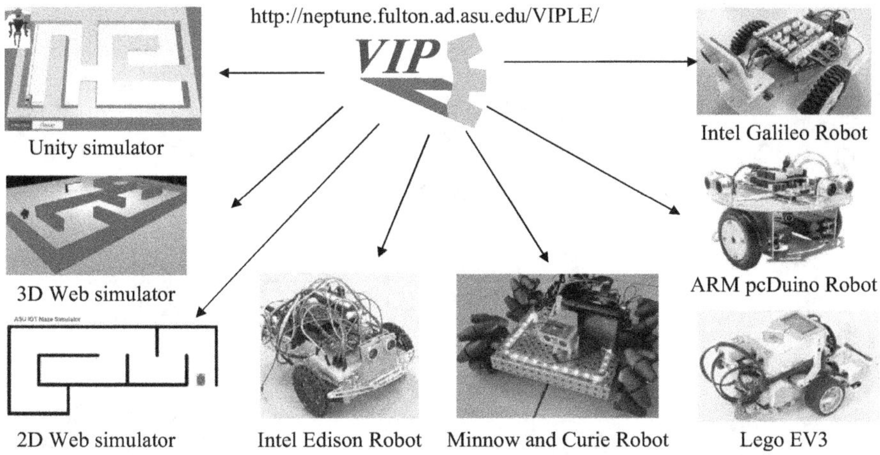

Figure 8.6 A variety of robot platforms supported by VIPLE

ASU VIPLE code. Four different hardware platforms have been developed in addition to the EV3. A specific set of services are developed to work with Lego EV3 as our open-source middleware cannot be installed on EV3. For all the other open-architecture platforms, a standard interface in JSON format has been defined and a middleware has been developed, which can be executed on all these platforms. A school can build the robots for the course based on the open-source middleware or purchase the prebuilt robots from a vendor that builds such robots. For example, the pcDuino robots based on ARM architecture and built to work with ASU VIPLE can be purchased from Link Sprite at http://store.cutedigi.com/pcduino-asu-viple-robot/.

The boards we use for building the RaaS units are IoT boards, with both processor and Arduino connectors, which makes the construction very easy. Figure 8.7 shows the hardware design and connection based on Intel's Edison board. No additional hardware or logic are needed, simply connect the sensors and motors to the board. We can have the Linux operating system and the driver library installed on the Edison board by following Intel's Edison's software installation guide at https://software.intel.com/en-us/articles/flash-tool-lite-user-manual.

We have developed a middleware for the Edison board, which can be downloaded from the ASU VIPLE site at http://neptune.fulton.ad.asu.edu/VIPLE/.

In addition to the simple robots shown in Figure 8.6, VIPLE can be used to control complex IoT and robotics systems. Figure 8.8 shows a more complex robot system which we are integrating into our RaaS and VIPLE environment. The humanoid and the dog robots are built from the same Robotis Biodloid Premium kits (http://www.robotis.us/robotis-premium/). There are up to 18 motors and a variety of different sensors on each robot. We are able to use the existing services in VIPLE to program the robots by calling the existing APIs available on the robots. We are also adding more specific services into VIPLE for make programming easier and more expressive. In addition to controlling the movement of the

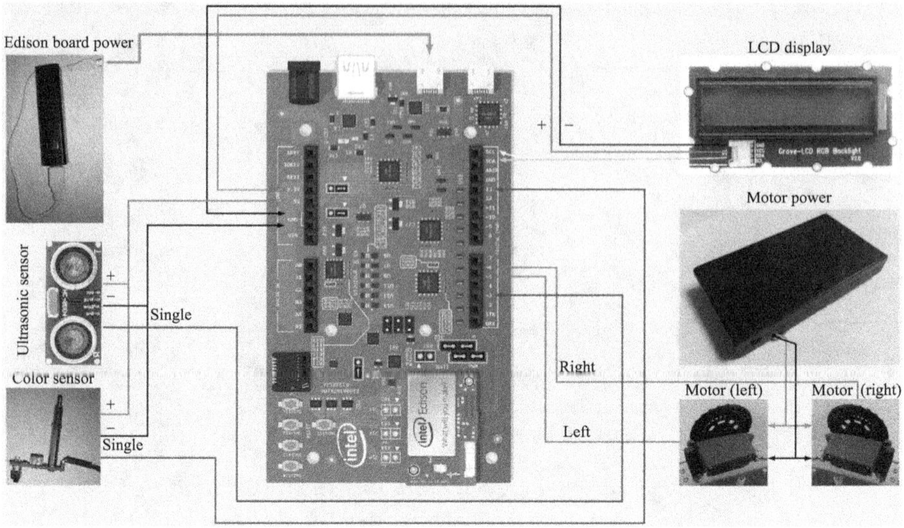

Figure 8.7 RaaS design based on Intel Edison board

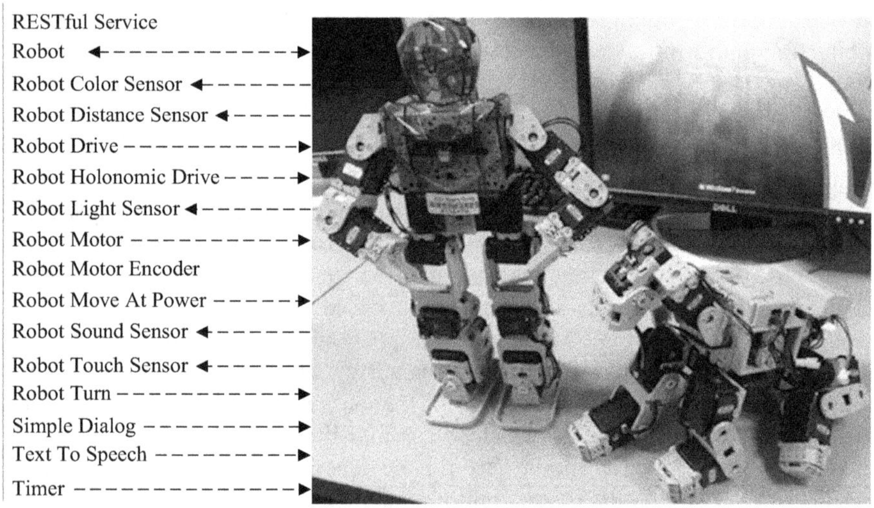

Figure 8.8 Integrating robotis biodloid premium robots into VIPLE

robots, we are adding human–robot interaction functions, such as voice recognition and video difference detection in patrolling. We are also adding services to support robot–robot interaction, so that the humanoid and the dog can communicate and work together between themselves.

The RaaS units can be physically simulated. Both a desktop simulator and Web simulators are developed. The desktop simulator is implemented using the

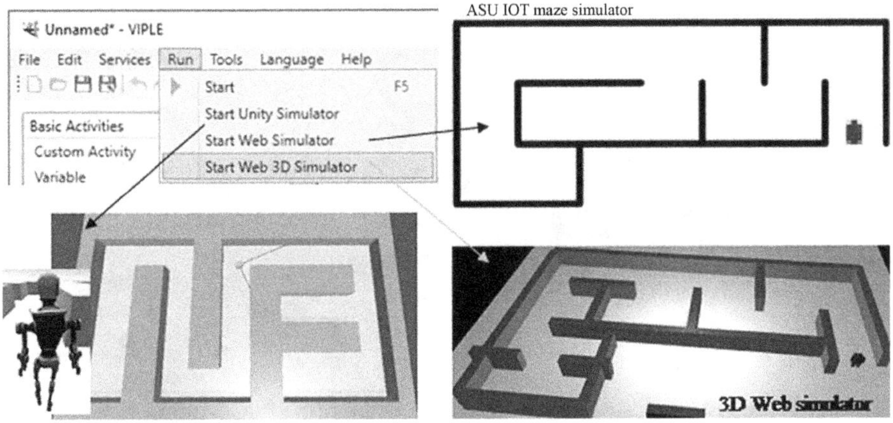

Figure 8.9 Different simulators in VIPLE environment

Unity 3D game engine, which runs on the same computer as the VIPLE program. The Web simulators are implemented in HTML/JavaScript, which runs in a Web browser on any computer or mobile device. One Web simulator is a 2D version, while the other is a 3D version, depending on the user's preference. Figure 8.9 shows the three simulators. The simulators consist of two parts: the robot and the environment. We can choose the robot to be a humanoid robot or a vehicle, and the environment is a maze that can be reconfigured dynamically while the robot is moving. We can create one robot or multiple robots in the environment. The 3D simulators can be changed dynamically, while the program is running and the robot is traversing the maze.

VIPLE programs can start with no simulator. In this case, either no device is needed, or the physical devices will be used. VIPLE programs can also start with one of the simulators to visualize the movement of the robot in the environment.

Different devices are identified in VIPLE through the robot, sensor, and motor configurations. Figure 8.10 shows the configuration of Robot Drive (left and right drive motors), Robot Distance Sensor, Robot Touch Sensor, and My Robot 0 (the main robot). When multiple robots are used, with My Robot 1, 2, ... , *n*, the devices can partner with different robots.

We can use the simulator for traffic simulation experiments. Figure 8.11 shows a traffic simulation experiment based on the Unity simulator.

The traffic simulator is responsible for spawning a number of intersections as well as a number of vehicles (rectangular blocks) each with their own origin and destination. Each green (lighter-colored) circular point represents a clear waypoint, a point on a virtual street where cars can pass through. However, the red (darker-colored) circular points shown on the figure represent points that are blocked and should not be passed. These assets together provide a means of simulating traffic where each car spawned in the environment has a starting point and destination point.

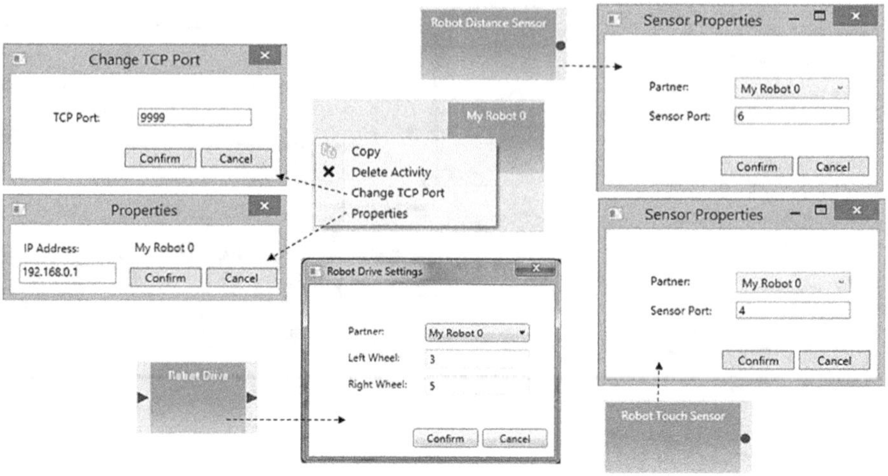

Figure 8.10 Configuration of IoT device ports

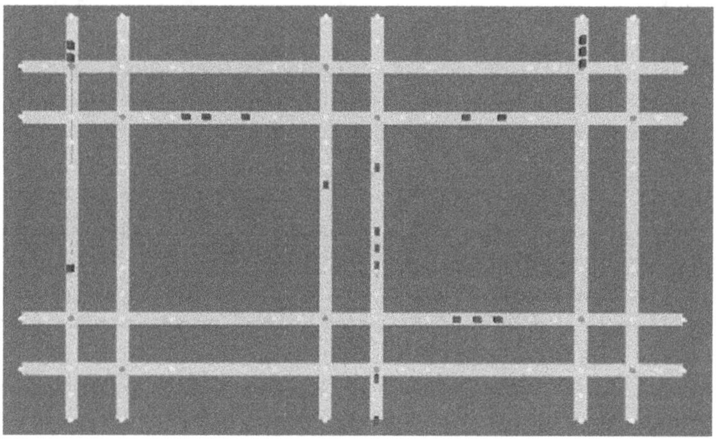

Figure 8.11 Traffic simulation in VIPLE and Unity simulator

The simulation makes it simple for users to program "self-driving" cars by emulating high-level behavior. The following car controls are implemented with the assumption that the car will complete the operation without crashing into other cars, or moving off the road.

- Changing lanes
- Accelerate/Decelerate
- Turn left or right at an intersection
- Follow a predefined route

In this way, the focus of programming is placed not on obstacle avoidance but instead on choosing the fastest route to get to the destination.

Applications developed for such simulators can then be used with physical devices. Once the middleware is installed on the physical robot, the RaaS unit can access the sensors and motors, and communicate with ASU VIPLE through Wi-Fi and the JSON object shown in Figure 8.3. This JSON object is standard across physical and simulated RaaS units and allows VIPLE to interact with either few to no changes.

The architecture of the middleware is shown in Figure 8.12. The sensor and motor controllers will access the device drivers installed in the operating systems to communicate with the devices. A JSON encoder will wrap the sensor values into the standard JSON object defined in Figure 8.3. The JSON object is then sent to the VIPLE code at the backend through the service interface. On the other hand, the control data from VIPLE is sent to the middleware through the service interface. It is decoded and sent to individual motor controllers to control the direction and speed of the motors. Two implementations of the middleware are available in JavaScript and in C++, and they can be downloaded as open source from ASU VIPLE site: http://neptune.fulton.ad.asu.edu/VIPLE/.

This middleware implementation not only standardizes control across simulated and physical RaaS units but also standardizes much of the underlying processing on the actual physical units. As Figure 8.12 shows, the middleware utilizes standard service interfaces and standard device drivers. Using such standards allows the middleware to be deployed to a variety of devices with little to no changes. Such reusability is made possible by installing an operating system with the same set of standard APIs (e.g., using Intel's Yocto Linux with their MRAA and UPM APIs requires only potential modification of sensor ports).

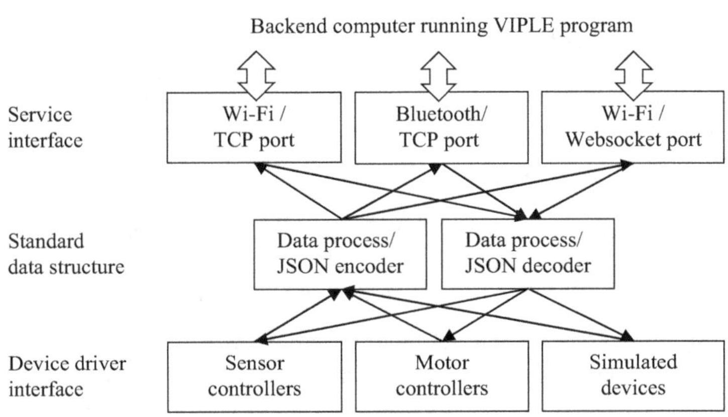

Figure 8.12 A variety of robot platforms supported by VIPLE

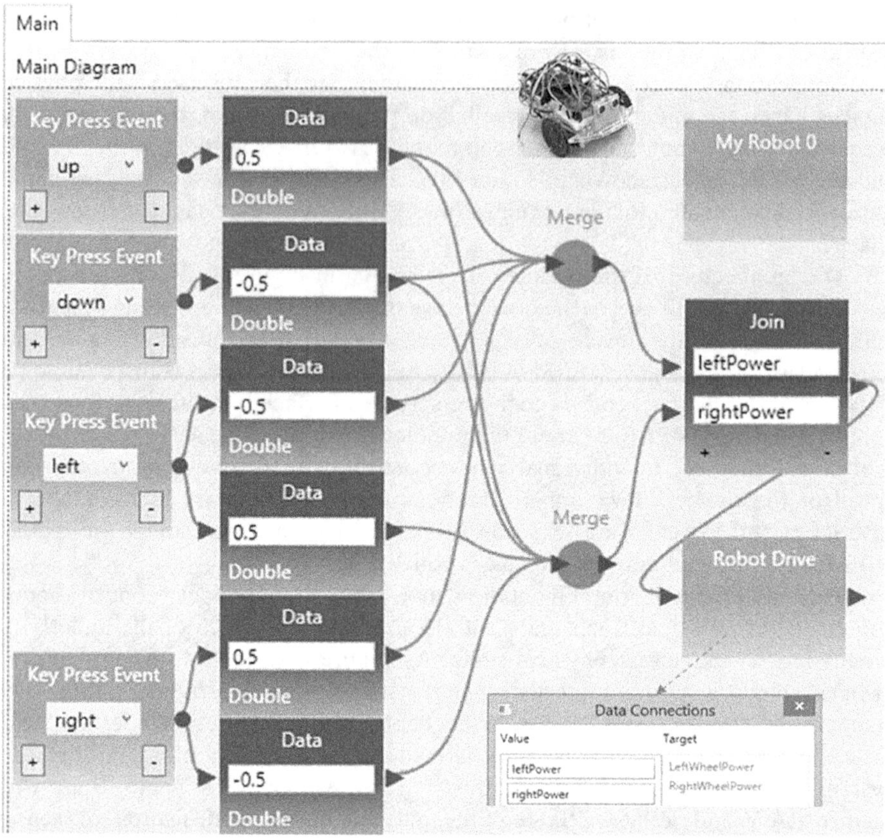

Figure 8.13 VIPLE code for drive-by-wire application

8.5 Robotics application development

Using VIPLE, high-level computational thinking can be directly applied to develop operational code to control robots. We illustrate the development through a few examples.

In the first example, we develop a drive-by-wire application, where we use a keyboard to remotely control the robot. To develop this application, students need to know only the following details.

The robot is driven by two wheels taking positive and negative powers. If we give

- two positive powers to both wheels, the robot moves forward;
- two negative powers to both wheels, the robot moves forward;
- left wheel negative power and right wheel positive power, the robot turns left; and
- right wheel negative power and left wheel positive power, the robot turns right.

Based on this high-level computational understand, the VIPLE code can be developed as shown in Figure 8.13.

The next example we want to develop is a right-wall-following application. We assume a robot has a distance sensor installed to face the right side and a distance or a touch sensor installed in the front. The computational thinking is as follows:

- The robot initially moves forward.
- If the right side distance suddenly increases, indicating an open road on the right, the robot turns 90 degrees to the right.
- If the front distance sensor reads a small value or the touch sensor is touched, indicating that there is no way to move forward further, the robot turns 90 degrees to the left.

Based on this algorithm, a simple VIPLE program can be constructed, as shown in Figure 8.14.

As another example, Figure 8.15 shows the program that controls the traffic of an intersection with an east–west road and a north–south road.

The program starts with a while loop with a true condition. A timer is used to control the duration of lights in each direction. The program also includes an emergency/ambulance signal detection. When such a signal (event) occurs, the traffic light will turn in favor of the signal's direction. The signal is simulated by key press events, 'e' for east–west direction and 'n' for north–south direction.

8.6 Conclusions

RaaS has been proven to be a useful concept for developing embedded systems that require Internet and cloud-computing support, such as ADSs and CPSs. RaaS defines the behavior model of the front-end devices and the standard interface between the backend computer and the devices. Each device can include a number of sensors and motors. The motors can be controlled individually or in groups. This section of the book applied the RaaS concept to design physical robots and simulated robots and their environments. The VIPLE development environment is developed for supporting RaaS concepts and offers a tool to ease the development and operation of RaaS units. VIPLE is defined based on π-calculus that bears complete behaviors and semantics. The basic activities and built-in services are explained and applied to develop different applications. Three basic applications are demonstrated.

Complete resource package is available in VIPLE Website (http://neptune.fulton.ad.asu.edu/VIPLE/) for downloading, including RaaS robot software and hardware design and installation guide, open-source middle code for downloading onto Intel Edison board, and free downloads of VIPLE Standard Edition and Developer Edition. Full documents and lecture PowerPoint slides are also available on the site for downloading.

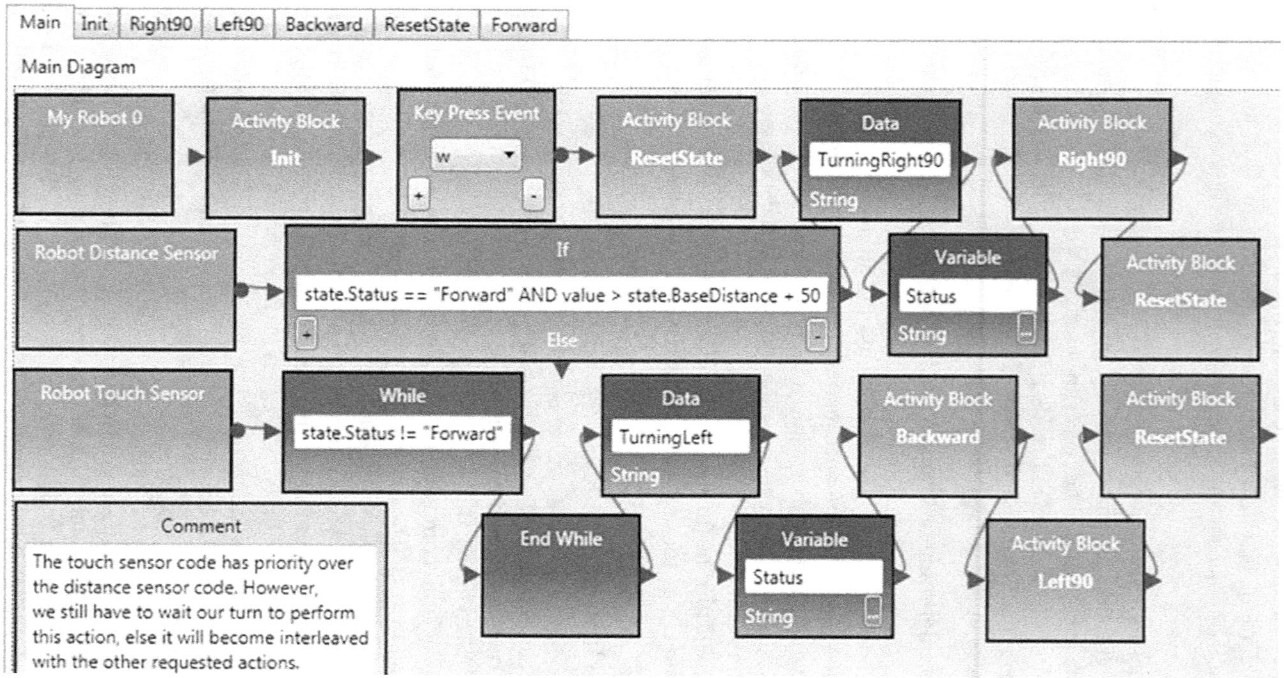

Figure 8.14 VIPLE code for right-wall-following application

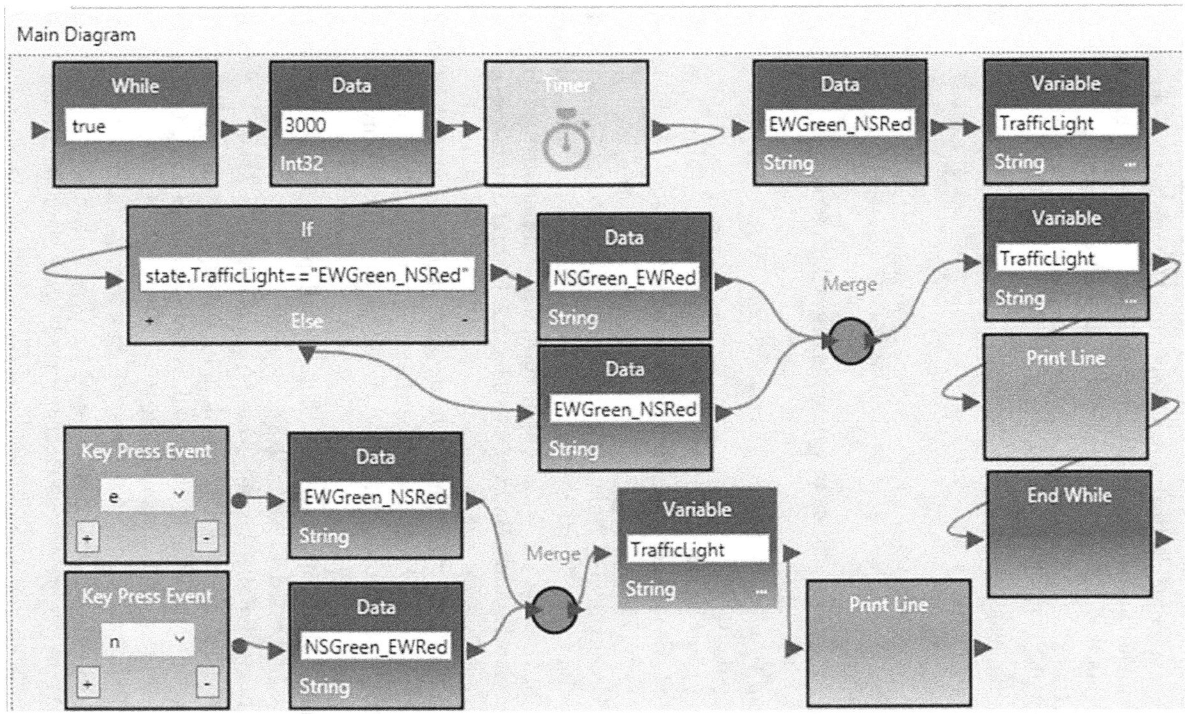

Figure 8.15 VIPLE code for a traffic light control

Acknowledgments

The research and development RaaS and VIPLE are conducted at IoT and Robotics Education Lab at Arizona State University. Multiple grants were used for the research and development efforts, including a US Department of Education FIPSE grant, an Arizona Science Foundation grant, and multiple Intel grants. Many faculty, students, and scholars have contributed a great deal of efforts in this project. Special thanks owe to Calvin Cheng, Yann-Hang Lee, Garrett Drown, Megan Plachecki, Mathew De Rosa, Sami Mian, Tyler Pavkov, Alexandra Porter, and Yufeng Ouyang, who made significant contribution in the development of RaaS and VIPLE.

References

[1] IoT in Wikipedia, http://en.wikipedia.org/wiki/Internet_of_Things. [Accessed 15 May 2018].

[2] Stephan Haller, http://services.future-internet.eu/images/1/16/A4_Things_Haller.pdf, 2009.

[3] Yinong Chen, Hualiang Hu, "Internet of Intelligent Things and Robot as a Service", Simulation Modelling Practice and Theory, Volume 34, 2013, Pages 159–171.

[4] Ton Steenman GM, Intel Intelligent Systems Group: "Accelerating the Transition to Intelligent Systems", Intel Embedded Research and Education Summit, February 2012, http://embedded.communities.intel.com/servlet/JiveServlet/downloadBody/7148-102-1-2394/Accelerating-the-Transition-to-Intelligent-Systems.pdf.

[5] Ragunathan (Raj) Rajkumar, Insup Lee, Lui Sha, and John Stankovic, "Cyber Physical Systems: The Next Computing Revolution", In 47th Design Automation Conference (DAC 2010), CPS Demystified Session, Anaheim, CA, 2010.

[6] Wikipia, Cyber-Physical System, http://en.wikipedia.org/wiki/Cyber-physical_system. [Accessed 15 May 2018].

[7] Jian Huang, Farokh Bastani, I-Ling Yen, and Wenke Zhang, "A Framework for Efficient Service Composition in Cyber-Physical Systems", In Proceedings of 5th IEEE International Symposium on Service Oriented System Engineering (SOSE), Nanjing, 2010.

[8] NSF Solicitation, "Cyber-Physical Systems", 2008, www.nsf.gov/pubs/2008/nsf08611/nsf08611.pdf.

[9] Kinji Mori, "Autonomous Decentralized System and Its Strategic Approach for Research and Development", IEICE Transactions on Information and Systems, Volume E-91-D, Number 9, 2008, pages 2227–2232.

[10] Autonomous Decentralized Systems (ADS) in Wikipedia: https://en.wikipedia.org/wiki/Autonomous_decentralized_system. [Accessed 15 May 2018].

[11] Yinong Chen, Zhihui Du, and Marcos Garcia-Acosta, "Robot as a Service in Cloud Computing", In Proceedings of the Fifth IEEE International Symposium on Service Oriented System Engineering (SOSE), Nanjing, 2010, pp. 151–158.
[12] Robot as Service in Wikipedia: https://en.wikipedia.org/wiki/Robot_as_a_service. [Accessed 15 May 2018].
[13] Ioannis A. Moschakis, Helen D. Karatza, "Evaluation of Gang Scheduling Performance and Cost in a Cloud Computing System", The Journal of Supercomputing, Volume 59, Number 2, 2012, pages 975–992.
[14] Zafeirios C. Papazachos, Helen D. Karatza, "Performance Evaluation of Bag of Gangs Scheduling in a Heterogeneous Distributed System", Journal of Systems and Software, Volume 83, Number 8, 2010, pages 1346–1354.
[15] Georgios L. Stavrinides, Helen D. Karatza, "Scheduling Multiple Task Graphs in Heterogeneous Distributed Real-Time Systems by Exploiting Schedule Holes with Bin Packing Techniques", Simulation Modelling Practice and Theory, Volume 19, Number 1, 2011, pages 540–552.
[16] Serverless computing in Wikipedia: https://en.wikipedia.org/wiki/Serverless_computing. [Accessed 15 May 2018].
[17] Yinong Chen, Service-Oriented Computing and Web Software, IoT, Big Data, and AI as Services, 6th edition, Kendall Hunt Publishing, Dubuque, 2018.
[18] App Inventor, http://ai2.appinventor.mit.edu/. [Accessed 15 May 2018].
[19] Alice, http://www.alice.org/. [Accessed 15 May 2018].
[20] Intel IoT Service Orchestration Layer: http://01org.github.io/intel-iot-services-orchestration-layer/. [Accessed 19 Sept 2016].
[21] Microsoft Robotics Developer Studio, https://msdn.microsoft.com/en-us/library/bb648760.aspx. [Accessed 1 Mar 2012].
[22] MRDS in Wikipedia https://en.wikipedia.org/wiki/Microsoft_Robotics_Developer_Studio. [Accessed 15 May 2018].
[23] ASU eRobotics Programming Environment, http://neptune.fulton.ad.asu.edu/WSRepository/erobotic/. [Accessed 15 May 2018].
[24] Robin Milner, Communicating and Mobile Systems: The π-Calculus. Cambridge: Cambridge UP, 1999.
[25] Frank Puhlmann, Mathias Weske, Using the π-Calculus for Formalizing Workflow Patterns. In: van der Aalst W.M.P., Benatallah B., Casati F., Curbera F. (eds) *Business Process Management. Lecture Notes in Computer Science*, vol. 3649. Springer, Berlin, Heidelberg, 2005.

Chapter 9

"JR EAST App" for customers' smartphones based on ICT

Takayuki Matsumoto[1] *and Takeshi Nakagawa*[2]

Abstract

East Japan Railway Company (JR East) is all-round and the biggest railway company in the world. Our business domains are railway, lifestyle and IT&Suica business, and we intended to grow continuously while meeting our social responsibilities as a trusted lifestyle service creating group. JR East released the "JR East App" which is an application (app) for smartphones in March 2014 aiming to improve the level of satisfaction with information service for passengers. This app aims to help those who use JR East to easily obtain necessary information by using their smartphones anytime and anywhere in a timely manner. In order to achieve this aim, JR East utilizes various kinds of location technologies such as GPS, Wi-Fi and ultrasonic beacons and provides content using real-time data unique to railway business operators such as operation statuses, location information and occupancy of trains. JR East also provides information on shops in its station buildings and around its stations and entertainment content consisting of e-books, games, etc. As of the end of December 2016, over 2.3 million smartphone users mainly within Japan have downloaded this app, and many people actively use this app. Access logs show that the content mainly used is real-time information related to the use of railways. In addition, the questionnaire survey revealed that around 80% of respondents are satisfied with this information service, and around 90% of respondents are willing to continue to use this app.

9.1 Introduction

East Japan Railway Company (JR East) has conducted customer-satisfaction surveys every year since 1997 to understand the trend of the general evaluation of the company. Figure 9.1 shows the results of the customer-satisfaction survey in 2013 before the release of the application (app). The vertical axis represents the level of

[1]Customer Service Quality Reformation Department, East Japan Railway Company, Japan
[2]Technology Innovation Headquarters, East Japan Railway Company, Japan

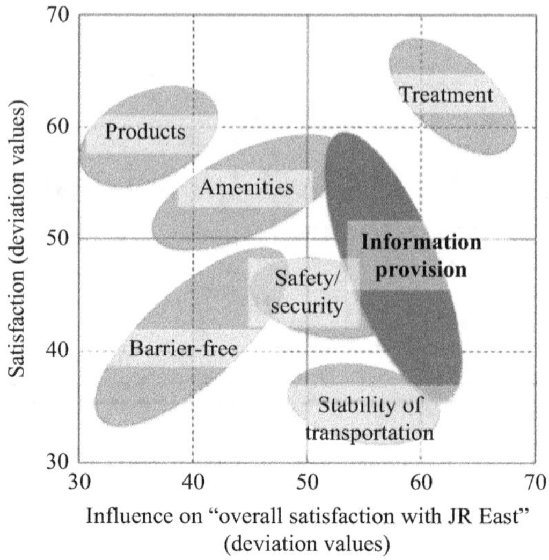

Figure 9.1 The results of the customer satisfaction survey

satisfaction with each item, and the horizontal axis represents deviation values of influence on overall satisfaction with JR East. Figure 9.1 shows that items related to "information service" which influence general satisfaction with JR East did not give complete satisfaction to the users of JR East's railways.

Apps for other companies' smartphones for railways are frequently used in Japan. For example, these are NAVITIME, EKI-TAN, Japan Transit Planner, etc. It is said that each app is downloaded about 20 million. As a feature, the route search from the station to the station when using the railway is the main function, depending on the app, there are many functions regarding railway use, such as supporting buses and walking. However, it can be said that there is a lack of context information that we aim for these apps, that is, dynamic information. In addition, contents linked with places indispensable for information provision in individuals are still insufficient. The JR EAST App made it possible to provide more personalized information by incorporating a mechanism such as dynamic information in the real-time, visualizing information on train presence and information service using train location on board into the app.

Therefore, improving satisfaction with "information service" is an urgent issue for JR East to deal with. JR East already has provided information on operations and other information through its website and displays at its stations and in trains of its lines. However, this information service was for many unspecified customers, which meant that satisfying the needs of individual customers was difficult. Therefore, JR East developed the "JR East App," an app for smartphones which enables information provision to each customer and released the app in March 2014. At the time of its release, this service was available only in Japanese.

9.2 Features of the app

The aim of the "JR East App" is to help those who use JR East to easily obtain necessary information through their smartphones anytime and anywhere in a timely manner. In order to achieve this aim, JR East has developed an app having the following features:

1. Information service according to contexts
   The opening page is the screen that summarizes information on specific stations of JR East to ensure provision of information according to locations, situations and other contexts of the users of the app. For example, Figure 9.2 shows the opening page that summarizes information on Tokyo Station. This screen also provides real-time information on weather and operation statuses. This is the aim of getting up in the morning and habitualizing to launch the JR EAST App. Also, by registering the nearest station and the route that you use every day on the top page, you can customize it according to the individual. In addition, on the opening page, JR East placed the buttons of content with the names such as "taking trains" and "using stations" based on users' behavior to help users have images on "when and what kind of information they are able to obtain."
2. Utilization of data unique to railway business operators
   JR East has created content for passengers by utilizing real-time data unique to railway business operators such as operation statuses, departure times and location information of trains and occupancy of each train car.
3. Actively adopting new technology
   JR East has actively adopted new technology such as location estimations utilizing signal strength of Wi-Fi, and identification of train cars of which passengers are on by using ultrasonic beacons installed in the train.

Figure 9.2 Opening page of JR East App

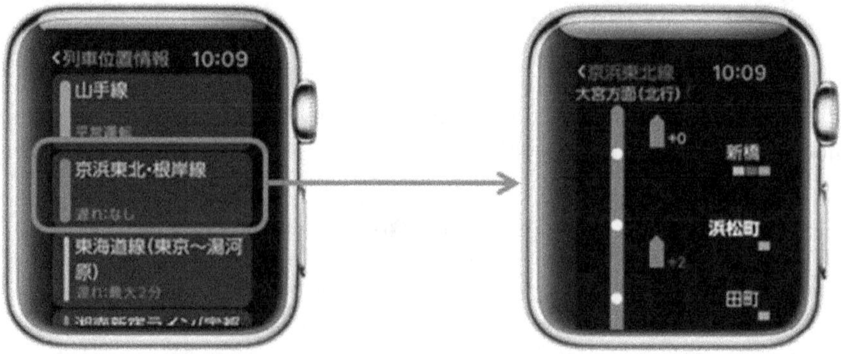

Figure 9.3 Train location on Apple Watch

4. Available for Apple Watch
   JR East App is also available for Apple Watch. You can see JR East's train operation statuses and location information of trains on small display of Apple Watch (Figure 9.3).

9.3 Content

The motivation of JR East to provide information for smartphones was to increase the overall satisfaction of users of JR East's railways by improving how to provide information as described above. In this context, "information" mainly means railway-related information. JR East considered that an information service for smartphones could realize the offer of useful added value to those who use railways under space and time constraints. JR East also believed that an information service utilizing smartphones had the possibility to create opportunities for new business from the viewpoints of business operators.

In addition to its railway business, JR East engages in various businesses related to travel by railways including management of commercial facilities in stations, station buildings and hotels, all of which have great potential. Therefore, JR East decided to provide not only content related to railways but also content related to marketing through the JR East App. Representative kinds of content included in this app are shown below.

9.3.1 Content related to railways

1. List of operation statuses
   Operation statuses of JR East's lines are displayed. With regard to the lines of Tokyo metropolitan district, the users of the app can see the same screen as that displayed near ticket gates of main stations through their smartphones (Figure 9.4).

Figure 9.4 Operation statuses

2. Location information of trains
   The location and time of delay of each train on the line that users wish to check are shown on this app (Figure 9.5). The content was created by using data of JR East's operation-management system. As of January 2016, JR East provides this service for 11 lines in the Tokyo metropolitan district and 39 lines in other areas and plans to continue increasing the number of lines to which this service is provided.
3. Yamanote Line Train Net
   The Yamanote Line is a loop line running in the center of Tokyo. Of JR East's lines, this line has the greatest number of passengers. JR East has started a special information service called "Yamanote Line Train Net" for this line. By detecting the train cars the users of this service are on, the service provides information on stops (transfer lines, platform maps, station maps) according to the car number and location of the train and the occupancy/temperature in those trains (Figure 9.6). The occupancy of coaches is calculated from the pressure of air suspensions of coaches. This value is usually used for controlling the strength of brake. We utilize these data for a passenger information service.

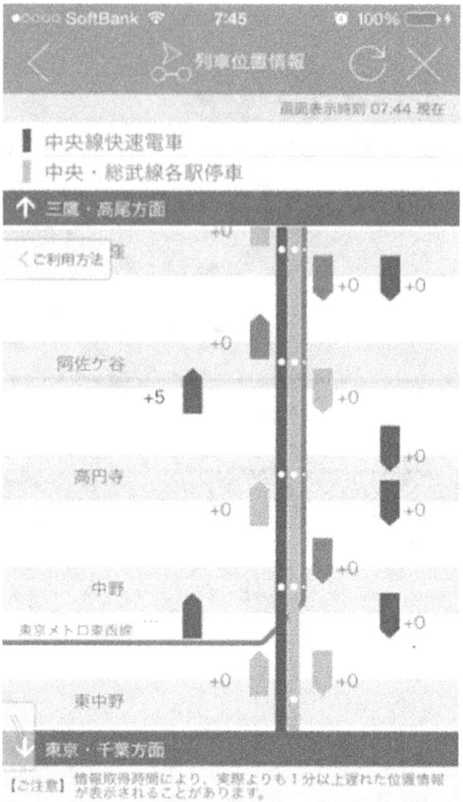

Figure 9.5 Location of trains

A device that sends ultrasonic waves (a beacon) is installed in each car of trains running on the Yamanote Line to provide this service. Microphones of passengers' smartphones detect ultrasonic waves sent from the beacon installed in each car of those trains to identify train numbers and car numbers the users of this service are on, and real-time data of the train (operation sections, temperature in the car, occupancy, etc.) is sent to their smartphones. The information described above is able to see even before getting on a train of the Yamanote Line.

9.3.2 Content related to marketing

1. Shop information
   The content offers information on shops and coupons including in-station shops operated by JR East Group and shopping malls and restaurants around stations. The list of in-station shops that shows only shops of the station is displayed on the smartphone screen (Figure 9.7).

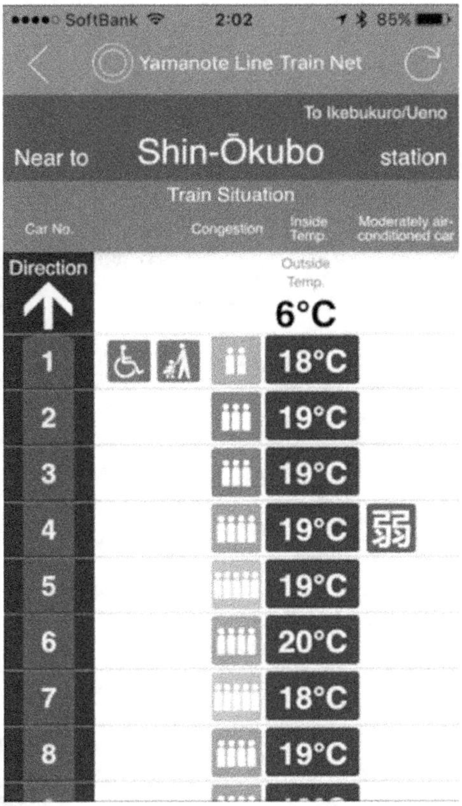

Figure 9.6 Yamanote Line Train Net

2. Entertainment information
   The app offers entertainment content designed to be accessed by the users who are under a time constraint while traveling by railways. For example, the content includes e-books and Sugoroku games (snakes and ladders) unique to this app utilizing location information (Figure 9.8).

9.4 System configuration

The content delivery system of the JR East App consists of the "smartphone app," the "ground system" and the "system on the train" and enables provision of various kinds of information by adding and linking external systems. Figure 9.9 shows the system configuration of the JR East App. For exchange of data between the external and the ground systems, some APIs are used to prepare for future extensibility. This is also a cornerstone for making open data in the future, and by actively incorporating Web technology from the design philosophy, JR EAST App was made to be a versatile and highly scalable system.

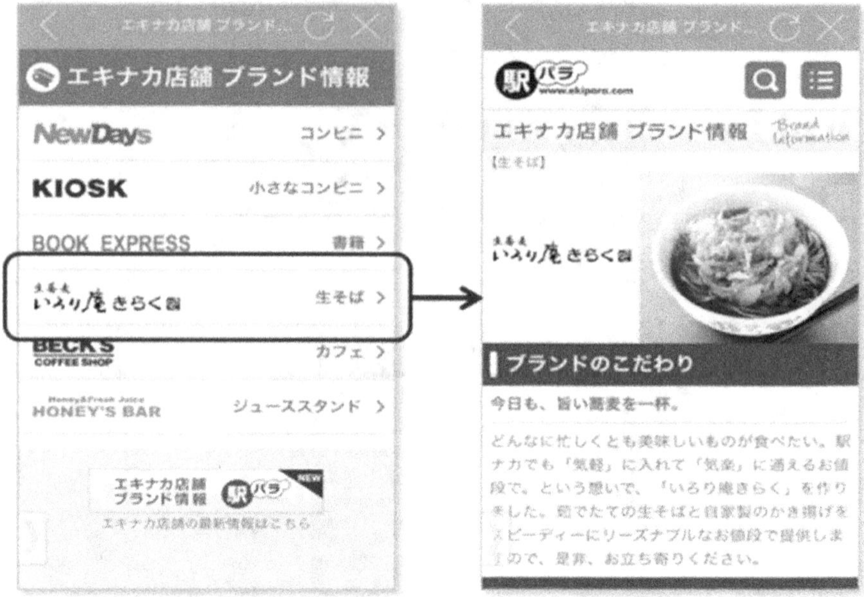

Figure 9.7 Shop information

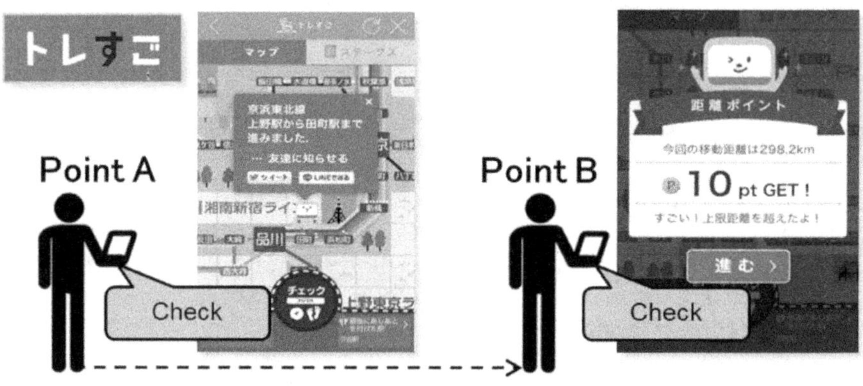

Figure 9.8 Entertainment information

JR East offers two types of apps—one for iOS and the other for Android. In order to display information quickly, some static data is included in the app. The ground system which consists of several servers mainly obtains and accumulates dynamic information from external systems and the system in trains, maintains static data and outputs information responding to requests from the smartphone app, etc. In order to provide dynamic information such as real-time information on departure of trains, information on train locations, operation statuses, weather at each station, etc., the ground system

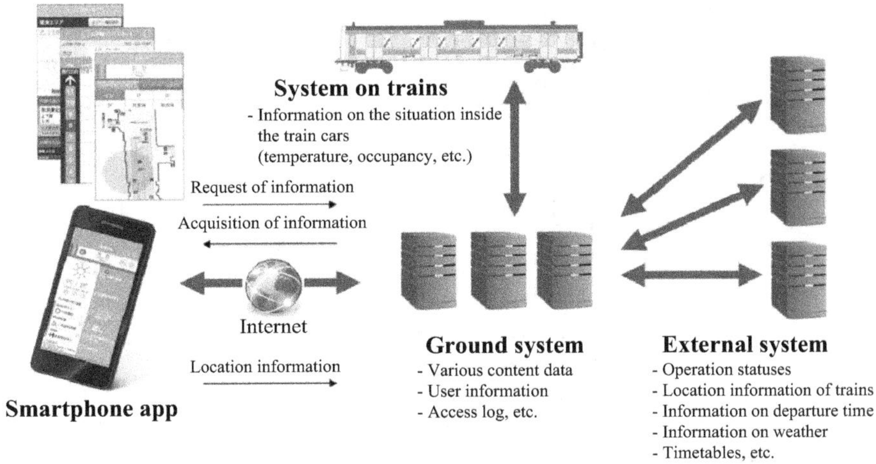

Figure 9.9 System configuration

obtains data from the Autonomous Decentralized Transport Operation Control System and other external systems. Then, a device to send data of cars and trains to the ground system on a regular basis and ultrasound beacon devices is installed in the train cars running on the Yamanote Line.

9.5 Use situations of the app

9.5.1 The number of users

As of December 2015, the apps for iOS and Android have been downloaded more than 1.5 million times in total, and the number of unique users a month is approximately 300,000.

9.5.2 Attribute of users

When starting the use of JR East App, the users are able to register their gender, date of birth and occupation on a voluntary basis. Table 9.1 shows the aggregate results (as of December 2015).

With regard to gender, males account for nearly 80% of all users. Of those who use railways more than once a week in the Tokyo metropolitan district of Japan, approximately 55% are men [1], which indicates the percentage of male users of this app is relatively higher than that of average railway users. With regard to age, although the percentage of users in their 40s is slightly higher, the percentage of users in their 20s, 30s, 40s and 50s are around 20%, respectively, and users in these age groups account for the majority of the users. In terms of occupation, over 70% of app users are workers including office workers, company executives and government employees.

Table 9.1 Attribute of app users

	Attribute	Number	Percentage
Gender (n = 1,182,418)	Male	915,638	77.4
	Female	266,780	22.6
Age (n = 1,182,441)	Under 20	94,110	8.0
	20–29	216,367	18.3
	30–39	249,238	21.1
	40–49	320,776	27.1
	50–59	229,939	19.4
	60+	72,011	6.1
Occupation (n = 1,102,218)	Junior or senior high school student	84,251	7.6
	University student	97,913	8.9
	Company or public employee	823,037	74.7
	Freelance professional or self-owned business	60,846	5.5
	Part-time employee	54,632	5.0
	Housewife	25,620	2.3
	Others	36,171	3.3

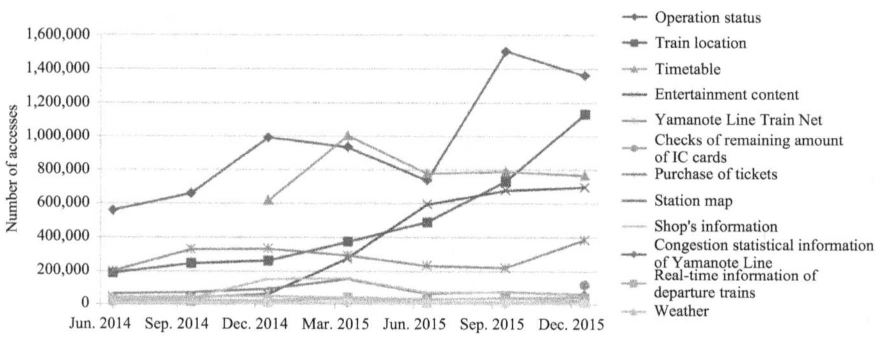

Figure 9.10 Transition of the total number of accesses to main content

9.5.3 Access logs

Figure 9.10 shows the transition of the total number of accesses to main content a month. The value plotted in June 2014 represents the total number of accesses to each type of content during the whole month of June 2014. For the period between June 2014 which is 3 months after the release of the app and December 2015, the number of accesses for a month was aggregated every 3 months. It is found that the types of contents which have been accessed by many users are operation statuses, location information of trains, information on the situation of the inside of the trains of the Yamanote Line and other real-time information related to the use of railways. "Congestion statistical information of Yamanote Line" is the content which shows the occupancy of coaches on Yamanote Line in the past. Those data are aggregated by each time zone and by each weekday.

Since the real-time information is important information in daily railway use by businessmen, it can be said that the high number of accesses of train operation information and train position information is as intended at the time of app development. Although the timetable is static information, it is easily guessed that it is the information that is frequently confirmed. Regarding entertainment, depending on the content, it is difficult to know how much information can be put in the future. Although it is somewhat unexpected that the information on the train departure point and the congestion information are low, it may be because the information seems not to change especially afterwards. We will continue to consider contents while looking at the log information.

9.6 Questionnaire survey

For approximately 10 months starting in June 2014 when 3 months had passed after the release of the JR East App and ending in March 2015, JR East conducted a questionnaire survey on the evaluation of content offered by the app and the app itself and the intention of continuous use of the app. The survey was included in the app as one of the "information" contents, and the users of the app responded to the survey on a voluntary basis. The survey obtained more than 10,000 responses even though it offered no incentives.

9.6.1 Attribute of respondents

Compared with the attributes of the actual users of this app, the percentage of male respondents is higher (84.9%), and the percentage of respondents in their 40s is higher (30.4%).

9.6.2 The level of satisfaction/intention of continuous use

The level of overall satisfaction with the JR East App and the intention of continuous use were asked. Respondents were asked to rate both questions on a scale of one to five. With regard to the level of satisfaction, 78.9% of all respondents gave a positive evaluation. However, the response of "very good" accounted for 32.8% and "good" accounted for 46.1%, suggesting the possible hope of the users for improvement. In addition, with regard to the intention of continuous use, 88.0% of all respondents answered positively. Of the positive answers, 66.1% answered "Yes," and 27.7% answered "Maybe, yes," showing a higher level.

9.6.3 Relationship between the level of overall satisfaction with the app and the level of satisfaction with each type of content

Figure 9.11 shows the relationship between the level of overall satisfaction with the app and the level of satisfaction with each type of content which is called CS portfolio. The vertical axis of the figure represents deviation values of satisfaction, and the horizontal axis of the figure represents correlation coefficients with overall

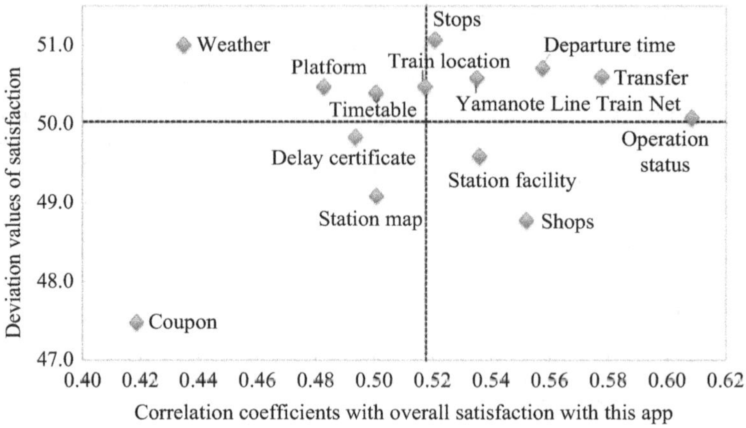

Figure 9.11 CS portfolio

satisfaction with this app. The group of contents in the upper right area of the figure highly contributes to the overall satisfaction with the app and gains a high level of satisfaction. In other words, this group of contents can be said to be an appealing point of this app. The contents in this area include operation status, location information of trains, information on the situation on the inside of trains of the Yamanote Line, the list of stops of each line, and transfer lines. In addition, the group of contents in the lower right area of the figure highly contributes to the overall satisfaction with the app but gains low level of satisfaction. In other words, improving this group of contents will attain a higher evaluation of the app on the whole. The contents in this area include information on station facilities and shop information.

9.7 Comparison with the English version of this app

So far, we have described the Japanese version of the app. JR East released the English version in March 2015, 1 year after the Japanese version was released. The English version of the app is for those who stay in Japan for a short term for the purpose of sightseeing, etc., and offers a part of the railway-related content of major stations (approximately 200 stations). This English version is also used by many visitors from Taiwan and Hong Kong. Based on the access logs, the content that gained the largest number of accesses was operation status, the same as the Japanese version. Station maps gained the second largest number of accesses followed by the list of stops of each line. We found that the number of users who are looking for basic information related to railways is larger in those who use the English version than in those who use the Japanese version although it is partly because the content of the English version does not include location information of trains and timetables.

9.8 Conclusion

This paper summarizes the motivation of development, the system configuration, the offered content, access logs and survey results of the "JR East App" developed by JR East with the purpose of providing information for individual smartphones of customers. Access logs and questionnaire surveys revealed that real-time information which is helpful for the use of railways and unique to railway business operators such as operation statuses, location information of trains and information on the situation on the inside of trains gains a large number of accesses and is highly evaluated. In addition, although information on facilities and shops in station buildings and around stations of JR East does not receive relatively high evaluation at present, they are found to have high correlation with satisfaction with the app on the whole and are important content with great promise. In addition, the customer satisfaction survey JR East conducted in September 2014, 6 months after the release of the app showed that satisfaction with "ease of finding out 'information on delay of trains' on the Internet" and "ease of finding out 'information on station facilities' on the Internet" increased by 2.3 and 2.0 points, respectively. From this result, we believe that this app has successfully increased the level of satisfaction with information service offered by JR East.

9.9 Future work

Technologies such as Internet of Things (IoT), big data and AI exponentially evolve, and these have a great impact on all industries, so it is called "4th Industrial Revolution." Regarding transportation industry, car-sharing system and autonomous car technology are about to break the boundaries between car and public transportation including railway.

In this background, we have just developed a technological innovation vision, "Mobility Revolution," which is based on IoT, big data and artificial intelligence (AI) targeting 20 years later. This vision consists of the following four pillars, "Safety & Secureness," "Service & Marketing," "Operation & Maintenance" and "Energy & Environment." Regarding "Service & Marketing," we are aiming to provide customers the value of "context-aware service," for instance, "door-to-door" service.

In recent years, it has been a trend to provide "door-to-door" service utilizing ICT mainly among railway business operators in Europe [2,3]. JR East is also aiming to provide the users of railways of JR East with "door-to-door" service by utilizing the JR East App. In order to achieve this aim, we believe that it is necessary to further improve content and add new functions. With regard to the content, JR East should provide information held by other railway business operators and information on other transport facilities including buses and taxies, very little of which are provided at present. In terms of functions, services for route search, online reservations and purchase of tickets should be provided. Ultimately, we will aim to create an environment to provide suitable information everywhere including in the train, at the station, in the town and at home by establishing an internal information platform, gathering various data within the company and sharing data with other companies and SNS.

References

[1] East Japan Marketing & Communications, Inc., Media Data (in Japanese), Retrieved 07/01/2016 http://www.jeki.co.jp/transit/mediaguide/pdf/jmg_md_013–037.pdf.

[2] SNCF, Door-to-Door Services, Retrieved 07/01/2016 World Wide Web, http://www.sncf.com/en/services/door-to-door.

[3] Deutsche Bahn, Overview of Mobile Service, Retrieved 07/01/2016 World Wide Web, http://www.bahn.de/i/view/GBR/en/prices/mobile/mobil-overview.shtml.

Chapter 10

Autonomous decentralised systems and society

Colin Harrison[1] *and Jeffrey Johnson*[2]

Abstract

The concepts of autonomous decentralised systems were developed for the control of large engineering systems such as high-speed trains with the goal of increasing their resilience where centralised control systems could not be trusted to be always operating and accessible. Historically, social organisations, both civic and commercial, were controlled or governed with a balance of centralised, top-down and decentralised, bottom-up mechanisms. We consider how the emergence of a 'hyper-connected' world, together with advances in education, political, and management science, is changing this balance. We begin by considering the emergence of the management of large-scale enterprises during the Industrial Revolution and contrast this with the roles of complex systems in the emergence of autonomous social structures. We go on to examine how this changing balance is becoming manifest in public and private social structures and conclude with thoughts on its evolution.

10.1 Introduction

While the political revolutions of the eighteenth and nineteenth centuries represented, to some degree, a devolution of centralised authority to a democratic franchise, the Industrial Revolution introduced strong systems of centralised, top-down management of employment and manufacturing. The former gave individual citizens a sense of freedom and individual accountability. The latter tethered workers and consumers to the large-scale employers and providers of goods and services, which engendered a widespread mindset that certain social and economic activities are best left to governments and large enterprises. In engineering, an autonomous decentralised system (ADS) is formed from parts that are designed to function independently but are also able to interact to meet the requirements. The concepts of ADS were developed for the control of large engineering systems [1]. Although the scale of ADS in public and private organisations is still small, it is

[1]School of Sustainability, Arizona State University Tempe, USA
[2]School of Engineering & Innovation, The Open University, UK

beginning to challenge the centralised mindset, and so it behoves us to study how such autonomous systems emerge, how they function, and how they grow. We refer to this area as Social ADS.

This centralised mindset began to weaken during the second half of the twentieth century due to rapid changes in education, social mores, methods of organising work and managing complex processes, methods for the customised production of goods and services, the availability in both wealthy and impoverished nations of finance for innovation, and, not least, the power of information and communication technology (ICT) and the Internet. These allowed individuals to access opportunities that were previously reserved to governments and large commercial enterprises and the emergence of smaller scale products and services based on autonomous, decentralised principles.

Prior to the Industrial Revolution that began in the United Kingdom in the late eighteenth century, the principal models for the management or control of large organisations were armies, royal courts, major trading companies, such as the Dutch East India Company, and major religions, such as the Roman Catholic Church. Compared to other countries, where the monarchy and church inhibited innovation by top-down control, the United Kingdom was more permissive and enabling due partly to the evolution of its parliamentary system, a nationally controlled church, a flourishing interest in science, and new methods of funding investment and insurance. Nonetheless, when the entrepreneurs of the Industrial Revolution were confronted with the need to create and manage large-scale enterprises, they relied heavily on such hierarchical models.

Until the mid-twentieth century, it seemed inevitable that both public and private organisations were destined to grow ever larger as they attempted to coordinate manufacturing, transportation, government, and other tasks at a global scale. The boundaries of such organisations were defined by access to primary resources such as raw materials, energy, and long-haul transportation. The behaviour of such organisations was studied in the 1920s by Coase [2], who concluded that the extent of these boundaries depended on the difference in cost between internal and external transactions. Hence, as the Internet and e-commerce reduced transaction costs with external suppliers, it was anticipated by some that this would lead to the deconstruction of large public and private organisations and their replacement by networks of smaller companies coordinated through standardised IT e-commerce protocols [3]. It was believed in the year 2000 that one could envisage ‘plug and play’ business process integration among independent parties to a global supply chain leveraging XML [4] and service-oriented architectures [5].

To some degree, this expectation has been fulfilled. Whereas the Ford Motor Company under Henry Ford I found it valuable to maximise vertical integration, even to the extent of owing rubber tree plantations, manufacturing in general has evolved into global networks of suppliers whose products or services are integrated by a core organisation. Equally, small social organisations or even individuals can leverage the Internet and business process standards to serve niche markets or deliver philanthropic services. But today it seems unlikely that large, centralised, top-down enterprise and social systems will completely disappear.

This chapter explores the evolution of this tension between autonomy and centralised control in social systems ranging from small community activities to national enterprises. We begin by considering the underlying theory of complex systems as it applies in social systems. In the section that follows, we look backwards to the Industrial Revolution and the evolution of the top-down–bottom-up tension during the nineteenth and twentieth centuries. We then consider how autonomous social systems emerge and develop and offer case studies of how existing social systems are evolving towards decentralised autonomy. Finally, we consider issues and opportunities raised for Social ADS by the continuing development of ICT.

10.2 Systems thinking, complex systems, and global systems science

We define a *system* as a set or collection of interacting elements where the behaviour of any element or component depends on its direct or indirect interactions with the other elements.

For social systems, this definition is more problematic. It is a common experience that different people see the same social situation in different ways, even when they are trying to reach a consensus. A common view in social science is that individuals construct their own subjective views of the social world [6], and that we have ways of interacting that enable us to overcome any differences and collaborate for our mutual benefit. Whether or not there is an objective social system underneath all this is an open question – if there is, none of us can see it objectively.

This matters because any theory or model of systems involves implicit or explicit decisions on which are the relevant elements and what are the relevant interactions. Guidance on this comes by extending the definition of system following Bignell and Fortune [7]. A system has phenomenology, i.e. it does something. For those wanting to design, engineer, or manage systems, some aspects of the assembly have been identified as being of particular interest. Now the inclusion of things as elements in the systems and the identification of relevant interactions depend on the purpose of the analysis. Thus, analysts with different purposes may indeed formulate different models or theories for the same situation.

Social systems have another problem that is less familiar in engineering: they are inherently unpredictable or indeterministic. For example, electronic systems are designed as assemblies of components connected in specified ways. The behaviour of the whole system *emerges* from the behaviours of the components and the way in which they are assembled. Although great creativity and imagination may go into designing a circuit, the electronic engineer *predicts* that by connecting the components in the given way it will have the emergent behaviour desired. Such predictions may be based on expertise and intuition gained by experience, but they can be tested by computer simulation or by experimenting with the working system. The theory of electronics is so well behaved that the models underlying the simulations can give *point predictions* that, given the initial conditions, the system will behave deterministically at least over a certain range of performance. Conventional control engineering is based on such theory and the predictions it enables.

Not all physical systems are point-predictable in the way that, for example, ballistics is. The weather is a counter-example. In 1963, the weather scientist Lorenz published a paper *Deterministic Nonperiodic Flow* [8] in which he argued that ‘Two states differing by imperceptible amounts may eventually evolve into two considerably different states’, i.e. that the evolution of some systems is *sensitive to initial conditions*. Since all measurements involve error, such systems cannot be predicted forever into the future. This work led to the field of dynamical systems and chaos.

Arguably, *all* social systems are sensitive to initial conditions. Wake up 1 min late, miss the bus, miss the train, miss the plane, and you may arrive at your meeting a day late. More generally, social systems are unpredictable in the sense of complex systems science [9]. Other reasons for social systems being unpredictable is that they are made up of large numbers of heterogeneous people with heterogeneous behaviours, that they have many interacting feedback loops, and that the available data are often incomplete and inconsistent. Nonetheless, complex systems science has developed computer-based methods to explore the behaviour of social systems. These include computer-simulated-agent-based modelling which enables the ‘space’ of future behaviours to be explored by running millions of simulations from many different sets of initial conditions. Such modelling does not give point predictions and, at best, can indicate how the system may evolve and the relative probabilities of various outcomes. However, it can generate unexpected and counter-intuitive behaviours, including sometimes making the analyst aware of ‘unknown unknowns’.

Policy is a major reason for being interested in the dynamics of social systems – will the proposed intervention have the desired outcome, and not have unintended consequences? Recently, global systems science [10] has emerged as a fusion between policy questions, complex systems science, policy informatics, and citizen engagements. The last two of these are particularly relevant to autonomous distributed systems.

10.3 Centralisation in industry

The invention of factory-based manufacturing and the harnessing of new energy sources lead to the emergence in the eighteenth century of enterprise-scale manufacturing and commerce. They were also causes of major social change as feudalism gave way to capitalism and workers moved from agricultural jobs in rural settings to production work in dense urban settings. While their new jobs were based on highly centralised, capitalist principles, the newly affluent working communities self-organised autonomous groups such as brass bands, choral societies, boys’ clubs, sports clubs, and many other cultural activities, not least trades unions. As we will show later, Social ADS is a ubiquitous pattern and in the twenty-first century; the capabilities of such organisations have been greatly extended through the Internet and ICT-based processes. In this section, we consider first the development of highly centralised manufacturing and services.

10.3.1 Emergence

The owners of the new factories took large financial risks and wanted to maximise the returns they could make on their investments. Such forces tended to centralise the activity to maximise the productivity of the investment with an initial focus on a narrow range of outputs that satisfied the customers' demands. This closed, centralised approach also served to preserve industrial secrets, which were great sources of wealth to the owners. Foremen were recruited from the ranks of non-commissioned army officers to direct the workers, and hierarchies of management were established to plan and track the factories' operations.

Distribution channels were set up to deliver manufactured products to customers. The commercial or private consumers of such novel products were likewise initially pleased to have access to materials or products that were previously unobtainable or unaffordable and hence were relatively undiscerning with respect to quality or specialised products. Moreover, the manufacturers had little need to market their products to these customers, since rising demand outstripped supply. Profits were high and, where possible, a manufacturer would attempt to establish a monopoly and extract high rents.

To be sure, the new capitalists had few choices as enterprises grew. Highly educated engineers who understood how these plants worked were rare as hens' teeth. Management was not yet recognised as a profession and was not taught in schools or universities. Hence, management skills were equally rare and little understood. Both technical and enterprise processes for such businesses were just emerging. So closed, centralised, top-down management was the best available solution.

Industrialisation also lead to rapid urbanisation. The rapid expansion of cities [11,12] required the provision of new services by municipal and central governments – housing, water heating gas, public health, police and emergency services, electricity, postal and later telecommunication services, radio and television broadcast services. No less than in the private sector, the public sector developed closed, centralised, hierarchical systems operating at maximum capacity and providing undifferentiated products to undiscerning and unconsulted customers for the provision of services such as public safety, water supply, transportation, and communications.

The ghosts of this centralised, hierarchical, highly integrated model still haunt many industries and public services. For example, heavy electrical engineering, which builds the massive machines employed in Gigawatt power stations, remains a dominant force, even though it is now cheaper to produce electricity in more decentralised and autonomous ways [13]. In some countries, centralised industries have collapsed, for example the steel and coal industries of Western Europe and North America gave way to competition from smaller, more versatile plants in India and China. However, the latter still have a tendency to grow towards the earlier, centralised model.

10.3.2 Evolution

While the complete deconstruction of enterprise and social systems imagined *ca* 2000 through the dramatic reduction in transaction costs, the adoption of standardised business processes and their representation in computer languages such as XML have not yet occurred, the wind is still blowing in that direction and the applications of Social ADS principles are increasing. Apart from the ICT-based factors just mentioned, other forces lean in this direction:

Returns to scale: Nineteenth-century industries were labour-intensive. In consequence the marginal cost of, say a ton of coal or a ton of steel, though lower than the average cost, was still relatively high. The scaling was roughly linear. Today many marginal costs decline almost exponentially with volume. The costs of developing, say a new computer chip or a sensor for a mobile telephone, are high and require specialised skills and tools, but when these costs are amortised over production runs of tens or hundreds of millions of pieces, the unit cost becomes minute. This also applies to larger systems, and hence, it becomes cheaper to make and install a million solar panels, each producing 1 kW, than to construct a conventional, 1-GW power plant. It is also faster, requiring far less time to plan and construct, and easier to locate, and it is therefore possible for small businesses to enter the utility industry.

Financial services: The global financial industry is in no danger of going out of business, although it may collapse occasionally under its own complexity, but it is no longer the only way to finance a new business. Decentralised, bottom-up methods such as Kickstarter [14] enable individual, small investors to raise seed money for new ventures by leveraging social media systems to assess the reputations of the innovators. Microfinance [15] has also proved effective in increasing the inclusiveness of financial services in regions lacking in banking services, notably areas of Africa, although it is also susceptible to abuse in the form of extortionate interest rates.

Embedded ICT: Developments in embedded intelligence, in miniature sensors, and in communications, especially wireless communications, make it possible to place the management of function, say demand-side energy management, at the point of delivery rather than requiring a centralised management system.

Educated consumers: The millennial generation that is now entering the global economy in large numbers has different expectations from previous generations. It has never known a world without the Internet, and hence, the concepts of decentralisation and autonomy are much more easily embraced.

There are however countervailing winds. Many of the same factors also lead to concentrations of functionality. Some examples are:

Returns to scale: Communication services work both ways, and so it becomes easier to aggregate small businesses and individuals into giant, online services such as Alibaba, Amazon, eBay, Facebook, and Google. In this world, economic power emerges from the capturing the attention of billions of people, from identifying and exploiting their interests and needs and from leveraging volume to drive down suppliers' prices.

Financial services: While Kickstarter enables the innovation of many startup companies by raising hundreds of thousands of dollars, it does not scale (yet) to the hundreds of millions required to launch a scalable business. So eventually the successful start-up companies need to turn to conventional venture capital for larger sums and for the lucrative business of initial public offerings.

Distributed ICT: The pervasiveness of intelligence, sensors, and communications permits the aggregation of the activities of individuals and their usage of infrastructure and services on a scale that was unimaginable even 15 years ago. There is still much further to travel on this road as the Internet of Things adds ever more sensors to public and private places, and the immense volumes of data collected daily become fodder for ever more sophisticated analytics on vast server farms. Public and private organisations that can capture and analyse these vast flows of information can acquire major political and economic power.

Stable employment: In highly developed countries, much of the period following World War II was a time of stability in enterprise and social systems. Employees of large enterprises and governments could expect lifelong employment. That stability was greatly reduced after 1990 as automation, new business management ideas, ICT, globalised industries, and other factors created a fast-moving world based on change and innovation. This has led to far more volatile forms of employment and the loss of good jobs in manufacturing.

We see here the Yin/Yang nature of decentralisation and autonomy in industry. In the following sections, we will examine how this has played out in a few major industries and government.

10.4 Autonomy in social systems

In the United Kingdom and many other countries, Social ADS is a ubiquitous pattern. People and communities spontaneously self-organise for many purposes including leisure, philanthropy, commerce, and political power. For example, a cluster of villages in the county of Buckinghamshire in the United Kingdom has a rich variety of self-organising leisure groups including a walking group, a cycling group, a gardening club, a wine tasting group, a choir, a book club, a bridge group, a brass band, bowls club, a cricket club, the British Legion for former soldiers, the boy scouts, an allotments society, various church groups, and many more.

Some of these local groups spring from the initiative of an energetic core and survive as long as that core group remains active. Other groups survive longer through committee structures that recruit new people to play active organisational roles. For example, the bowls club was established in 1922 [16], while the brass band was granted charitable status in 1973 [17]. Some local groups are actively engaged in contributing to the social good. For example, a voluntary organisation helps to maintain a community-owned green space by clearing brambles, thistles, and nettles; maintaining the barn, fences, and gates; litter picking; path maintenance; orchard mowing, pruning, and mulching; improving drainage; hay-raking; installing nesting boxes and hedge-laying [18].

Sometimes the local organisations self-organise to form higher level structures. For example, the Woburn Area Arts Society [19] is the local branch of The Arts Society (previously called the National Association of Decorative and Fine Arts (NADFAS)). The history of NADFAS illustrates the evolution of an autonomous distributed social system. In 1965, Patricia Fay founded the Chiltern Antiques Group, and in 1968, she was instrumental in uniting eleven arts societies to form NADFAS. In the 1970s, NADFAS had a period of rapid expansion and in 1972 became a registered charity. A major initiative was launched in 1973 to record the buildings, interiors, and history of churches and places of worship. In 1994, central London premises were purchased, and in 2002, NADFAS was legally incorporated as a UK company limited by guarantee. A scholarship scheme was established in 2006, and in 2010, NADFAS Heritage Volunteers achieved the milestone of contributing 300,000 hours, saving museums and galleries £5 million. Over the last 50 years, NADFAS has grown from a small local organisation to a major, national Social ADS, supporting the arts at local and national levels [20].

The National Trust, which cares for historic properties and areas of countryside, has a similar history. It was founded and registered under the Companies Act on 12 January 1895 by Octavia Hill, Robert Hunter, and Hardwicke Rawnsley. The Trust rapidly acquired many historic houses and estates. In 1886, it bought the clergy house in Alfriston as its first building; in 1899, it acquired its first nature reserve at Wicken Fen; in 1900, it was given Kanturk Castle [21]. Today it owns over 775 miles of coastline, over 247,000 ha of land, over 350 historic houses, gardens, parks, ancient monuments, and nature reserves. From its bottom-up, self-organising beginnings in the nineteenth century, the trust has become one of the United Kingdom's largest charities with an annual expenditure of half a billion pounds [22].

By its nature, commerce depends on the evolution of autonomous distributed systems. For example, on a small green at the centre of the same Buckinghamshire village, there is a Sunday farmers' market organised by one of the local traders. At one of the rustic stalls, the head brewer and owner sells craft beer made by his new microbrewery at the back of a house in a nearby village [23]. What started as a part-time enterprise became a commercial organisation in 2015 with the installation of a 6-barrel brewery. This company now supplies beer to many local outlets and some as far away as Manchester. This kind of autonomous entrepreneurship is an essential element of the capitalist system. As with The Arts Society in the voluntary sector, a small autonomous distributed system like this begins with a small local initiative that may grow to become a national or international organisation.

A current example is the emergence of the Maker Movement [24], in which communities of artisans establish communal workshops, sometimes offering highly sophisticated tools and materials, and jointly develop and share designs and production techniques. This movement, whose origins are traceable to the early twentieth century and the advent of radio as a medium of entertainment, spreads organically as clusters of artisans discover one another and expand from a corner in an individual's workshop to many thousands of square feet of space, often in a former manufacturing building.

10.4.1 The formation dynamics of autonomous distributed systems in society

Such Social ADS thus often emerge spontaneously from social interactions and go on to evolve into complex interdependent systems. The following illustrates the dynamics of such evolution:

1. **Phase 1: Getting started**
   (i) One or more people are motivated to create a system to do something. The motivation may come from a desire to do something pleasurable, e.g. play tennis, go walking, go cycling, sing in a choir, attend interesting lectures. It may come from a sense of concern or social justice, e.g. 'making things better', being a Samaritan [25], becoming a scout leader, becoming a first aider, listening to children reading at school, being a mountain rescuer. It may be motivated by seeing a commercial opportunity, e.g. starting a brewery, starting a window cleaning business.
   (ii) There are no insuperable obstacles to stop them doing so. That is, there are no legal or social impediments forbidding the activity, unless the purpose is deliberately criminal, challenges the law, or contravenes strongly held social norms. There are no other groups able to oppose and halt the initiative.
   (iii) The interested parties provide or have access to the resources necessary for the enterprise, e.g. money, time, premises, knowledge, and experience.
2. **Phase 2: Implementation**
   (i) The founders work together to form the system and succeed or fail at the micro level.
   (ii) They become certificated, e.g. take legally required examinations, join professional bodies, or establish their own formal qualifications for accreditation.
3. **Phase 3: Expand locally**
   (i) The level of activity expands, e.g. involving more people, hiring employees, making more product, selling more product, and increasing the range of services.
   (ii) The activity federates with other organisations, e.g. by forming a group of synergistic organisations, joining a federation of sports clubs, or making agreements with other commercial organisations.
4. **Phase 4: Expansion to national and global scales**
   (i) The activity establishes a network of quasi-autonomous organisations that operate under the aegis of the original, parent organisation, e.g. by opening new branches in other towns or franchising the activity.
   (ii) The activity may develop a shared ICT platform that provides services to the national or global community. Such services may include education and training, provision of shared knowledge or other resources, reputation management, referrals, commercial and financial operations, user and customer support, and so forth.

The initial, localised activity thus has completely autonomous origins, but it may grow to regional, national, or even international scale through the emergence of a network of quasi-autonomous branch organisations at the local level. However, to ensure cohesion of these distributed activities, it is likely to acquire some set of rules, principles of behaviour, or standards that are replicated to the branches and become conditions for affiliation to the founding organisation. In this way, small bottom-up ADS may evolve into large, multilevel systems which combine bottom-up and top-down organisational structure and control [26].

10.4.2 Social ADS in large organisations

As sketched above, Social ADS begin as small enterprises. Can such innovation occur within larger organisations such as private and state run industries, large institutions such as universities and the military, and government at local and national scales?

Entrepreneurial individuals in large organisations face the problem that other people within their organisation may oppose and attempt to halt an initiative. The reasons may include lack of imagination or being rule-bound, jealousy or spite, and risk aversion. The reasons may also include the proposed innovation being demonstrably unworkable, that it undermines another initiative, there being insufficient resource available, or there being no 'market' for its activities.

A further distinction may be that the growth of a new activity must in many cases take place within the existing organisation, which may be constrained by space, capital, willingness to hire new employees, and so forth. An exception would be an organisation based on a franchise business model.

Although it may be difficult for individuals at intermediate levels in large commercial organisations to implement initiatives, the board is expected to organise autonomously to run the business. A lesson from the science of complex systems is that strong top-down management may be inappropriate for some systems. One reason for this is that board level managers cannot know the details of what happens in the organisation at the micro level. For this reason, board-level managers may best conceive of their organisations as distributed autonomous systems and managing them as such, by, for example mandating (financial) outcomes rather than production or marketing processes. Compare this to laws and regulations imposed by governments, which usually mandate or exclude specific behaviours in order to achieve socially desirable outcomes.

10.5 Social control systems

We have illustrated above that collective social behaviours of all kinds are governed and regulated through a mixture of centralised and decentralised mechanisms. Culture and laws are the control mechanisms that we have developed over several thousand years [27] to create desired outcomes such as safe and stable societies. They work by creating frames of reference that are known to all members of the society and can be compared *post hoc* to the actual behaviour of an individual who can then be judged on whether he or she has violated a law, a custom or a regulation.

Any control system must choose whether to attempt to define permitted or forbidden inputs (a law, a custom or a regulation) or desired ranges of outputs (outcomes). Government and societies are mainly motivated by outcomes – safe transportation, food, a well-educated workforce, a balanced budget – which cannot be directly achieved. Instead governments enact laws and regulations – speed limits, years of primary and secondary educations, taxes – that are intended to ensure the desired outcomes.

The challenge for such input-based control systems is how to deal with unforeseen circumstances. Speed limits may ensure safe travel under normal conditions, but may be dangerously inappropriate when the roads are icy, and drivers must apply their own judgement. Hence, centralised laws and regulations serve to control individual behaviour, while national, local, and personal culture provides decentralised guidance for what each individual may judge to be acceptable behaviour. Decentralised autonomy is valuable for a society by enabling the individual to respond appropriately when confronted by exceptional circumstances and thus contributes to the society's resilience.

The balance between freedom to innovate and constraints on doing so is illustrated by the expectations of engineers when working on safety-critical systems. The professional codes that protect engineers are drafted in a general way in the context of particular technologies, but do not specify what any particular system must or must not do. Engineers are expected to apply professional judgement as to what is safe when they create new systems. They do this in the knowledge that if things go wrong, there will be an enquiry that will indeed analyse the particular system and if necessary hold to account those individuals that do not meet the standards set.

Such decentralised, cultural control of behaviour emerges and evolves over long periods to sustain desirable social outcomes. In tolerant societies, cultural control allows the individual considerable autonomy. We generally trust our fellow citizens to behave in way that supports our desirable social outcomes, even if their behaviour occasionally contravenes laws or regulations. Whether through centralised or decentralised control, few societies wish to invest in a police force that can monitor and command every individual's behaviour, although some societies may encourage citizens to act as enforcers of policies or faiths. Hence, citizens may assert autonomy within and even beyond legal boundaries. Depending on the scale of enforcement, societies exhibit varying attitudes to laws and regulations.

Present-day technological innovation may perturb this balance between centralised and decentralised social control mechanisms as we consider in the following section.

10.5.1 ICT and Social ADS

While ICT has had major impacts on personal systems (email, Skype, Facebook, audio and video digital streaming, and so forth) and on enterprise systems (global banking, global manufacturing, air travel, and so forth), its impact on social control systems has been marginal. Governments have applied ICT to existing bureaucratic processes (identity registration, permitting, taxation, and so forth), but it is only recently that transformational applications have emerged. One of the strongest and most contentious examples is the ability of national security agencies to monitor personal communications and personal movement at hitherto unprecedented levels.

More relevant here are the citizen-to-government communication channels that have been opened up. For example, consider the web-based open data processes [28], through which governments publish non-confidential information at the national, regional, and city levels ranging from committee reports, tax and spending by district, crime and health statistics by district, real-time traffic information, and public works job tickets. In the other direction, ICT can also be the means of enabling policy makers to engage with citizens and stakeholders.

For example, in the United Kingdom, the government runs a website that allows citizens to start petitions. If 10,000 people sign the petition online, the government promises to respond [29]. A successful example was a petition to end the tax on tampons that gathered more than 300,000 signatures. Policy informatics can also aid communications between policy makers and citizens by making available the models and data used in policy formulation and allowing others to investigate possible policy outcomes [10].

The emergence of a highly instrumented world [30] with machine-based reasoning capable of matching or exceeding some aspects of human cognition may introduce further transformational change. In such an environment, Casey and Niblett [31,32] have suggested that, instead of defining desirable behaviour via laws and regulations that can be applied *post hoc* to assess an incident, an individual or a device operated by an individual, notably a self-driving car, may be given real-time guidance on whether an intended action would be considered 'legal'. They call these advisories 'micro-directives'.

Micro-directives are issued by a centralised governance or control system that assesses the individual's situation and determines permissible actions that may be taken without putting at risk the goal of a safe and stable society. An example that is currently debated in the ethics of autonomous systems such as self-driving cars concerns the problem of what such a vehicle should do if a pedestrian unexpectedly steps into the roadway. The vehicle is posited to have the option of quickly moving into the opposite lane, but would then collide with an oncoming vehicle that it knows to contain no human being. Since the centralised system is fully informed about the situation of every autonomous vehicle, it can assess this problem and issue a micro-directive that authorises or instructs the vehicle to change lanes and unavoidably collide with the on-coming passengerless vehicle.

In fact in most countries, individuals routinely give themselves such permissions to make small violations of the laws that go undetected or that would not cause a policeman to intervene, e.g. minor parking violations. Societies tolerate such autonomous judgements for the resilience and agility they enable at fine-grained spatial and temporal scales. Creating a centralised system that automates such judgements is a troubling step. But the scenario is real. The advent of autonomous vehicles however requires either an embedded system in each vehicle for analysing such an ethical dilemma or a centralised system that can perform such instantaneous judgements. This will no doubt require extensive ethical and legal analysis and debate before such vehicles become widely permitted.

We may use the term 'Policy Informatics' for these processes of, on the one hand, engaging citizens in policy development and, on the other hand, of generating hard real-time, spatially specific commands that are derived from centralised policies. Policy Informatics is central to Global Systems Science. It is the means of coordinating Complex Systems Science with policymakers and with citizens. Computer graphics and visual analytics can be very important in this [10].

Against this rather dystopian view, we note that ICT also greatly favours the formation dynamics of Social ADS. Small numbers of individuals can reach very large numbers of people through websites, Google email and document services, Facebook, Twitter, YouTube, Flickr, and the many other free services available and thereby create decentralised, autonomous organisations that may disrupt existing centralised organisations.

10.6 Case studies of social and enterprise systems in transition

In recent times, the advent of the Internet, the Internet of Things, Big Data, and artificial intelligence technologies begin to exert enormous impacts on the traditional social control methods described above and in the operating models of large enterprises. In this section, we explore two examples of these impacts with the aim of identifying those characteristics that favour autonomous, decentralised systems in society and those that favour centralised, top-down systems in the private and the public sectors.

10.6.1 Telecommunications

Telecommunication systems have followed a very similar trajectory to electricity utilities. A flowering in the late nineteenth century of many, small local or regional telephone networks was followed by their aggregation into national monopolies based on the control of the analogue copper network. In this network, bandwidth was a scarce, but highly valuable resource and consumers would pay high tariffs by the minute for its use.

The advent of digital technologies and of fibre-optic cables transformed scarcity into abundance. However, the national monopolies had a technological control point – all of the intelligence required for the operation and management of these systems was embedded in centralised switches. Innovation, in the form of the Intelligent Network principles [33] took the form of increased intelligence in the switches, while the user's terminal, the basic telephone, remained dumb. This control point became threatened in around 1990 by the popularity of personal computers and the public emergence of the Internet, which is based on the opposite principle – as little intelligence as possible in the network and as much as possible in the hands of the users at the periphery. The realisation that this transition was imminent led to an infamous white paper by David Isenberg entitled 'The Rise of the Stupid Network' [34].

Also unlike the former telephone networks, the Internet famously has no centralised management system. As the name implies, it is formed from the interconnection of many subnetworks that are managed by their individual owners. In the mid-1990s, telco engineers still believed that the Internet must be inherently unstable and would 'melt down' sooner or later.

The development in the early 1990s of asynchronous transfer mode [35] enabled all forms of telecommunications to share the same pipes. Data communication services have long been based on tiered, flat-rate tariffs, but as the cost of fibre-optic bandwidth fell dramatically in the 1990s, the perceived value of voice bandwidth declined. The notion of paying by the minute has today almost completely disappeared. *Ca* 1996 'Voice over IP', that is voice telephony on a data network, took off. The introduction of the asymmetric digital subscriber loop [36] in the late 1990s removed the last-mile bottleneck and sounded the death knell of the historic model of telecommunications. In 1998, a major US consulting company published a report entitled 'Telecommunications in the Age of Zero Tariffs' and Cairncross published 'The Death of Distance' [37]. Today many services [38,39] exist to provide free, worldwide voice communications based on network capacities that were unthinkable even 20 years ago.

We see here that a major, global industry can be violently transformed against its own will by technological innovation in less than 30 years from one based on centralised, top-down services to one that provides a multitude of autonomous, decentralised services based on a common, open platform – the Internet. In some countries, operators may still hold monopoly power over last-mile communications, but revenues from fixed networks, as opposed to mobile networks, have declined strongly, and the utilities have little interest in investing in this area for the primary purpose of data communication on tiered flat-rate tariffs. Instead, fixed network operators seek new revenues streams from streaming entertainment media and from new services such as home or building automation [40].

10.6.2 Local government

Local governments, particularly municipal governments, have proven strongly resistant to transitions towards autonomy and decentralisation at various levels. Their organisational models resemble fiefdoms descended from eighteenth century royal courts and manifest closed, centralised, top-down control. Apart from tradition, there may be legitimate reasons for this stasis. Government agencies deal with large amounts of personal information such as financial status, tax and criminal records, property ownership, educational records, and so forth whose confidentiality must be protected and which therefore precludes the easier sharing of information in the private sector. Officials are accountable for results in ways that private-sector managers are not and are thus cautious of losing their own independence. In the United States, strong public-sector unions impede changes in operational practices. These legitimate reasons can, however, become justifications for a lack of transparency and accessibility that isolates agencies from the public they are intended to serve.

While these issues remain, the recent decade has shown some progress in opening up municipal government and providing new roles for grass-roots civic organisations. One innovation is Open Data [41], which has its roots in the publication of reports on municipal websites in the 1990s and in the open-source movement of the same period [42]. This has since evolved from static information published as documents to real-time data, enabling greater accountability, transparency and action. Today cities such as Amsterdam [43], Barcelona [44], London [45], and San Francisco [46] make available hundreds of datasets on topics ranging from public health inspections, the geographic distribution of budget spending, public works work orders or the current positions of municipal buses. Such data are highly valuable to civic organisations in assessing the effectiveness of policy and in developing arguments for policy changes. It is also valuable to individual citizens who want to know when their streets will be repaired or when the bus is going to arrive at a given bus stop. It even provides a viable mechanism for agencies to share data among themselves.

Many of the principles of Open Data derive from previous work on open-source software that created concepts such as the Creative Commons [47], which ensures free access to communally developed works and the right to republish or to prepare derivative works. These roots ensured that civic organisations had access to the ICT skills needed to develop tools for visualising and understanding the Big Data now being published by municipal governments. These movements coalesced in the last decade into the Open Cities [48] movement in which civic organisations and individuals are able to apply their innovation to grass-roots problems through 'hackathons' [49].

10.6.3 National government

As an example of the potential for Social ADS, consider what is called the bed-blocking problem in the United Kingdom. Simplifying the problem considerably, bed-blocking occurs when a person has completed their clinical treatment but cannot leave hospital because they cannot care for themselves at home, and there is no place in a residential home. This problem reflects the health and social care system in the United Kingdom in which health care is provided free of charge by the National Health Service and social care such as residential accommodation for the old and infirm is provided by local councils. These two subsystems are funded differently and have different incentives.

The National Audit Office report *Discharging older patients from hospital* (2016) summarises this as follows:

> Unnecessary delay in discharging older patients ... from hospital is a known and longstanding issue. For older people [this] can lead to worse health outcomes and can increase their long-term care needs [and] is an additional and avoidable pressure on the financial sustainability of the NHS and local government. ... Older people are cared for in hospital by the NHS, but once discharged some may need short- or long-term support from their local authority or community health services. ... The number of recorded delayed transfers of care has increased substantially over the past two years [a 31% increase between 2013 and 2015]. The main drivers for this increase are the number of days spent waiting for a package of home care (which more than doubled between 2013 and 2015, from 89,000 to 182,000) and waiting for a nursing home placement or availability (which increased by 63%). [50]

This problem is long standing and very complex, reflecting social and administrative systems at many levels from the individual patient stuck in hospital to local councils being unable to provide them with a suitable care package to the Ministry of Health trying to manage a system it only partially controls. However, it illustrates the great opportunities for ADS.

One approach to this problem is to analyse it top down, design a new 'integrated' health and social care system, and implement it. Such reorganisation has been tried before, and, even on a smaller scale, the complexity has resulted in disaster. Instead of this top-down approach, consider the possibilities for bottom-up self-organisation. One thing is clear. Whatever solution is implemented eventually, it will be highly dependent on ICTs.

To illustrate this, consider an elderly lady living alone in a rural area. In order to stay in her own home, suppose she needs three half-hour visits a day from carers who help her dress and make breakfast in the morning, help her prepare lunch and dinner, and help her tidy up her get ready for bed in the evening. One solution to providing her care is through a specialist company, which in this case would be very expensive, due in part to a lot of travelling being involved.

Another solution is to have the care provided more locally. Suppose there are people in the village who would be glad to provide care, paid or voluntary, but not every day and not always at the same time. For example, a young mother might be pleased to earn a little extra money on the day her child is at nursery. Thus, we hypothesise a heterogeneous set of, say, 20 people, willing to provide care from time to time. Being realistic, some of these people will be more reliable than others. Despite this, it is essential that the care system put in place is robust. It is not acceptable for care visits to be missed. How might such a system self-organise?

First, it is assumed that each of the twenty people is recruited into the ADS. This could be done through social media such as Facebook or local authority websites. Second, it is necessary that the individuals providing care are coordinated with those receiving it. This can be done by mobile telephony and specialised apps that contact all the 20 carers, who may be unaware of one another, and coordinate the times that they are free with the visits to be made. These apps can form a 'community' on the 20 mobile phones – they don't need to be on a server. Furthermore, they can store and share data. With this information system, the Social ADS can know if a carer is in the wrong place at the wrong time, likely to miss their visit, and take remedial action be bringing in someone else from the ADS and so on.

Is this kind of system possible? We think it is, indeed such an approach is used in the village of Brookfield, Connecticut, to manage the resettlement of Syrian refugee families using a team of 20 volunteers. Much of the necessary technology already exists. In the case of bed–blocking, the national and local governments can provide the incentives for the necessary self-organisation. One can imagine ecologies of apps evolving as social needs change. We think this is a very exciting and plausible opportunity for engineers now and in the future.

As final, brief examples, consider first Amazon's Mechanical Turk [51], an online service that allows an organisation or an individual to compose an *ad hoc* group of skilled workers to execute simple or complex tasks. This extends as far as developing a company's annual report. Second, note a recent newspaper article [52] commenting on how small, Internet-based companies providing healthy foods are having a significant impact on revenues of companies such as Nestle in the mainstream food industry.

We see here the possible coordination of the closed, centralised, top-down methods of government with the autonomous, decentralised methods of civic and commercial organisations.

10.7 Conclusions

In the World Until Yesterday, Diamond [53] discusses the spectrum of governance mechanisms across social groups ranging from families, to clans, to tribes, and to nations, and the need for such mechanisms to expand with the scale of the group. Harrison [54] suggests that as such groups expand, an individual leader eventually needs to delegate to a hierarchy of trusted lieutenants. In modern times, such human delegation has been complemented by rapid and pervasive telecommunications and sensing technologies.

The essential element here though is trust. In hierarchical governance, the individual leader of the social structure is trusted to look after the overall well-being of the group. As the group and its territory expand, the appointed delegates inherit from the leader some of the trust placed in him or her by the members of the group. The average member of the group has some degree of trust and respect for these delegates or mechanisms even though he or she may not directly know them.

In an autonomous system, this hierarchical chain of trust may be largely absent. The average member of the group may have direct knowledge of the people or mechanisms managing the system if it is localised. But if the system is spatially distributed, perhaps even globally distributed, such direct knowledge may not be feasible. In its place, indirect knowledge has until recently relied on proxies such as brands, limited word-of-mouth, or consumer review magazines. The advent of the Internet and the emergence of social interactions via the Internet in the last 20 years provide new mechanisms for establishing reputation and trust, albeit more limited than trust based on direct interaction.

Examples of the roles of such indirect trust are seen in services such as the open-source movement, the Maker Movement, Mechanical Turk, eBay, Uber, Airbnb, and social media sites such as Facebook and LinkedIn. These services enable small, localised groups or individuals, for example Uber drivers, to benefit from opportunities that are accessible through a large, complex platform that also serves to establish trustworthy engagements between providers of goods or services and clients. While, as we have shown, the scaling of modern technology enables small Social ADS groups to engage in activities formerly reserved to large public or private organisations, we believe that it is this ability to develop indirect trust that is central to the growth of such ADS. While this indirect trust brings many advantages, it must still complemented by direct, hierarchical trust for many of the critical functions provided by government.

In this article, we have seen that social and technological changes are leading to rebalancing of the tension between centralised, top-down systems and ADS. ADS principles have gained and may continue to gain ground, but the adoption of

Social ADS has not so far led to the wholesale deconstruction of large, centralised, top-down systems. While Social ADS approaches offer channels for human independence and innovation, these are countered by limitations to distributed control and management as well as financial and social forces.

In the end, Coase's Law [2] still applies. Some tasks, for example large-scale manufacturing of cars or governing a city, are simply too technically, financially, legally, and socially complex to be decentralised by current Social ADS methods. It is, however, not beyond imagining that platforms similar to eBay or the Mechanical Turk might emerge that can address such complexities, making Social ADS a predominant organisational force in the twenty-first century.

Acknowledgements

We gratefully acknowledge many discussions with Prof. Kinji Mori, Dr. Ralph Dum, and our colleagues in the Global Systems Science and Policy community.

References

[1] Mori, Kinji (1993), *Autonomous decentralized systems: concept, data field architecture and future trends*, Proceedings of the IEEE International Symposium on Autonomous Decentralized Systems (ISADS), 1993, pp. 28–34.

[2] Coase, Ronald (1937), *The Nature of the Firm*, Economica, **4** (16): 386–405. doi:10.1111/j.1468-0335.1937.tb00002.x. JSTOR 2626876.

[3] Boh, Waifong; Soh, Cristina; Yeo, Steven (2007), *Standards Development and Diffusion: A Case Study of RosettaNet*, Communications of the ACM, **50** (12): 5762. doi:10.1145/1323688.1323695.

[4] Bray, Tim; Paoli, Jean; Sperberg-McQueen, Michael; Maler, Eve; Yergeau, François (2008), Extensible Markup Language (XML) 1.0. https://www.w3.org/TR/REC-xml/ Retrieved 2018-10-29. World Wide Web Consortium.

[5] Arsanjani, Ali; Booch, Grady; Boubez, Toufic; *et al.* (2009), SOA Manifesto. www.soa-manifesto.org Retrieved 2018-10-29.

[6] Nowak, Andrzej; Kacprzyk-Murawska, Marta; Serwotka, Ewa (2017), *Social psychology and the narrative economy*, in Johnson, J.; Nowak, A.; Ormerod, P.; Rosewell, B.; Zhang, Y-C., (eds), Non-Equilibrium Social Science and Policy, Springer (Berlin), DOI 10.1007/978-3-319-42424-8_3.

[7] Bignell, Victor; Fortune Joyce (1984), *Understanding Systems Failure*, Manchester University Press (Manchester), 1984.

[8] Lorenz, Edward (1963), *Deterministic Nonperiodic Flow*, Journal of the Atmospheric Sciences, **20** (2): 130–141.

[9] Hayek, Friedrich (1978), *The results of human action but not of human design*, in New Studies in Philosophy, Politics, Economics, University of Chicago Press (Chicago), pp. 96–105.

[10] Dum, Ralph; Johnson, Jeffrey (2017), *Global systems science and policy*, in Johnson, J., Nowak, A., Orerod, P., Rosewell, B., Zhang, Y-C., (eds), Non-Equilibrium Social Science and Policy, Springer (Cham), pp. 209–225.

[11] Batty, Michael (2006), *Rank Clocks*, Nature, 444: 592–596 (30 November 2006). doi:10.1038/nature05302.

[12] Weber, Adna (1899), *The Growth of Cities in the Nineteenth Century: A Study in Statistics*, Macmillan (New York), pp. 218–222, 439–442, 444–445.

[13] Donohoo-Vallet, Paul (2016), *Revolution … Now*, U.S. Department of Energy, https://www.energy.gov/eere/downloads/revolutionnow-2016-update Retrieved 2018-10-29.

[14] Kickstarter, https://en.wikipedia.org/wiki/Kickstarter Retrieved 2018-10-29.

[15] Microfinance, https://en.wikipedia.org/wiki/Microfinance Retrieved 2018-10-29.

[16] The Woburn Sands Bowls Club, https://sites.google.com/site/woburnsands-bowlsclub/ Retrieved 2018-10-29.

[17] The Woburn Sands Band, http://www.woburnsandsband.co.uk Retrieved 2018-10-29.

[18] Edgewick Farm, http://www.edgewickfarm.btik.com/AboutUs Retrieved 2018-10-29.

[19] The Arts Society Woburn Area, http://the artssociety.org/woburn-area Retrieved 2018-10-29.

[20] NADFAS History, https://theartssociety.org/about-us Retrieved 2018-10-29.

[21] The National Trust History, https://www.nationaltrust.org.uk/lists/our-history-1884-1945.

[22] The National Trust, https://www.nationaltrust.org.uk/documents/annual-report-2014-15.pdf.

[23] Hornes Brewery, http://www.hornesbrewery.co.uk Retrieved 2018-10-29.

[24] Anderson, Chris (2013), *20 Years of Wired: Maker movement*, http://www.wired.co.uk/article/maker-movement Retrieved 2018-10-29.

[25] Samaritans, http://www.samaritans.org/about-us/history-samaritans Retrieved 2018-10-29.

[26] Johnson, Jeffrey; Fortune, Joyce; Bromley, Jane (2018), *Multilevel systems and policy*, in Mitleton-Kelly, Eve; Paraskevas, Alexandros; Day, Christopher (eds) Handbook of Research Methods in Complexity Science and their Applications, Edward Elgar (London), pp. 363–387.

[27] Code of Hamurabi, https://en.wikipedia.org/wiki/Code_of_Hammurabi Retrieved 2018-10-29.

[28] Open Data, https://en.wikipedia.org/wiki/Open_data#In_government Retrieved 2018-10-29.

[29] Petitions to the UK Government, https://petition.parliament.uk Retrieved 2018-10-29.

[30] Chen-Ritzo, Ching-Hau; Harrison, Colin; Paraszczak, Jurij; Parr, Francis (2009), *Instrumenting the Planet*, IBM Journal of Research and Development, **53** (3): 1:1–1:16.

[31] Casey, Anthony; Niblett, Anthony (2016), *Self-Driving Laws*, University of Toronto Law Journal, **66** (4), http://dx.doi.org/10.3138/UTLJ.4006.

[32] Casey, Anthony; Niblett, Anthony (2015), *The Death of Rules and Standards*, Available at SSRN 2693826.

[33] Stallings, William (1992). *ISDN and Broadband ISDN*. Macmillan Publishing Co., Inc. (London).

[34] Isenberg, David (1997), *Rise of the Stupid Network*, Computer Telephony **5** (8): 16–26.

[35] Handel, Rainer; Huber, Manfred; Schroder, Stefan (1998). *ATM Networks: Concepts, Protocols, Applications*. Addison-Wesley Longman Ltd (Essex).

[36] ITU (1999), *G. 992.1: Asymmetrical Digital Subscriber Line (ADSL) Transceivers*, ITU (Geneva).

[37] Cairncross, Frances (2001), *The Death of Distance: How the Communications Revolution Is Changing Our Lives*, Harvard Business Press (Boston, MA).

[38] Skype, https://en.wikipedia.org/wiki/Skype Retrieved 2018-10-29.

[39] WhatsApp, https://en.wikipedia.org/wiki/WhatsApp Retrieved 2018-10-29.

[40] Home Automation, https://en.wikipedia.org/wiki/Home_automation Retrieved 2018-10-29.

[41] Manyika, James; Chui, Michael; Farrell, Diana; Van Kuiken, Steve; Groves, Peter; Almasi Doshi, Elizabeth (2013), *Open Data: Unlocking Innovation and Performance with Liquid Information*, McKinsey Global Institute, pp. 21.

[42] Open Source Model, https://en.wikipedia.org/wiki/Open-source_model Retrieved 2018-10-29.

[43] Open Data, Amsterdam, https://data.amsterdam.nl/ Retrieved 2018-10-29.

[44] Open Data, Barcelona, http://opendata.bcn.cat/opendata/en Retrieved 2018-10-29.

[45] Open Data, London, https://data.london.gov.uk/ Retrieved 2018-10-29.

[46] Open Data, San Francisco, https://data.sfgov.org/ Retrieved 2018-10-29.

[47] Creative Commons, https://creativecommons.org/licenses/ Retrieved 2017-01-05.

[48] Open Data Institute, https://theodi.org/ Retrieved 2018-10-29.

[49] Code for America, https://www.codeforamerica.org/ Retrieved 2018-10-29.

[50] National Audit Office (2016), *Discharging Older Patients from Hospital*, National Audit Office, ISBN 978-1-78604-05.2-7, 26 May 2016, https://www.nao.org.uk/wp-content/uploads/2015/12/Discharging-older-patients-from-hospital.pdf Retrieved 2018-10-29.

[51] Mechanical Turk, https://www.mturk.com/mturk/welcome Retrieved 2018-10-29.

[52] *A Life Less Sweet – Nestlé Looks for Ways to Boost Stale Growth as Consumers Snub Unhealthy Food*, The Economist, 7 January 2017, https://www.economist.com/business/2017/01/07/nestle-looks-for-ways-to-boost-stale-growth-as-consumers-snub-unhealthy-food Retrieved 2018-09-11.

[53] Diamond, Jared (2012), *The World Until Yesterday: What Can We Learn from Traditional Societies?* Allen Lane (London).

[54] Harrison, Colin (2018), *Augmented Intelligence and Society*, in Araya, Daniel (ed) Augmented Intelligence, Peter Lang, Inc. (Bern).

Chapter 11

Internet of Simulation: building smart autonomous decentralised systems

David McKee[1], Stephen Clement[1], and Jie Xu[1]

Abstract

The increasing autonomy, complexity, and decentralisation of modern systems require an increasing reliance on large-scale simulation to cope with the social, technical, economic, and environmental challenges of the world we live in. These large-scale simulations aid in the understanding of processes and design of complex systems. The Internet of Simulation (IoS) is an emerging trend towards a decentralised ecosystem of geographical disparate simulations that are readily combined to form more complex simulations. This chapter explores some of the challenges in implementing this distributed simulation system and its use in design, maintenance, analysis, and training of complex systems. These include the management of complex interactions between integrated simulation components and wider operational challenges. The contextualised applications of IoS and some of the relevant issues' post-deployment are discussed.

11.1 Internet of Simulation characteristics

Simulation is used in almost every industry and affects nearly every aspect of society. Virtually, every product in use has been designed and tested using some level of simulation, whether this be office furniture, vehicles, or complex safety systems such as air-traffic control. As cyberphysical systems become autonomous and decentralised, the interactions between sub-systems and components become more complex. Co-simulation has become a vital element of design and virtual prototyping in order to handle these complexities. Co-simulation allows multiple simulations, of different components, to be combined together into a single larger simulation. This allows for virtual prototyping and discovery of emergent behaviours earlier in the engineering life cycle. Simulation also facilitates *what–if* analysis allowing the exploration of scenarios that are either infeasible or too costly

[1]School of Computing, University of Leeds, UK

to be physically tested. Being able to therefore pull together numerous, potentially hundreds of, simulations into a single co-simulation provides an essential and powerful capability for designing evermore complex and integrated cyberphysical systems and particularly ensure those aspects of automation within those systems are appropriately tested.

This gives rise to the concept of an Internet of Simulation (IoS) which derives from an ever-growing need to design larger and more complex system which are themselves deployed into complex massive-scale cyberphysical systems [1]. IoS extends the concepts of IoT with virtual things and then bridges the virtual and physical worlds. The core characteristics of IoS break down into three areas: (1) the relationship to IoT as a network of virtual things, (2) the iterative and hierarchical concepts of Simulation as a Service (SIMaaS) and Workflow as a Service (WFaaS) for massive-scale co-simulation, and (3) the interconnectivity between IoS and devices within the physical world of IoT.

Internet of simulation characteristics:

- A specialism of the Internet of Things (IoT) comprises interconnected virtual system components, agents, or virtual environments defined by cross-domain collections of network-enabled, variable fidelity, and heterogeneous models and simulations.
- Through composing multiple virtual entities by defining their interactivity, a system simulation can be constructed and distributed.
- The *simulated things* contained in the IoS can be connected to the IoT via a real-time bridge.

11.1.1 Simulation as a Service

The IoS paradigm is built first and foremost on providing simulations as services over a network. Doing so requires

1. interface management using standards including the IEEE High-Level Architecture (HLA), Web service (WS) standards, and others so as to ensure simulation interoperability and
2. managed computational infrastructure from specialist hardware through virtualisation in the Cloud or on high-performance computing (HPC) platforms.

The underpinning principle of SIMaaS is ensuring the interoperability of simulations, simulations tools, and other systems. In order to guarantee this interoperability, three aspects must be managed; standardisation, timing and synchronisation, as well as internal and external states. In terms of standardisation, there are two starting points from the disciplines of service computing and co-simulation. In the first instance, there are the Software as a Service commonly used standards including WSs, SOAP, and REST. These provide the basis for the service aspects of SIMaaS providing interfaces to allow remote access to hosted simulations. For example, WSs typically have a definition specified using WSDL which specifies the methods that can be called as well as inputs, outputs, and parameters, and these can generated using almost any technology.

11.1.1.1 Simulation interoperability

SIMaaS facilitates simulation publication and integration. This requires

1. standardisation of both service and simulation interfaces;
2. causal consistency, including timing and synchronisation; and
3. internal and external state management.

The simulation community have however typically used official standards by either OMG, IEEE, or SISO. These include the Data Distribution Service (DDS), the HLA, and Functional Mock-up Interface (FMI). However, of the numerous commercial simulation tool providers, very few adhere to these standards and those that do typically provide only partial support for an individual standard and add vendor specific features. As a result, the method for exposing simulations becomes particularly challenging and requires expert engineering to understand how each of the underlying technologies function and also their interdependencies. Moreover, the current standards only manage the interoperability in terms of data and do not manage timings, synchronisation, or aspects such as state.

Given the wide range of tools and technologies that could be involved in exposing simulation as services, there are two general patterns (shown in Figure 11.1) that can be used for the publication of simulations as services.

1. **Embedded function block** where a tool-specific library or plug-in is created that encapsulates the information for network communication and interactivity. This then allows the dropping of a block into the diagrammatic representation of the simulation, within the tool. The block can then be wired up as a component or function within the simulation. Such a function block must support three modes of operation: *pass-through*, *simplified*, and *co-simulation*. The first allows the simulation to run independently for the purposes of testing. In doing so, the block must pass all data through from input to output without any modification or time delay. In the *simplified* mode, the block must provide some level of data transformation that represents the behaviour that would be expected

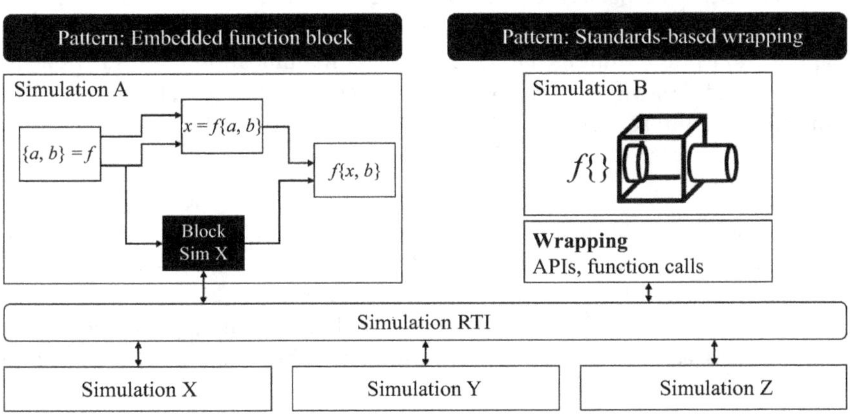

Figure 11.1 SIMaaS creation patterns 1 and 2

during a co-simulation. This particular mode would be typically implemented using data-sets. In the *co-simulation* configuration, the block must hide the complexities of time integration as well as network management. An example of this type of approach in a specific tool is the use of Simulink® s-functions.

2. **Standards-based interface wrapping** based on formal standards by the likes of IEEE, OMG, or SISO. This pattern moves a step further than the function blocks by fully encapsulating the simulation tool or model within a wrapping that is compliant with the distributed co-simulation or service standard. There is currently one main standard, IEEE HLA 1516, which is supported by a runtime infrastructure (RTI). In this fashion, a simulator is wrapped as federate which appropriate controls for information exchange, synchronisation, and coordination. These federates are then combined into a federation which describes the overall connectivity between the simulations as well as data and logic constraints. Underpinning the simulation structure is an RTI which is responsible for managing all communication. There are two main technologies that are currently adopted. The first is OMG DDS which provides a publish/subscribe model with low-level management of the network to provide soft real-time guarantees. The second is WSs which provides greater level of abstraction and an ability to create co-simulations across more wide spread, possibly global, networks but does not provide the same level real-time guarantees. Adopting this approach typically requires custom code generation for a given simulation model. Another technology is the FMI standard which facilitates model exchange and co-simulation using a mixture of XML files and compiled C-code.

At this point, these patterns assume the existence of an RTI. They do not capture issues around deployment of SIMaaS in environments such as Cloud computing or HPC. Therefore, a third pattern can be considered which extends the *standards-based interface wrapping* with virtualisation:

3. **Virtualised SIMaaS** In this fashion, the wrapping process should create an agent which runs as a WS hosted within either a virtual machine (VM) or container. The packaged VM, or container, includes the simulation and its constituent parts (tool and model) as well as the SIMaaS agent providing an application programming interface (API), communication, and function management. In this way, a simulation can be easily deployed across a wide range of infrastructure as well as deployed several times for different co-simulations.

11.1.2 Workflow as a Service

Once simulations are published using the SIMaaS paradigm, the next step is to facilitate co-simulation. Using today's standards and technologies, co-simulation is normally limited to connecting only two simulations together. There are very few scenarios which combine more, and those that do typically sit within the training domain with live, virtual, and constructive simulation. Live, virtual, and constructive simulation refers to a simulation involving humans-in-the-loop with virtual models being used for training (constructive) purposes such as military scenarios. The reason for this is the lack of standards and technologies to facilitate general simulation integration and the rapid integration of numerous simulations into larger more complex system-level simulations.

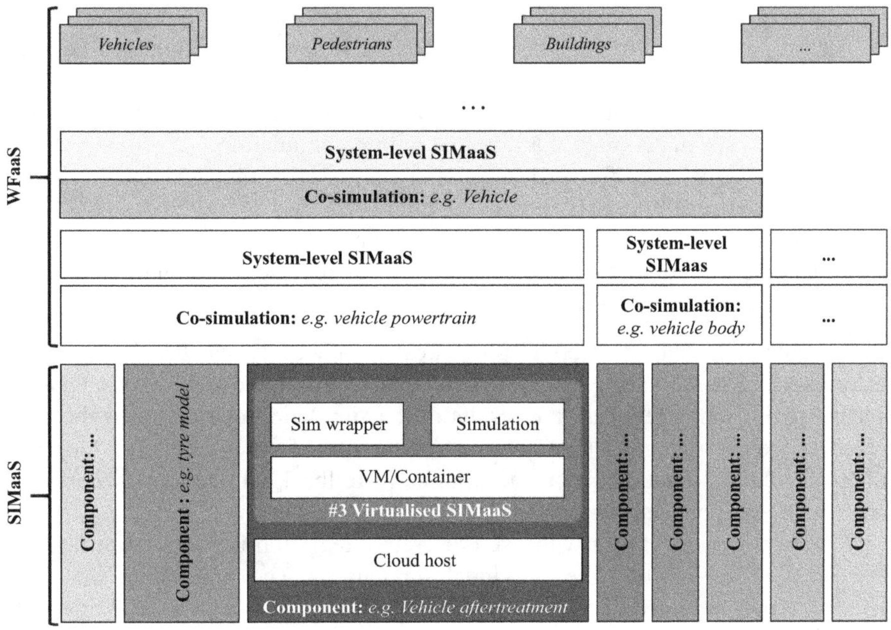

Figure 11.2 Iterative nature of WFaaS and SIMaaS with an automotive example

However, building specifically on pattern 3 for SIMaaS provision, it is possible to build massively scalable and autonomously managed co-simulation platforms. As shown Figure 11.2, component-level simulations can be exposed using the SIMaaS paradigm, integrated into system-level simulation using WFaaS, and the re-exposed using the SIMaaS paradigm as a system simulation.

In terms of creating co-simulations, the technologies described as part of the first two SIMaaS patterns provide the starting point. HLA, for example, can facilitate co-simulation with multiple discrete-event simulations. However, there are several key challenges that must be addressed when building a co-simulation ecosystem, the most prominent being causality, instability, and validation.

11.1.2.1 Causality

The temporal nature of the simulations could be either discrete event, discrete time, or continuous. This results in several possible configurations of a co-simulation. In the first two cases where all simulations are discrete, the integration is relatively straightforward. The management system must, however, guarantee the temporal consistency of events and signals. Temporal consistency refers to the order and visibility of updates, and guaranteeing it is one of the major and most important challenges in co-simulation. These are of several types:

- *Strict* – Any change to a property or value must be instantaneously seen by all other simulations. In other words, if x is written at time t, then x is immediately available to be read before time $t+1$. This approach has a very heavy performance cost.

- *Sequential* – Does not require the instantaneous visibility of values, but the ordering must remain the same. This approach does not guarantee a deterministic result.
- *Atomic* – Is sequential consistency with a real-time constraint and the use of semaphores. This is often not possible with co-simulation.
- *Causal* – A weaker form of sequential consistency that guarantees that all events are causally related and are in order.
- *Weak* – Does not enforce the synchronisation.
- *Optimistic* – Similar to weak consistency but reverses computation when an out-of-order event is detected.

Most simulation is either strictly, optimistically, or weakly synchronised with the performance cost decreasing as the level of synchronisation decreases. Conversely, as synchronisation is reduced, the results of the simulation can no longer be guaranteed to the same degree. One approach is the use of time warp [2] which uses either reversible computation or checkpoints to rollback simulation to a previous state to correct inconsistencies.

One of the major issues with even discrete integration is potential differences between time-steps within the individual simulations. The simulations may have differing representations of time in terms of both logical and execution time. The former being modelled by the simulation, and the latter representing the time taken to run the simulation. If these differ, techniques such as zero-order and first-order hold should be considered.

If however one or more of the simulations are continuous, strict attention must be paid to the control and data-flow through the co-simulation. If a continuous simulation is only downstream of others, then the temporal issues are negligible. However, if a continuous simulation is part of a loop in a co-simulation, whether it be with discrete or other continuous simulations, there are several issues. Specifically, continuous simulations are represented algebraically, and the tools use solvers implementing methods such as Runge–Kutta or Dorman–Prince to iteratively solve for a given time-step. To facilitate co-simulation with continuous simulations, the eco-system must either facilitate some global solver across the entire system or provide a mechanism to interface with the relevant points during the iterative solver cycles.

11.1.2.2 Instability and validation

Within engineering disciplines, most simulations are not currently built with co-simulation in mind. Therefore, any given simulation will have underlying assumptions including internal constants and values that may not be compatible with other simulations. Therefore, when performing co-simulation at any scale, it is essential that there are mechanisms for validating both the independent simulation results and the combined co-simulation results. One of the challenges with this validation is distinguishing between emergent system behaviours and simulation incompatibilities due to interfaces, data types, etc.

These issues mean that performing large-scale co-simulation is challenging and must be carefully managed. However, successful simulation integration and deployment of an IoS has huge implications across and wide range of applications and industries.

11.1.3 Relationship to IoT

The IoT describes the use of digital technologies and the interconnection of components, devices, and services at a massive-scale across networks [3]. IoT differs from concepts such as sensor networks; in that, it incorporates intelligence, either through context-aware computation or smart connectivity [4]. IoT extends across multiple domains and covers a wide range of technologies. These technologies provide the basis for various applications such as smart cities and autonomous vehicles. The domains and applications shown in the figure as well as the many others which form part of the concept of IoT each have specific requirements with regards to aspects such as infrastructure, security, interconnectivity, and standardisation. As a result, depending of what the specific *things* are, the specification of an IoT can be very different.

IoS provides an extension of IoT that is of specific interest to communities from the domains of manufacturing, government organisations, aerospace and defence [1]. IoS provides additional virtual elements and technologies which can be either used as a *Virtual IoT* or in conjunction with IoT using a real-time communication bridge. IoS can therefore be applied across application areas including product design, smart cities, as well as command and control systems.

11.2 Engineering applications

Simulations are used across disciplines from engineering and manufacturing, industrial applications, through machine learning and artificial intelligence (AI). The remainder of this chapter looks at the benefits of co-simulation across these particular areas.

11.2.1 Design and virtual prototyping

Simulation is used at every stage of product design and analysis with the concepts of virtual prototyping becoming commonplace to reduce the cost of product development and reduce time to market without compromising product quality. For example, in designing a transmission system of a vehicle, it consists of several components including engine and transmission, transmission control unit, and vehicle dynamics. Within the automotive industry, each of these operates within a different tool sets and are traditionally not suited for co-simulation [5]. Additionally, even within a single organisation, these simulations will physically exist in different locations introducing challenges for communication. This can be achieved using a combination of the SIMaaS patterns described previously along with careful management of the underlying communication infrastructure.

Given co-simulation during the design and analysis manufacturing phases, the concept of IoS can be extended further to look at integration across the entire engineering life cycle and through product life as shown in Figure 11.3 [6]. This

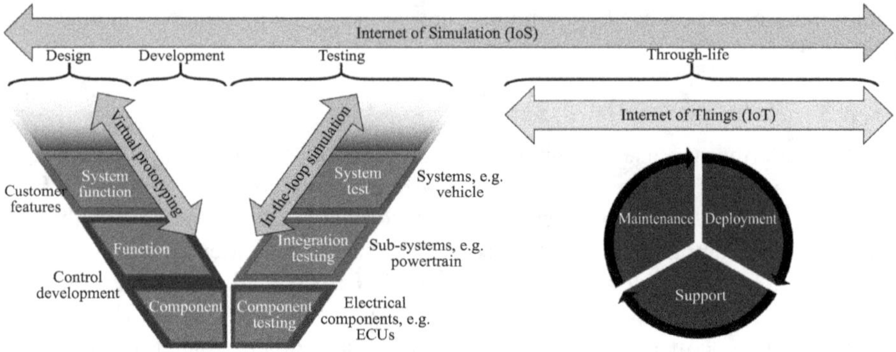

Figure 11.3 Engineer life cycle from concept to deployment and maintenance

moves beyond the engineering processes during the design and testing phases of products to the maintenance and analysis of deployed products.

11.2.2 Industry 4.0 (Industrial IoT)

Originally arising out of the aerospace domain, the concept of a digital twin is now central to the growing use of industrial IoT and Industry 4.0. The digital twin is a virtual construct that mirrors all information embedded in a physical component or system linked throughout life cycle [7]. Simulation is a vital component of the digital twin; it encompasses an ultra-high fidelity simulation integrating the data and functionality of the real system [8].

It has been previously noted that there is difficulty in generating digital twins given the large amount of engineering they require [9]. However, IoS is designed to integrate multiple smaller component simulation using co-simulation to provide a larger system simulation. This allows for the digital twin to be constructed in the same way as the system itself, from smaller system components, and has the benefit of every system component having a digital twin as well as the systems as a whole. This can be used to improve the development time and support the system through the engineering life cycle [6]. The same simulations used for design can be used as digital twins as in Figure 11.3.

11.3 Artificial intelligence and machine learning

There is a large trend today towards automation using AI and machine learning, so called ambient intelligence [10] which is currently being propelled by advances in big data which allow for the collection and processing of large amounts of data. This is relevant to IoT systems where sensors can collect large amount of real-world data. This data can typically be applied in machine-learning applications for classification or regression.

Typically, these applications utilise supervised learning techniques where labelled training data is used to train the desired model. But this presents a number of downsides as the complexity of the problems, machine learning is applied to, increases. Usually, an increase in the complexity of the problem requires an increase in the amount of data required for training. This presents two problems, collecting this data and labelling it. Labelling is usually performed manually, so an increase in data requires an equal increase in manual effort to prepare the data set. Additionally, the collection of data may not be possible or may be difficult. This is especially true if the problem involves systems which are dynamic or not yet built. In answer to these problems, there is increased interest in unsupervised and semi-supervised [11] techniques where implicit cost of labelling is mitigated.

In particular, techniques such as generative adversarial networks (GANs) and deep reinforcement learning are particularly suited to learning using simulation. If an AI is used to control aspects of a system, simulation can form a vital role in its training. This is an example of transfer learning [12] where a simulation is used to train a model before it is applied to the real-world system. In the case of reinforcement learning, the model is only guided by metrics that measure its performance. It can provide inputs to a simulation and evaluate the simulated effect of those inputs on the system, adjusting the model as necessary. In the case of GANs, one network (the generative network) might provide environmental inputs to a simulated system, while the other network attempts to maintain a level of system performance. In both these cases, being able to construct and execute elaborate simulations from individual component simulations allows for more complex problems to be solved.

Finally, another essential application in the world with increasing automation is safety assurance. Before AI's are allowed to control safety critical systems, they should be proved to be safe, especially as they become more prevalent and each AI is interacting with more dynamic systems. In this application, IoS would allow for a fully realised simulation of a dynamic system, and its environment can be used to test the underlying models, presenting it with many different scenarios and measuring its performance empirically.

11.4 Conclusion

Simulation is vital for the development and application of autonomous, decentralised, cyberphysical systems. IoS captures one approach to this problem, capturing behaviours in component simulations and facilitating large-scale simulation by integrating these components in a distributed environment. The concepts of SIMaaS and WFaaS capture this integration and can be provided in Cloud-based environments. We have seen that the application of this approach to engineering, and design challenges can dramatically improve the design process and provide life cycle support using digital twins. We have also seen that this large-scale simulation can be useful to improve autonomy by providing ready access to detailed simulations which facilitate unsupervised learning for increasingly autonomous systems.

Acknowledgements

This work has been supported by the jointly funded UK-EPSRC and Jaguar Land Rover grant EP/K014226/1.

References

[1] Mckee D, Clement S, Ouyang X, *et al.* The Internet of Simulation, a Specialisation of the Internet of Things with Simulation and Workflow as a Service (SIM/WFaaS). In: 11th IEEE International Symposium on Service-Oriented System Engineering (SOSE 2017). IEEE; 2017.

[2] Jefferson DR. Virtual Time. ACM Trans Program Lang Syst. 1985; 7(3):404–425.

[3] Gubbi J, Buyya R, Marusic S, *et al.* Internet of Things (IoT): A Vision, Architectural Elements, and Future Directions. Future Gener Comput Syst. 2013; 29(7):1645–1660.

[4] Buyya R, Dastjerdi AV. Internet of Things: Principles and Paradigms. Cambridge, MA: Elsevier; 2016.

[5] Mckee D, Webster D, Xu J, *et al.* DIVIDER: Modelling and Evaluating Real-Time Service-Oriented Cyberphysical Co-Simulations. In: IEEE International Symposium on Real-Time Distributed Computing (ISORC); 2015.

[6] Clement S, McKee D, Romano R, *et al.* The Internet of Simulation: Enabling Agile Model Based Systems Engineering for Cyber-Physical Systems. In: IEEE International System of Systems Engineering Conference (SoSE); 2017.

[7] Grieves M, Vickers J. In: Kahlen FJ, Flumerfelt S, Alves A, editors. Digital Twin: Mitigating Unpredictable, Undesirable Emergent Behavior in Complex Systems. Cham: Springer International Publishing; 2017. p. 85–113.

[8] Boschert S, Rosen R. In: Hehenberger P, Bradley D, editors. Digital Twin – The Simulation Aspect. Cham: Springer International Publishing; 2016. p. 59–74.

[9] Datta SPA. Emergence of Digital Twins. CoRR. 2016. Available from: http://arxiv.org/abs/1610.06467.

[10] Sadri F. Ambient Intelligence: A Survey. ACM Comput Surv. 2011 Oct; 43(4):66, 1–36.

[11] Kingma DP, Mohamed S, Jimenez Rezende D, *et al.* Semi-Supervised Learning with Deep Generative Models. In: Ghahramani Z, Welling M, Cortes C, *et al.*, editors. Advances in Neural Information Processing Systems 27. Curran Associates, Inc.; 2014. p. 3581–3589.

[12] Weiss K, Khoshgoftaar TM, Wang D. A Survey of Transfer Learning. J Big Data. 2016;3(1):9.

Part 4

Concept-oriented business and services (business model) new business model inspired by ADS

Overview

In recent years, the society has experienced several changes in its ways and methods of consuming. Nowadays, the diversity and customization of products and services have provoked that the consumer needs be very diverse and frequently changing. As a result, cooperation among multiple companies, and formation of strategic alliances, for providing better and quicker services to consumers has increased.

Moreover, with the wide spread of the Internet and the massive use of mobile communication, service providers can offer global services and customers are empowered by global accessibility. As a result, customers have become more demanding, and the conditions of service provision are becoming more and more stringent. Due to the wide choice range, the users' desires have become highly dynamic and unpredictable.

Because the environment and the market trends have become so diversified and dynamic, new business models are required by heterogeneous e-business entities for cooperation to enhance individual capabilities regarding diversified customer's trends. Considering the dynamic environment, intense competition and possible business losses, these new business models must cope effectively with continuously changing heterogeneous requirements.

Since its conception in the 1970s, the autonomous decentralized systems (ADS) approach has been successfully applied in a wide range of application areas, such as transportation, financial, economic and information systems. During these years, ADS has enabled novel systems architectures that can cope with heterogeneous requirements in very dynamic and continuously changing environments. However, the current business situation in a globalized and massively interconnected world has forced ADS to evolve and innovate. In this sense, ADS has led and inspired not only new systems, but also new business models.

This part presents some of the business models inspired by ADS.

The first case study in this part is on mission critical information systems. The dependency and use of computer systems have incremented exponentially in the last years. It is necessary to change the way we design critical information systems since a failure in one of these systems could have very serious consequences.

The authors propose a new concept, architecture and technologies called ADS-oriented architecture (ADSOA) that provides uninterrupted services ADSOA has been inspired on ADS since it takes as an analogy the survival mechanisms of living organisms, as well as, the backbone of the specialization of functions through the genetic code that cells provide to a living thing.

The second case study in this part is on the blockchain. The author discusses the similarities of this technology with ADS, specifically the functional software architecture that provides secure and uninterruptible service. This chapter also covers the evolution from its origin as a subsystem of so-called cryptocurrency to an autonomous decentralized form that enables the general-purpose capability of smart contracts. The author mentions the application of blockchain technology in various industries, such as finance, health care, publishing, software, etc.

The third case study in this part presents a practical application of ADS in England. The authors observe that railway infrastructure is becoming more diverse and variable over time, and that these requirements are different from country to country. They suggest that a key requirement to keep ahead of structural changes in the market is to transform the railway business into a single enterprise flexibly and incrementally based on the new concept that they call "autonomous decentralization." The main focus of this chapter is to report the transformation of Hitachi's railway business when it entered into the UK railway market.

The fourth case study of this part discusses the development of a concept model of sustainable business through alliances based on railway infrastructure management and technology. The authors note that nowadays customer needs are so diversified that companies are finding it difficult to offer highly satisfied customer service. In this sense, new value should be created by alliances among different companies. The authors describe the alliance strategy that JR East has developed by a vertically integrated business model in railway business. They also describe how ADS has inspired the technology that supports these business alliances.

The final case study of this part presents the current efforts for developing cyber physical systems (CPS). The authors also identify the most relevant and major concepts regarding to this new research field. Security, safety and privacy are some of the major concerns that must be considered in the design of CPS since they operate in worldwide open environments. They also mention that governance and new business models are critical in the operation of CPS.

Chapter 12
Autonomous decentralized service-oriented architecture

Carlos Perez-Leguizamo[1]

Abstract

In a globalized market, technology evolution is neither an option nor an election but a necessity. The future of both humans and systems points toward a singularity within a short timeframe. We are converging into a symbiosis, which is why we have been evolving most traditional systems architectures and technologies into critical and high availability systems, so they can provide services available 24/7 in a precise and uninterruptible way. As a response to the necessities previously mentioned, there currently exist some efforts focused on solving said problems. A remarkable option among them is the autonomous decentralized system-oriented architecture (ADSOA), a novel architecture that takes as an analogy the survival mechanisms of living organisms, as well as the backbone of the specialization of functions through the genetic code that cells provide to a living thing, in order to deal with constant requirements that exist under a high demand system that provides uninterrupted services. Thus, ADSOA becomes a solid bridge between hitherto known architectures to business architectures that demand 24/7 services for internal and external business partners.

12.1 Introduction

A Mission Critical System is a system that requires uninterrupted operability and timely response in order to provide services to meet current business demands. The lack of these characteristics can result in serious financial losses, damage to the integrity of the enterprise and, in the worst-case scenario, loss of human life.

By the end of the twentieth century, our use of and dependency upon computer systems had incremented to the point of propagating everywhere, from the most complex banking systems or decision-making systems to the simplest like watches or kitchen appliances. Since then, and because a wide variety of solutions are available in the market, users' experience has established a paradigm about how

[1]Central Bank Operating Systems Development Division, Central Bank of Mexico, Mexico

systems should respond to their daily challenges, not only in their web applications but also in their electronic devices [1].

This idea has changed the way we should think and design our critical systems, especially in hospitals and nuclear plants, because a failure in one of these systems could have very serious consequences.

The problems previously mentioned make it necessary to conceptualize and design new ways of facing them.

12.2 Autonomous decentralized systems requirements

Nature has always been a source of inspiration, and in this sense, we have taken the autonomous decentralized system (ADS) systems as our primary idea and starting point for the autonomous decentralized system-oriented architecture (ADSOA) concept.

Living matter and cells are concepts that come from biological sciences. Both share the characteristic of metabolizing and self-perpetuating themselves. In a strict sense, any being that includes these features in its organism is considered alive.

We can generalize the universe of living organisms and attach to them three fundamental principles:

- All living organisms are made of one or more cells. Even the cell itself is a living organism.
- Cells are the basic unit of structure and function in living things. They are able to transmit information themselves by sending energy and matter through communication channels. One of the most important channels is the blood stream.
- Cells are created from existing cells. This last characteristic allows cells to self-perpetuate and transmits knowledge about how to carry out their functions, but still any cell will be unique and unrepeatable.

Figure 12.1 we relate the concepts of cell theory with an informatics system, which in this case will be the functioning of a calculator.

At the beginning, we have an organism that carries out four functions: division, addition, multiplication and subtraction. In step 2, we divide it into its individual functions that together comprise the first seen organism. Once divided, they are ready to be decentralized.

Following the diagram's flow, step 3 shows that the individual operations need to be multiplied. As a result of this stage, we obtain replicas of the same operation. We need to follow a special replication algorithm depending on the critical level of the functions, because it is possible that one operation is more important than the others. Therefore, for the most critical operations we will have more replicas.

In step 4, the products of the multiplication are disaggregated into different ecosystems that allow communication between them.

Then in step 5, the products in the same ecosystem and foreign ecosystems communicate with each other using molecules sent through the bloodstream. In terms of technology, we consider the bloodstream as the data field (DF) and the molecules as the messages.

At this point, the whole system acquires a form equivalent to a living organism.

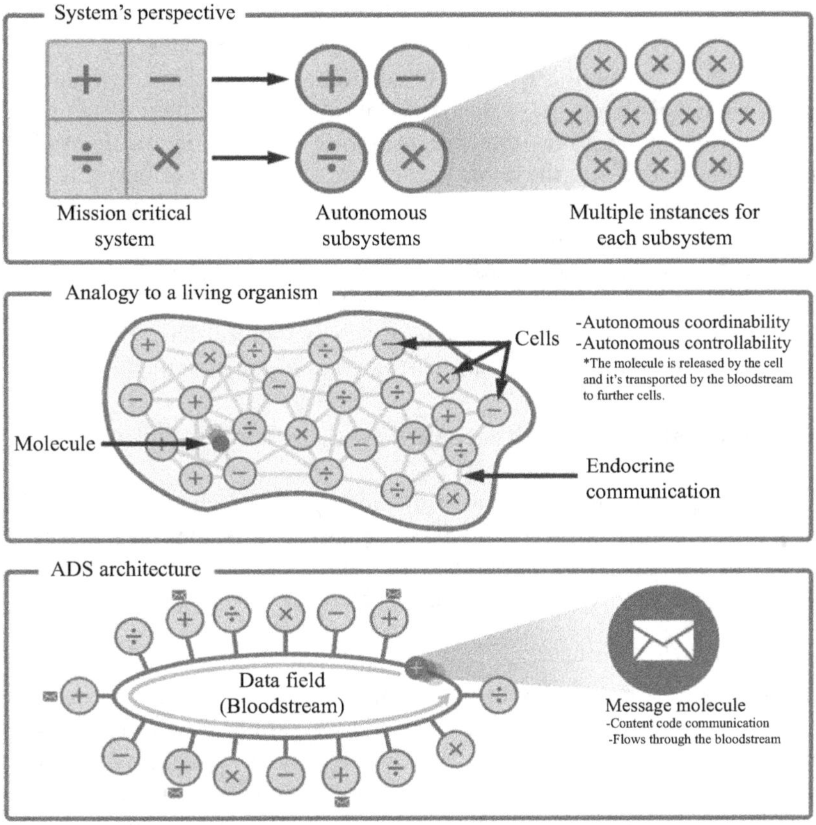

Figure 12.1 Autonomous decentralized systems (concept and architecture)

Finally, we have to give some intelligence to cells, so they can self-control and self-coordinate. It means that upon encountering a failure, cells that remain alive must have the capacity to continue operating themselves.

The advantages of this architecture are varied. The most remarkable advantage is that systems are able to distribute the load of operations into separate subsystems, as well as monitor the flow of operations and status of each individual "cell." Thus, if a part of the system crashes, it could be repaired, starting from the moment that the operation blacked out.

12.3 Service-oriented architecture requirements

Another modular technology that was used to build the concept paradigm of this paper is that of a service-oriented architecture (SOA).

SOA is an architecture used to transmit information among decoupled systems without the need to integrate them. This is possible because it uses standardized means to transmit information using typical network protocols.

Systems included in SOA architecture could be upgraded and integrated to the backbone of a company's systems without stopping or damaging the current operation required by online systems. In addition, the way that communication is carried out after applying SOA standards to your infrastructure allows for automatic managing of messages while transmitting information. In the end, these services will be coded into the deoxyribonucleic acid (DNA) of our cells.

SOA is based on services. You can represent any system with these services. A service is nothing more than a small program that, at the time of execution, is loaded, executed and ultimately destroyed. This is the basic principle we are taking in this analogy: the services are coded into the DNA of the cells. The objectives of this technology are to make computing resources more efficient and to manage online systems without having to stop the services of the system.

12.4 Concept and architecture based on biological analogy

This section of the paper will explain the concept and technologies of ADSOA, merging both paradigms of ADS and SOA. Because of this, the concept of ADSOA should not be explained as a computer system process, or merely a program, but a mix of both concepts based on a biological reference.

12.4.1 Complex systems

The human body is a complex uber system made of macro subsystems, like the cardiovascular, skeletal, muscular, nervous and endocrine systems. Likewise, these systems are divided into organs that allow the macro subsystems to work in a fine and perfect manner. At this point, it seems that we will get nothing more but, if we look into the essential components that give the perfect cohesion to organs, we will find cells [2].

Cells are little gears; without them the body would not be able to have diversification. They give the properties to some specific parts of the body, and they have an amazing communication system. Between them, they can notice if someone of kin is dying or is already dead. They can notice if the minimum number necessary for a population of cells to survive had been reached. To summarize, cells keep control of the unreliable resources.

Finally, what allows the whole body's organs and subsystems to keep in communication with each other is the blood stream, which carries information from one system to another and nourishes the whole body with essential information that will keep it synchronized.

Now that we have the background, it will be easier to understand the complexity of ADSOA by setting a simple example: a calculator that provides basic arithmetic operations.

Figure 12.2 shows an example of the human body represented as a calculator system.

Calculators have four fundamental operations that are addition, multiplication, subtraction and division. We can think of them as the macro subsystems that make up the body. They could be divided into organs like subtraction organ, multiplication organ and so on. Then we can also divide them and replicate the

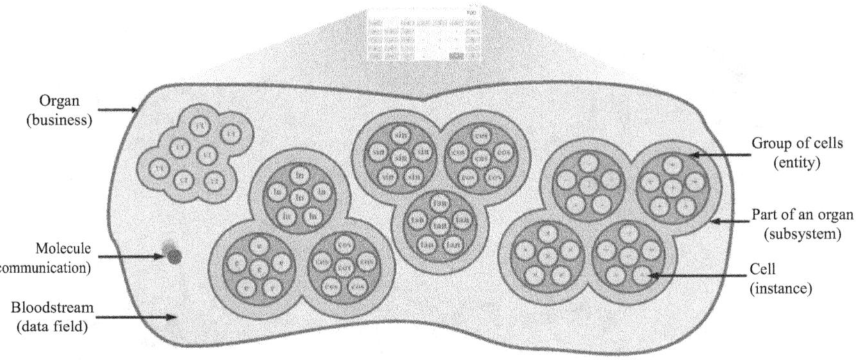

Figure 12.2 System biological analogy

divided operations into micro parts like the cells do: addition cells, subtraction cells and so on. We could set this individual group of cells into different servers, so now they could be separated as the human body's macro subsystems are.

Cells' fundamental operations can communicate as the blood stream does, via a socket from one server to another. The multiplied cells communicate between them like a server communication, adapting them as a service.

Thus, we are able to communicate and share status data, allowing us to notice if something goes wrong with an individual service, even if it is from a different kind of operation.

Based on both SOA and ADS paradigms, ADSOA follows their fundamental principles, which include fault tolerance, real-time and service-oriented computing.

ADSOA has uncountable applications in more than one working sector, but for practical purposes, we limited this paper exclusively to Fintech opportunities. For example, following Figure 12.3, you will find the flow of a company's processes and the philosophical application of ADSOA. For this standardized sector, we always have the business as the first parameter to attend to the requirements of the organization and implement the backbone of ADSOA in order to implement a perfect integrator of decoupled systems, which is adequate to include more and more systems in a scalable and safe way.

Once the process of the company is understood, the systems are divided into their essential functions. They are transformed into services, and thereby all services can be divided.

Before the division, every single part is multiplied into a number depending on the performance analysis and hardware resources. At this point, we put the remains of this treatment into different servers and we give them the name of entities, so they can keep a standard of ADS.

Communication among entities must be established, so that they may transmit information among each other.

At this point, we almost have completed the minimum requirements for designing an ADSOA system, but we still need the backbone. The backbone in SOA architectures is the enterprise service bus (ESB). An ESB is so important that without it SOA does not exist.

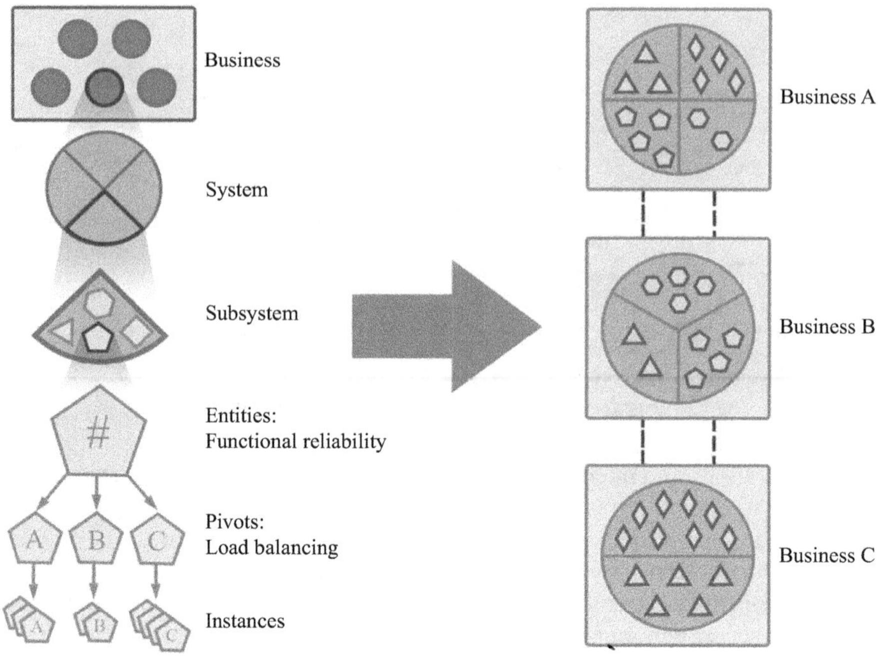

Figure 12.3 ADSOA design process

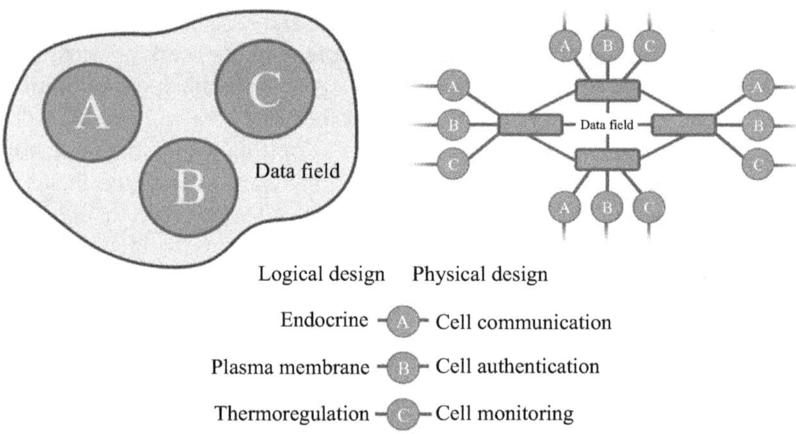

Figure 12.4 Cellular signalling (software-based DF)

12.4.2 Cellular signalling (software-based DF)

The DF, as shown in Figure 12.4, is the core of the architecture, becoming the communication deliverer and receiver; it works as the blood stream, nourished by the information broadcasted from the services. The cells communicate with each other by sharing molecules through the blood stream. In our case, messages identified by the content code are delivered through the DF.

The DF is designed with a topology mesh of interconnected processes that bind the communication between nodes, in such a way that when an entity requires communication with another, it puts the message into a DF, and the DF takes care of doing a multicast.

In logical design, there are three important entities. One of them is the communication system or endocrine system as it is called for the cells, it is the responsible of delivering the message to the cells that require it.

Another one is the cells monitoring also called thermoregulation, which allows us to know the state of the communication channel. It allows us to see what size and at what speed the messages are travelling.

Finally, we have the authentication system also called plasma membrane. This is in charge of determining which cells can participate in this communication, depending on the health of the cell. If it is damaged, the decision of repairing or replacing it is taken instead of keeping tracking of the communication.

12.4.3 Cell-oriented design (autonomous processing entity)

The autonomous processing entity (APE), as shown in Figure 12.5, is the modular entity of this whole infrastructure called ADSOA. When designing a system for a specific business, a methodology of designing the individual pieces of the final product is needed. Remember that the most complex part of any development is the design process. This paper does not intend to state a preferred pattern design, but rather the design and structure of ADSOA.

Here, in a logical way, the modules it contains are described. It is in charge of processing the functionality that lives in the DNA.

The diagram on the left shows the cell, and to its right the APE. The cell contains a module that allows communication with other cells. This module communicates through the DFs to the APE and sends messages through it.

The nucleus is in charge of processing service requests or functionality that this cell has programmed in its DNA. It takes a request, executes the operation, sends the message resulting from the operation and finally destroys the service.

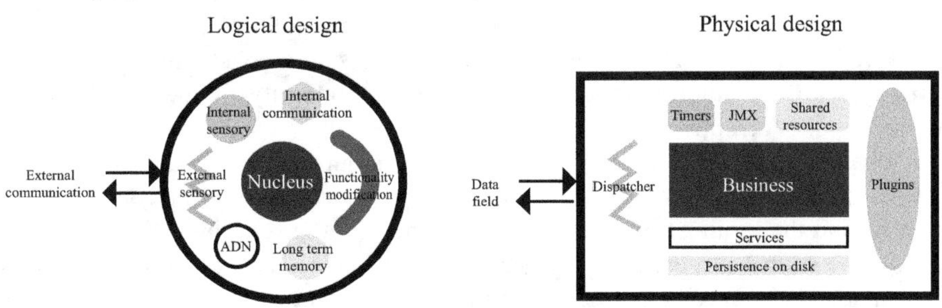

Figure 12.5 Cell-oriented design (autonomous processing entity)

The cell also has a memory module where the state of the cell will persist (long-term memory). Everything is stored on disc; there is nothing stored in memory. This gives us certain benefits, such as reducing memory costs.

The cell also contains a module called autonomous behaviour that is in control of its self-recovery. In the case of biological cells, this is known as cellular mitosis, and it is in charge of generating another cell from itself.

Another subsystem of the cell is the internal communication. It supports communication among the different modules inside the cells.

The sensory flagella are a system of sensors that exists within the cell and allows observation of the behaviour of nearby cells to determine if those cells are healthy or not. The sensors take decisions based on the status of nearby cells and determine if they are to be cloned or not.

The foundations of ADSOA are based on the DF and the ADS's and SOA's fundamental pieces.

12.5 ADSOA technologies

The best way to explain ADSOA is comparing the way it deals with fault tolerance with the way the human brain does. The brain is the primary receiver, distributor and coordinator of information in our bodies; it is the master control. Traumatic brain injury may lead to damage, destruction or even loss of certain parts of the brain, but fortunately the surviving parts can take over the missing functions. Remaining stable neurons near the damaged areas learn the information the damaged neurons had. Thus, the brain can resume its functions and preserve information.

Besides neurons, there are cells that make up the body organs. Through differentiation, specialized types of cells are created and used in different parts of the human body. The essence of the cell is that it can be replicated by mitosis, generating a new cell with the same characteristics of the first.

12.5.1 Fault tolerance

Cells communicate with each other. The messages that are transmitted among them operate in the same way, because they perform the same function. Cells will process the messages that are only of their kind. If a message becomes damaged or broken, it will still be transmitted as healthy data. The DF will then make a trace so as to find the point of origin of the problem and stop further generation of broken messages.

ADSOA retrieves information about the data status constantly. It has a DF property with a unique identifier that verifies the integrity of the data and will advise of data corruption.

When a message is damaged, the status of the inquiry will not match to the status that the other cells received, and the flow for that operation will be rolled back in order to find the turning point. Two scenarios are visualized at this point in order to recover the data: keep the integrity, save the checkpoint and keep transmitting the information between cells, or find the point before the data was corrupted and replace it with the temporary data which was created during the session of the data.

12.5.2 Self-recovery

Our body is able to recover from injury. When you exercise throughout your life, your muscles acquire the necessary knowledge to increase in size quicker than those persons whose muscles do not have that information, product of years of training.

What gives this ability to humans? The answer to that question is that the cells of your body store information about actions you made in the past, and they are able not only to reproduce by mitosis but also to transmit the information stored inside their DNA to the newborn cell. Thus, the new cells know how to replay the scenarios conducted by their predecessors.

Internal parts are discorporated from the global system, so the essential functions or cells that conform the body can be visualized. Each cell is drawn in a different colour, because they belong to different parts of the body. For example, heart cells are not the same as lung cells or tooth cells. In the case of ADSOA systems, the same principle applies. For example, bringing back the calculator analogy, the addition operation is not the same as the multiplication and so on.

Finally, cells are multiplied and divided, but what happens if some of them fail? ADSOA will fix the problem by tracing the route of the information to the point of failure.

If a problem has no fix, the cell is destroyed, and healthy cells will recover the missing one just like a living body works with its damaged parts.

12.5.3 Online services management

Our brain is capable of adding knowledge and learning about behaviour, experiences or senses without stopping the operability of the body. Neurons are always improving, creating binds to reach other neurons and hardening the old connections. They also are able to die during daily operation or improve while others are dying. This is the skill that makes the brain the most important organ in our body. We could live without a kidney, but not without a brain.

In the past, trying to upgrade, update or implement a new system meant a loss of client service time in production environments, a loss of test time in pre-production and error correction environments, or a loss of development time in a development environment.

ADSOA allows a company's systems to be updated, upgraded or re-implemented during daily operation without stopping or affecting the productivity of any system that is already online.

12.6 Summary

The constant changes inside the market are derived from the technological advances that have carried humankind to change paradigms with the same speed as the technology changes.

If a company intends to prevail in a stable state in this globalized world, adapting must be in a strict sense, natural, not only adaptation in the processes'

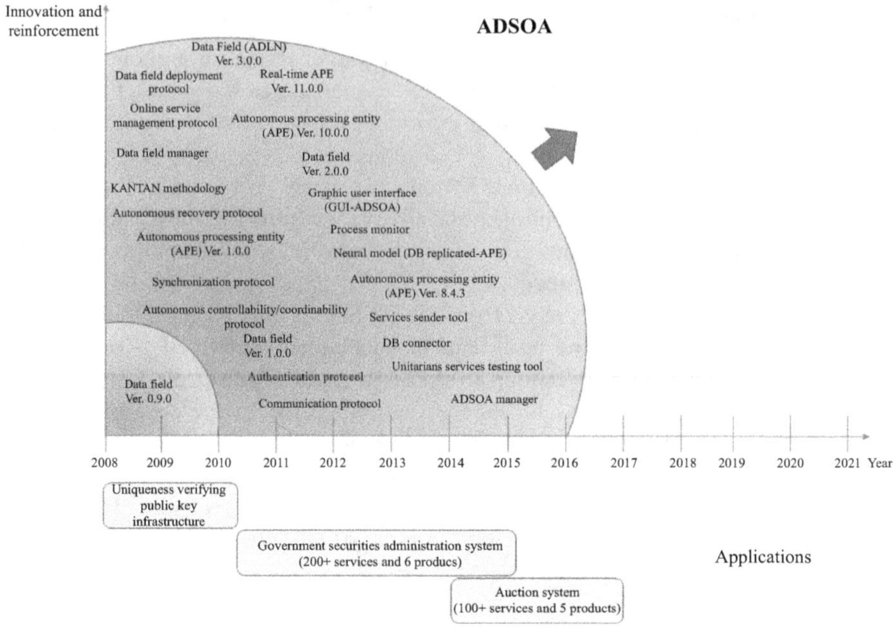

Figure 12.6 ADSOA roadmap

maintenance during the daily operation inside the enterprise, but also adapting to the clients' demands.

ADSOA consists of a set of technologies proposed to adhere to the standards of high availability, stability, scalability and speed of implementation. Its architecture is based on cellular theory as a central idea and in the ADS and SOA fundamentals as the backbone of the system.

Finally, ADSOA has been implemented in architecture of high availability systems, where data loss during the process of communication among applications results in human losses, severe infrastructure catastrophes and extreme loss of money. Thus, a company with an infrastructure of this magnitude must be able to face its internal and external clients' demands, as well as react with anticipation against the variables in the globalized environment. Figure 12.6 shows the roadmap of ADSOA since its conception in 2008 until now. During this period of time, more than 20 technologies have been designed and implemented. In addition, three mission critical information banking systems have been constructed and operated with ADSOA.

12.7 The future

In a not distant future, we will find ourselves in the era of the singularity or also called the Internet of Information, where the technology will advance to such a level that the ordinary businesses and tools as we know them will coexist in an interconnected and completely decentralized way through billions of computers,

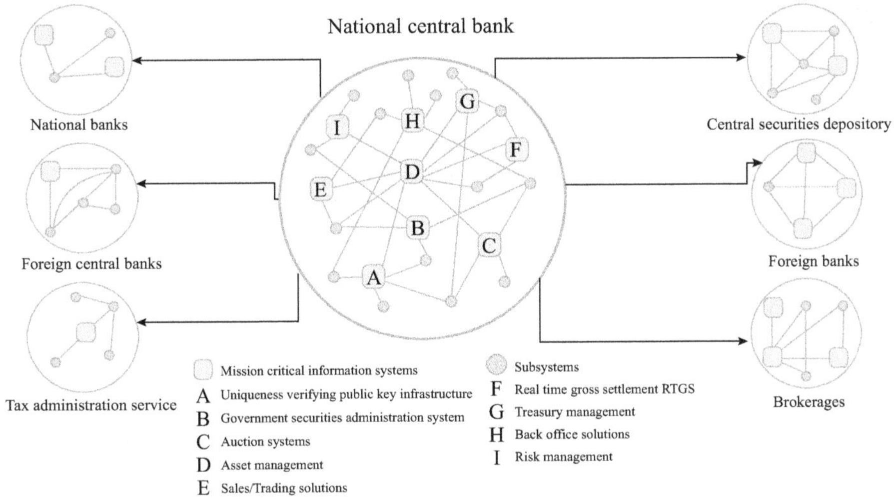

Figure 12.7 ADSOA future

smartphones and any other 'smart' devices that exist nowadays [3]. This paradigm shift will allow that the quotidian and specialized tools can be accessible anywhere and anytime in a trusted way.

When that time comes, technology will have such a high complexity that it will become similar to a living organism, where each connected piece of hardware will act as a node that will interact in the whole system as a cell within a big organism. All these components will work steadily, with the goal of prevailing within the body and reproducing when necessary.

Since its conception in 2008, ADSOA has been designed to work with critical systems in the banking area. The main idea considered since the design of its backbone always was to keep the systems operating as interconnected bodies and to maintain the integrity of messages among its individual parts. In this sense, ADSOA has become a key tool to accompany the evolution of technology to the horizon predicted in this text. A conception of this world can be appreciated in Figure 12.7.

Acknowledgements

I would like to express my deepest gratitude to Dr Kinji Mori for his continuous support and advice during the whole development of this work.

Josué Hérnandez, Sebastián Gódinez and Víctor Tapia, thanks for being not only one of the founders and best supporters of this life project called ADSOA but also for being a true friend. Andrés Villagomez, thanks for your unconditional dedication and effort to this project.

I would like to express a special word of appreciation for Banco de México and José Antonio Hernández who have supported us unconditionally during the implementations and operation of ADSOA infrastructure and based systems.

References

[1] Tapscott D. and Williams A.D., *Macrowikinomics: New Solutions for a Connected Planet*, Portfolio, New York, 2010.

[2] Meyer C. and Davis S., *It's Alive: The Coming Convergence of Information, Biology, and Business*, Crown Business, New York, 2003.

[3] Shirky C., *Here Comes Everybody: The Power of Organizing Without Organizations*, Penguin Books, New York, 2008.

Chapter 13

The role of blockchain in autonomous distributed business services

Doug McDavid[1]

Abstract

This chapter discusses the form of autonomous decentralized software known as the blockchain. It covers the original usage of blockchain distributed ledger technology, a part of the Bitcoin cryptocurrency. It summarizes the functional software architecture of blockchain that provides secure and uninterruptible service. It briefly discusses the growth and proliferation of cryptocurrencies throughout the economy. The main focus of the section is the application of blockchain technology beyond currencies, as a disintermediating force in various industries, such as finance, healthcare, publishing, and software, including the potential for widespread distributed, autonomous organizations, based on smart contracts built on the blockchain technology.

> While the technology is complicated and the word blockchain isn't exactly sonorous, the main idea is simple. Blockchains enable us to send money directly and safely from me to you, without going through a bank, a credit card company, or PayPal. Rather than the Internet of Information, it's the Internet of Value or of Money. It's also a platform for everyone to know what is true—at least with regard to structured recorded information [1].

13.1 A question to pursue

This section addresses a technological opportunity that has been emerging as blockchain technology has gained prominence. We start by posing a preliminary question for the reader, which will frame the rest of this section. Let's start by reviewing the autonomous decentralized system (ADS) point of view:

> An autonomous decentralized system (or ADS) is a decentralized system composed of modules or components that are designed to operate independently but are capable of interacting with each other to meet the overall goal of the system.

[1]International Society of Service Innovation Professionals

'Dynamic changes in social and economic situations demand next-generation transport systems to be based on adaptive and reusable technologies and applications. Such systems are expected to have the characteristics of living systems composed of largely autonomous and decentralized components ... ' This situation drives the ADS imperatives of 'on-line expansion, on-line maintenance, and fault-tolerance [2].'

This design paradigm enables the system to continue to function in the event of component failures. It also enables maintenance and repair to be carried out while the system remains operational. ADSs have a number of applications including industrial production lines, railway signalling, and robotics [3].

With the introduction of blockchain-based systems, the list of applications grows to include banking, financial, legal, and various commercial and governmental enterprises. Technologists and executives with an ADS mindset will find new opportunities among socio-technological systems. However, these new opportunities will inevitably be very complicated, quite malleable over the course of doing business, and subject to the sophistication and nuance of human language.

ADS is applied to build systems that mimic living systems. The living-system approach makes even more sense when it is applied to systems that support living human organizations [4].

So here's the promised question. The question is as follows: 'If blockchain experts wanted to listen and learn, what could ADS experts teach this new generation of technologists?'

The rest of this section will provide background for our question by describing blockchain attributes, problem domains, economics, history of experience, and foreseeable challenges.

13.2 Why this matters?

We continue to observe the rapid penetration of technologies and devices throughout the built environment. No one knows better than the readers of this volume, how much the physical infrastructure has been overlaid and interpenetrated by networks of computing and communication.

At the same time, computing has advanced in areas such as natural language processing, pattern matching, and deductive reasoning. This opens the path for applications to penetrate deeper into the fabric of organizations, institutions, and the professions. As Susskind suggests, it may turn out that 'increasingly capable machines will transform the work of professionals, giving rise to new ways of sharing practical expertise in society. This ... can be characterized in many ways: as the industrialization and digitization of the professions; as the routinization and commoditization of professional work; as the disintermediation and demystification of professionals. Whatever terminology is preferred, we foresee that, in the end, the traditional professions will be dismantled, leaving most (but not all) professionals to be replaced by less expert people and high-performing systems' [5].

From another angle, the future of the professions looks much like the vision of the 'enterprise of one', each individual interoperating as an autonomous agent within a world of distributed work, that professional work includes problem solving, aesthetic enhancement, application of knowledge, etc. The positive view of this evolution includes a race to the *top* because that's where the most obvious value gets created. In order to stay on the high road to greater value creation and acknowledgement, a grand juncture is needed among service providers and service consumers in a knowledge-based marketplace of problem-solving and fulfilment of desires. These service providers are the very people who are redefining the nature of work. Emerging technologies, such as artificial intelligence (AI) and natural language processing, provide the capability for the negative and the positive scenarios of race to top or bottom.

SI Hayakawa taught the following premise, on which much of modern linguistic thought is based: 'No word ever has exactly the same meaning twice [6].' This motto recognizes the importance of context in human interaction, and the level of complexity this introduces into our communication webs. The domain of human interaction is exactly where natural language words, legal and contractual terms, value negotiations, and blockchain-based smart contracts exist.

ADS has long been involved in dealing with nuance and continua of meaning and implications of variables that can be sensed. The deeper we get into domains of human interaction, the more valuable and widespread the ability to handle nuance and discontinuities will be.

13.3 What is blockchain?

It is in this context that a new, multipurpose technology known as the blockchain has recently made an appearance. In the discussion that follows, you will see that I've let the participants in this technology emergence largely speak for themselves. My primary hope is to weave a somewhat coherent story!

> The blockchain is basically a distributed database. Think of a giant, global spreadsheet that runs on millions and millions of computers ... It uses state-of-the-art cryptography, so if we have a global, distributed database that can record the fact that we've done this transaction... it [can also] record any structured information, not just who paid whom but also who married whom or who owns what land or what light bought power from what power source ... So this is an extraordinary thing – an immutable, unhackable distributed database of digital assets. This is a platform for truth and it's a platform for trust [7].

This makes blockchain 'a general ledger system that records transactions. You have a recorded and accessible electronic record of every exchange made by everyone. This can be used for any contractual exchange electronically [8].'

Blockchain is often equated with the term 'distributed ledger technology'. 'Although these terms are often used interchangeably, they are not strictly equal to each other. A DLT can be defined as a replicated, shared, and synchronized database. A blockchain shares these characteristics but also introduces unchangeable, digitally recorded and signed data in separate … blocks. Blocks are stored in an append-only chain, and each block is linked to a previous block through a cryptographic signature called a hash' [9].

The following specification appears in the original Satoshi Nakamoto paper on Bitcoin as an electronic cash system. This provides an excellent description of the blockchain mechanism (without mentioning the word 'blockchain'): 'The solution we propose begins with a timestamp server. A timestamp server works by taking a hash of a block of items to be timestamped and widely publishing the hash, such as in a newspaper or Usenet post. The timestamp proves that the data must have existed at the time, obviously, in order to get into the hash. Each timestamp includes the previous timestamp in its hash, forming a chain, with each additional timestamp reinforcing the ones before it' [10].

13.4 Problems addressed by blockchain technology

That specification formed original implementation of blockchain as part of the architecture of Bitcoin, the cryptocurrency. In the original Bitcoin paper, Satoshi Nakamoto proposed a solution to a problem that had plagued all previous attempts to create a viable form of electronic 'cash'—the possibility of spending the same 'coin' or token more than once—the famous double spend problem. 'Commerce on the Internet has come to rely almost exclusively on financial institutions serving as trusted third parties to process electronic payments. While the system works well enough for most transactions, it still suffers from the inherent weaknesses of the trust based model … What is needed is an electronic payment system based on cryptographic proof instead of trust, allowing any two willing parties to transact directly with each other without the need for a trusted third party' [11].

The solution to the double spend problem requires the ability to guard against tampering with the historical ledger of transactions. In other words, to spend some amount of currency, and still appear to own it, one must alter the historical record of it already being spent. 'In this paper, we propose a solution to the double-spending problem using a peer-to-peer distributed timestamp server to generate computational proof of the chronological order of transactions. The system is secure as long as honest nodes collectively control more CPU power than any cooperating group of attacker nodes [11]'.

Another problem addressed by the blockchain technology consists of need to defend against a variety of security breaches and attacks, including denial of service attacks, 51% attack, nothing-at-stake attack, long-range attack, bribery attack, etc. [12]. The emphasis on defence against attacks is understandable, given that blockchain as quickly been adapted to the general problem of validating interests (ownership) of items of value. The ownership of conspicuous levels of value tends to attract nefarious motivations.

One key theme that runs through blockchain discussions is the issue of trust and the social infrastructure that supports the institution of the trusted 'third party'. Blockchain provides an approach to putting technology in the place of trusted individuals and institutions. However, '"trustless" is largely a misnomer – and too much of it is not necessarily a good thing. Even the most ardent advocate of Bitcoin, the most prominent "trustless" network, will extend trust regularly to a certain extent, whether to the banks and payment processors which process their fiat currency transactions when they purchase Bitcoin, to the manufacturers of their computers and operating systems on which they run the Bitcoin client, and to the Bitcoin Core developers themselves ... It is not as if trustworthiness is a new problem. Entire industries have been developed to handle the boundaries and rules of various points along the trust-to-trustlessness spectrum – not the least of which is the legal system, which operates as a potent safeguard for consumers and commercial entities alike [13].'

13.5 Implementations of blockchain

There are three areas where both blockchain networks and blockchain clients stand out:

- Blockchain clients and networks validate everything (great for compliance and auditors; not so great for fraudsters).
- Blockchain clients are highly independent (and thereby fault tolerant).
- Blockchain clients and networks are automated [14].

Most modern databases store the world state of the data and keep the logs of transactions with the database as a separate 'thing'. Blockchain clients, on the other hand, build the world state of the data from the blocks of authenticated transactions that are 'chained' together over time. Thus, it is always immediately possible to tell if something is valid, as it must have come from a valid history, and everyone agrees completely on the sequence of valid history [14].

At the time of writing, blockchain implementations are evolving rapidly. Single monolithic 'blockchain' technologies are being reframed and refactored into building blocks at four levels of the stack [15]:

1. Applications;
2. Decentralized computing platforms ('blockchain platforms');
3. Decentralized processing ('smart contracts') and decentralized storage (file systems, databases), and decentralized communication; and
4. Cryptographic primitives, consensus protocols, and other algorithms.

13.6 Classifications of blockchain implementations

Blockchain has evolved to the level of maturity that certain key defining characteristics differentiate variant forms of blockchain implementation. For example, Monax.io (formerly Eris Industries) provides two useful classifications in terms of what they refer to as the 'permissioned spectrum' and the 'optimization spectrum' [16].

The permissioned spectrum provides two options:

- On the one hand, we have access to the technology of permissioned blockchain networks. These networks are based on nodes that can be made fully public or use whitelisting for control. These nodes will reject blocks from not-whitelisted nodes. Permissioned blockchain networks only deal with the functionality required for that particular network rather than all larger, unpermissioned networks.
- On the other hand, we have unpermissioned blockchain networks. These networks carry forward the original Bitcoin idea of an open network that accepts anyone who can plug a computer into the internet. 'These open blockchain networks lack an access control layer and as such handle anti-spam and consensus via purely economic mechanisms ... These blockchain networks are the best solution for censorship resistance ... They also have public governance mechanisms, and have been designed to provide the data management backbone for a variety of applications. That means that they were probably not well suited for any one type of application. Depending on what application one is seeking to build this may be a benefit or a detriment.'[1]

The optimization spectrum recognizes more of a continuum than the binary choice of permissioning. The optimization spectrum runs from systems that optimize logic toward ones that optimize transactions:

- Blockchain clients on the logic-optimization end of the scale focus on verifiably tracking title transfers in a distributed environment. They emphasize property transfer mechanisms, clearing and settlement, and provenance validation. These property audit blockchain clients often do 'provide some limited logic capabilities (Bitcoin's reference client, famously, has its multi-signature capacity which operates in a similar area of logic). However, they have been optimized to track movement of title over "property" from one node on the network to another'.

[1]There is a major and originally unanticipated consequence of the popularity of Bitcoin, coupled with the computational intensity of the permissionless protocol. "According to VICE Motherboard, in June 2015, "One Bitcoin transaction required the same amount of electricity as powering 1.57 American households for one day, of which the average sized home is approximately 2700 square feet. To give some global context, 1.57 American homes roughly equals 20 Hong Kong homes or 11 mainland China homes. As the Bitcoin network grows so does electricity demand and today the average Bitcoin transaction costs as much as powering 3.67 average American homes for one day, or 46.75 Hong Kong homes and 25.7 mainland China homes, respectively. In June 2015, the Bitcoin network was consuming enough energy to power 173,000 American homes and today that figure has grown to 1,018,762 which represents a $\sim 6\times$ increase. If we assume it's still early days of this ecosystem where Bitcoin can process seven transactions per second, then just how energy efficient and sustainable will it be as we begin to imagine scaling out Bitcoin so transactional throughput and volume can compete with the likes of VISA and its 24,000 transactions per second?

"To give some more comparative context with respect to the electricity cost of Bitcoin, CERN uses 1.3 terawatt hours of electricity annually to power the Hadron Collider particle accelerator versus the 11 terawatt hours (and rapidly growing) of electricity annually that Bitcoin consumes." —http://blockgeeks.com/bitcoins-energy-consumption/

- Blockchain clients on the other end of the scale 'give application developers a clear and efficient way to verifiably track business and governance process logic in a distributed environment. They are a good fit for application developers seeking to build complicated business process automation applications'.

13.7 Validation and consensus options

Another very fundamental architectural split in the blockchain world is unfolding at the time of writing. This other dimension refers to the difference between blockchain systems that provide different consensus and validation mechanisms.

- A proof-of-work (POW) validation protocol that uses an economic measure to deter denial-of-service attacks and service abuses on a network by requiring some work from the service requester. A POW protocol can be compared to a puzzle. A puzzle must be moderately hard to solve (but feasible) on the requester's side but easy to check for the service provider. "Easy" in this sense means with a small amount of computation; "hard" means with a lot of computation power required. One of the first uses of the POW protocol was to stop spam in mailboxes. In this situation, a sender of a message would have to complete a proof of work in order for a server receiving the message to accept it'.
- A proof of stake (POS) protocol, on the other hand, 'is a consensus algorithm for public blockchains that is intended to serve as an alternative to proof of work. While the POW method asks users to repeatedly run hashing algorithms (solving puzzles) to validate electronic transactions, POS asks users to prove ownership of a certain number of assets (their "stake" in the assets), which are exchanged in a network (kind of currency). Some public blockchain based applications use a hybrid approach for distributed consensus by combining POW and POS elements'.
- A third consensus mechanism is provided by practical byzantine fault tolerance (PBFT) and 'is used widely in private blockchain implementations. Unlike POS, trust is entirely decoupled from resource ownership. Each participant in a network publishes a public key. A message coming through the node is signed by the node to verify its format. Once enough identical responses are received, then a consensus is reached that the message is a valid transaction. PBFT is devised for low-latency storage systems. This mechanism is applicable to digital asset based platforms that do not require a large amount of throughput, yet demand many transactions. Consensus can be reached rapidly and efficiently' [17].

13.8 Blockchain and environs

The architecture of the blockchain and related functionality requires and provides a set of capabilities that can be incorporated into the architectures of virtually any kind of enterprise.

In one way or another, a functioning blockchain system needs to be able to sense the environmental and marketplace state of affairs, as well as relevant events, initiated by known or unknown parties. Any blockchain system needs to be able to sense the environmental or marketplace state of affairs, as well as relevant events initiated by known or unknown parties.

When these parties need to be known for the system to work, identifiers allow the system to record and retrieve access and transaction permissions.

Certain relevant aspects of events (such as a payment) are recorded and loaded into a variable payload, and a hash is made for the record. This hash plays the role of an immutable index link to that record.

A block for a chain concatenates some number of records, and a hash is also generated for each block. The chain itself is maintained by chaining the blocks, each block containing the hash of the previous one.

A potentially very large number of nodes in the network must be spent whatever time it takes to periodically (continuously) catch up with the full history for that chain, sometimes by recapitulating all the way back to the so-called genesis node at the start of the chain.

Certain relevant aspects of events (such as a payment) are recorded and loaded into a variable payload, and a hash is made for the record. The payload for a record may include any form of data, smart contract provisions.

The payload for a record may itself include actionable data, or it may point to some off-chain data store, where details of the recorded state of affairs stored for full access (like a patent, book, or provisions of smart contracts) are recorded.

Sometimes, the system needs to identify participating parties. A white list of known and vetted participants allows the system to record and retrieve access to transaction permissions.

Various smart contracts may be fulfilled by people or by things. If the latter, we are into a powerful form of the Internet of Things (IoT).

Smart contracts influence the state of affairs as they are fulfilled by people or by the IoT. A feedback loop is set up to:

- guide behavior;
- activate a stream of behavior that impacts the state of affairs; and
- sense, evaluate, and record the results on the blockchain.

A key point is the feedback loop that is set up to activate and then evaluate some stream of behaviour, being directed by smart contracts and recorded on the blockchain.

Tokens accumulated through blockchain activity are assigned to wallets. This currency can be spent on goods and services. Currencies can also be traded for other cryptocurrencies and moneys, such as national fiat money.

A fundamental purpose of blockchain is to provide a distributed ledger. Queries to the ledger are designed to validate claims of purported stakeholders, as illustrated by the following simulated dialogs:

- 'My contract should say I made the following contribution'.
- 'Yes sir, we see that contribution on March 20, 2017'.

or

- 'No, there's no evidence of that claim. Sorry sir'.

13.9 Problem domains

Application of blockchain technology can exert a disintermediating force in various industries and situations. Melanie Swan writes, 'Bitcoin is just one example of something that uses a blockchain. Cryptocurrencies are just one example of decentralized technologies. And now that the Internet is big enough and diverse enough, I think we will see different flavours of decentralized technologies and blockchains. I think decentralized networks will be the next huge wave in technology. The blockchain allows our smart devices to speak to each other better and faster' [18].

Domains and industries where we might expect to find smart contract use cases include digital identity, records, securities, trade finance, derivatives, financial data recording, mortgage, land title recording, supply chain, auto insurance, clinical trials cancer research, and many others [19].

It's not too surprising that interest has been steadily growing in the application of blockchain technology in banking and other financial institutions. Some of the most well-known financial institutions are getting on board the blockchain bandwagon including BNP Paribas, Société Générale (SocGen), Citi Ban, UBS, and Barclays [20].

The Bank of England (BoE) received a flood of publicity recently with a study of the feasibility of the bank issuing a cryptocurrency of its own. Such a currency might be called, as BoE does, a Central Bank Issued Digital Currency (CBDC). The bank addressed the fact that 'the technology already exists but this (central bank issuance) has never been done. It should be more surprising that more central banks are not considering the same. The rise of private digital currencies has already pulled a certain amount of transactions outside the purview of state banks. The BoE document reviews both the pros and cons of a CBDC reviewing structural issues, price, output stability, and financial stability [21]'.

In fields outside of banking, a few illustrative examples of possible smart contracts include the following:

- Insurance contracts, in which the parameters of an insurance policy are written into smart-contract code and enforced automatically;

- Escrow, in which a smart contract protocol sequesters messages or funds held on a distributed ledger until the occurrence of some event when the smart contract automatically performs or instigates a stated contractual action; and
- Royalty distribution to artists and other associated individuals.

13.10 Distributed autonomous organizations

This brief survey of blockchain technology would not be complete without a brief mention of the idea of a distributed autonomous organization (DAO). The basic idea is to build entire business models based primarily on blockchain capabilities.

This idea played out in a very public initiative simply called 'The DAO', a decentralized investment fund, based on the Ethereum version of cryptocurrency, blockchain, and smart contracts. The DAO constituted a complete enterprise implemented in smart contracts. In order to fund this idea, crowdfunding raised the equivalent of $130M—the largest crowdfunding effort in history, at that point.

The DAO is the first iteration on the Ethereum network of an idea that has been floating around the cryptocurrency space for a few years now, which is that you could take all the functions of an investment vehicle—fund storage, project vetting and approval, fund disbursement, and profit allocation—and handle it on a blockchain, thereby creating what is effectively a corporation without jurisdictional anchors. Equally attractive to some is the fact that a blockchain-enabled organization is completely transparent and does not rely on a managerial class with high salaries to complete its functions. Everything is done by the code, which anyone can see and audit. What investors who jump on board do rely on, however, is the expertise of the people who write and audit the code. They have to trust not only that the software is secured but also that the governance models work the way they are intended [22].

The DAO project allowed people to pool their funds and then vote on decisions on what the fund would invest in. Shortly after the initial funds were raised, however, the fund was attacked, and about $50M was drained into the account owned by the attacker before the process was stopped. In spite of disclaimers that the code running the DAO was the final authority, Ethereum ultimately underwent a 'hard fork', which resulted in further confusion and unexpected consequences.

13.11 The state of play

The world of blockchain started out as a kind of libertarian model of distributed disintermediated finance and commerce. This started out as the exclusive playground of hackers and coders. To a large extent, this is still true for this emerging technology development. On the other hand, this scene is attracting some very big and serious players, as we've seen from the involvement of, for instance, the BoE, among others.

IBM has been developing their 'Hyperledger' project, and Microsoft has released an Ethereum-based product called 'Blockchain as a Service' under their Azure Cloud offering. Slock.it has made progress in the effort to combine blockchain with the IoT. Eris Industries (renamed Monax.io) has taken a position as experts in Ethereum-based smart contracts. New business and initiatives are appearing almost every day.

An unmistakable sign of the times and trends has recently been signalled by patent activity in the blockchain space. 'Recently, some of the biggest names in business, from Goldman Sachs to Bank of America and MasterCard, have quietly patented some of the most promising blockchain technologies for themselves. Through mid-November, the number of patents that companies have obtained or said they have applied for has roughly doubled since the start of the year, according to law firm Reed Smith [23]'.

13.12 Conclusion

The question that was proposed at the outset reads 'If blockchain experts wanted to listen and learn, what could ADS experts teach this new generation of technologists?' This question, of course, needs to be answered by ADS experts themselves, as well as by the blockchain experts who may benefit from ADS experience and expertise.

From the perspective of the enterprise, opportunities to apply the blockchain-based smart contract approach can appear bewilderingly numerous and complex. Every interaction that an enterprise pursues in the course of business can be subject to some level of smart contract guidance and control. Every enterprise has relationships in four dimensions: demand (customers, etc.), suppliers of all types, reward (sources of funding and revenue), and control (regulators, media, and the public) [24]. These billions of interorganizational relationships all provide opportunities for improvement with smarter technology.

In 'the DAO', we see a very direct attempt to interface code to culture—to literally create an enterprise that itself entirely consists of code, and the human participants who hope to manipulate and benefit from the capabilities of that code. This may seem extreme, and of course, it is an example of technology in the hands of aficionados. Lessons learned from such early promoters may be able to carry the technology forward to the next wave of adopters. It's inevitable that there will be many hybrids of human and code-based DAOs.

In financial, commercial, and governmental enterprises system, faults imply breach of contracts, with often rippling consequences. Expansion and maintenance (E&M) of these organizational systems can be directed by smart contracts running on blockchains.

The smart contract will not replace financial experts, although integrated with predictive analytics will provide powerful services. The best bankers will be the ones who become the best smart contract users.

The smart contract will not replace lawyers, although couple with machine-learning formidable tools will emerge. The best lawyers will be the ones who become the best smart contract users.

This phenomenon, replicated across entire professions, will unleash a torrent of value-creation opportunities for those who become cognizant of how to harness these developments as they roll out.

I'd like to end with a statement I made several years ago, which has since gained in wider relevance beyond the original context of a discussion about systems engineering for autonomous weapons systems: 'When technology is increasingly inserted into the fabric of our lives, there is an accountability that includes both the creator and the installer of the technology. This linkage is not always understood as an explicit responsibility, and is a dangerous point of potential abdication of responsibility. Clarifying and making explicit the accountabilities, which are evolving at the combined rate of evolution of technology and prevailing social preferences, falls into the domain of the systems engineer. ... [who] must develop the methods and tools to keep increasingly autonomous and complex techno-institutional systems within acceptable envelopes of determinism. This [requires] ... aligning the desires of participants and beneficiaries with institutional as well as technological architectures' [25].

References

[1] Tapscott, D and Tapscott, A. Blockchain Revolution: How the Technology Behind Bitcoin is Changing Money, Business, and the World, NY: Penguin Publishing Group. Kindle Edition, 2016.

[2] Introduction to this volume: Autonomous Decentralized Systems (ADS) and their applications in Intelligent Infrastructure.

[3] Wikipedia. Available from https://en.wikipedia.org/wiki/Autonomous_decentralized_system.

[4] McDavid, D. All Services All the Time: How Business Services Serve Your Business, NY: Business Expert Press, 2016.

[5] Susskind, R and Susskind, D. The Future of the Professions: How Technology Will Transform the Work of Human Experts, Oxford: Oxford University Press. Kindle Edition, 2015.

[6] Hayakawa, S. I. and Alan, R. Language in Thought and Action, NY: Harcourt Brace, First Harvest Edition, 1990.

[7] Zampfir, V. The History of Casper – Chapter 2, NY: Media, 2016. Available from https://medium.com/@Vlad_Zamfir/the-history-of-casper-chapter-2-8e09b9d3b780#.94uwo9brz, Accessed Feb 2017.

[8] Skinner, C. VALUEWEB: How Fintech Firms Are Using Bitcoin Blockchain and Mobile Technologies to Create the Internet of Value, Singapore: Marshall Cavendish International (Asia) Pte Ltd. Kindle Edition, 2016.

[9] Marzantowicz, K. and Jedrzejczyk, M. Distributed Ledger, Distributed Consensus, and Their Impact on the Financial Services Market, Cutter Business Technology Journal, NY, 2016.

[10] Nakamoto, S. Bitcoin: A Peer-to-Peer Electronic Cash System, 2009. https://bitcoin.org/bitcoin.pdf.
[11] Nakamoto, ibid.
[12] Zampfir, op cit.
[13] Monax. What is a Permissioned Blockchain?, NY: Monax, 2016. Available from https://monax.io/explainers/permissioned_blockchains/, Accessed Feb 2017.
[14] Monax. What is a Blockchain? NY: Monax, 2016. Available from https://monax.io/explainers/blockchains/, Accessed Feb 2017.
[15] Pon, B. What is BigChainDB?, NY: Media, 2016. Available from https://blog.bigchaindb.com/what-is-bigchaindb-38aff031bf51#.7sn263gjw, Accessed Feb 2017.
[16] Monax, op cit.
[17] Roberts, D. The Bitcoin Book Boom, NY: Fortune, 2015. Available from http://fortune.com/2015/03/06/bitcoin-book-boom/, Accessed Feb 2017.
[18] Crowdfund Insider. Here is the Bank of England Document on the Central Bank Issuing Digital Currency. London Crowdfund Insider, 2016. Available from https://www.crowdfundinsider.com/2016/10/90761-bank-england-document-central-bank-issuing-digital-currency/, Accessed Feb 2017.
[19] Digital Chamber of Commerce. Smart Contracts – 12 Use Cases for Business and Beyond, Washington, D.C.: Digital Chamber of Commerce, 2015. Available from http://www.the-blockchain.com/docs/Smart%20Contracts%20-%2012%20Use%20Cases%20for%20Business%20and%20Beyond%20-%20Chamber%20of%20Digital%20Commerce.pdf, Accessed Feb 2017.
[20] Pouliot, J-M. Do You Think Big Banks Will End Up Supporting Bitcoin in the Near Future?, NY: Blockgeeks, 2016. Available from http://blockgeeks.com/questions/do-you-think-the-big-banks-will-end-up-supporting-bitcoins-in-the-near-future/, Accessed Feb 2017.
[21] Crowdfund Insider. op cit.
[22] Peck, M. Ethereum's $150-Million Blockchain-Powered Fund Opens Just as Researchers Call for a Halt, IEEE Spectrum, May 28, 2016. Available from http://spectrum.ieee.org/techtalk/computing/networks/ethereums-150-million-dollar-dao-opens-for-business-just-as-researchers-call-for-a-moratorium, Accessed Feb 2017.
[23] Kharif, O. Big Banks are Stocking up on Blockchain Patents, NY: Bloomberg, 2016. Available from https://www.bloomberg.com/news/articles/2016-12-21/who-owns-blockchain-goldman-bofa-amass-patents-for-coming-wars, Accessed Feb 2017.
[24] McDavid, D. Enterprise Stakeholder Types. Mt. View, CA: LinkedIn, 2016. Available from https://www.linkedin.com/pulse/21-enterprise-stakeholder-types-doug-mcdavid?trk=prof-post, Accessed Feb 2017.
[25] McDavid, D. Determinism and Determination in Socio-Technological Systems, INCOSE INSIGHT, 2012.

Chapter 14

Change and expansion of business structure using ADS concept in railway market

Yuichi Yagawa[1], Hiromitsu Kato[1], Shuichiro Sakikawa[1], and Gaku Suzuki[2]

Abstract

Due to global changes in social structure, the needs for social infrastructure are diversified and continually changing over time. To capture the structural changes in the market and to sustainably and continuously improve the quality of life of users, innovative manufacturing is required to flexibly and incrementally innovate their business in finance, organizations, technology, and operations. In this chapter, we explain our incremental expansion of social infrastructure business to the UK railway market using the autonomous decentralized system concept.

14.1 Changes in value structure

14.1.1 Changes in value structure for railway infrastructure

Social structure is globally changing. In the developed countries, social infrastructures, like railways which are part of a country's urbanization process, are now becoming too old. Furthermore, the growth of industrial competitiveness has stunted. In contrast, in the developing countries, construction of social infrastructures is increasing rapidly as urbanization advances.

Changes in social structure lead to changes in value structure. To create a sustainable society, it is expected that services in the developed countries bring about not only simple replace of aging infrastructures but also new added value. For example, the UK Class 395 high-speed rolling stock, launched in December 2009, shortened the required time between Ashford and London from 83 to 37 min and added value to society by improving the quality of life (QoL) of users. In the developing countries, on the other hand, changes which stimulate economic growth are occurring. In these countries, domestic demand for social infrastructures is

[1]Research & Development Group, Hitachi, Ltd., Japan
[2]Yashima & Co., Ltd., Japan

rising rapidly as urbanization increases. Triggered by these changes, cultivation of a domestic industry, active development of technology, and low-cost construction of infrastructures are starting.

For global deployment of social infrastructures that capture these value structural changes, it is important to establish a business scheme that matches the business structure of the market. In the developed countries, manufacturers are required to collaborate with companies in other fields for operations, maintenance, and finance as well as manufacturing. On the other hand, it is necessary for the developing countries to plan for local businesses and support a local consumption society by actively using external consulting companies or starting their own consulting business.

14.1.2 ADS business architecture

The needs for social infrastructure are diversified globally and are continually changing over time. To capture these structural changes in the market and sustain improvements to users' QoL, manufacturers will need to flexibly and incrementally innovate their business into a single company that includes finance, organizations, technology, and operations. Figure 14.1 shows a business architecture based on an autonomous decentralized system (ADS) concept. A prevalent business architecture has a structure in which a value chain connects each pair of companies, and a maximization of business value is reached by optimizing each process. For example, in the UK railway business, railway facility management, train operation, and rolling stock management companies have value chains which connect another company. In contrast, the ADS business structure connects those companies through a value network and optimizes the entire value chain.

To optimize the entire value network, manufacturers are required to manufacture the product to get maximum operation value on top of the typical approach of finding value in pursuing product performance. In addition, the operator is required to realize the operations for getting the maximum service value. The ADS business

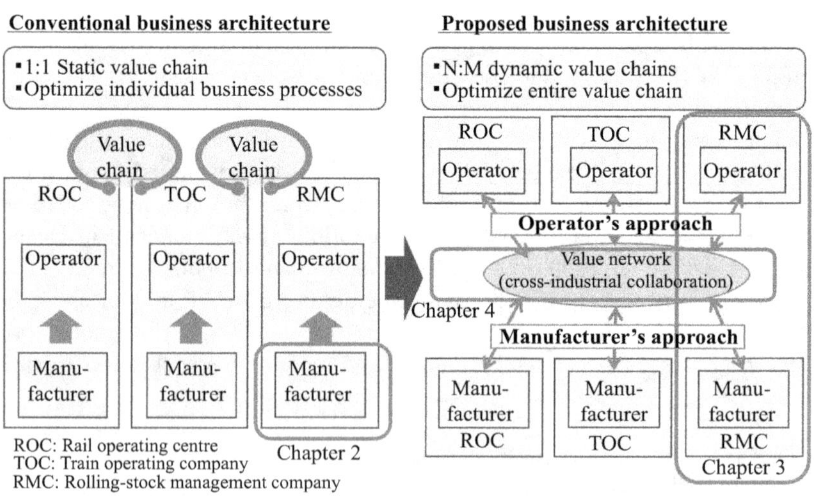

Figure 14.1 Conceptual scheme of ADS business architecture

architecture enables manufacturers and operators to reconstruct a value network regardless of its conventional business field. Furthermore, connecting between companies or manufacturers and operators indirectly, this architecture has the flexibility to enter a business or withdraw from it. To deploy a social infrastructure business, it is necessary to promote incremental business development, which adopts a business entity as a structural unit and completes optimization in correspondence to mutual business needs with a symbiotic relationship between existing companies.

In this chapter, a business transformation from manufacturer to maintenance operator, and finance entities based on an ADS business architecture is described with Hitachi's experience in expanding its business to the UK railway market as an example. In Chapter 2, a business development approach as a rolling-stock manufacturer based on a conventional business architecture is introduced. In Chapters 3 and 4, a business development approach based on the ADS business architecture is shown. Chapter 3 is devoted to describing the transformation from rolling-stock manufacturer to maintenance operator. Transformation to a finance-business entity and the reconstruction of a value network caused by establishing a new business entity are described in Chapter 4.

14.2 Efforts for global expansion

14.2.1 Features of railway market in the United Kingdom

After the privatization of national railway in United Kingdom in 1993, the railway market there had adopted a "vertical separation" structure in which operation was collectively managed by multiple train-operating companies (TOCs) and railway facilities throughout the country by the network rail. The basic structure of the UK railway market after 1993 is shown in Figure 14.2. TOCs were granted franchise authority for 7–10 years by the UK Department for Transport (DfT) and operated in about 25 business zones of each line. Since the franchise period was short, it was difficult for TOCs to make large-scale investment in assets like rolling stock. Under those circumstances, the rolling stock was sold to three bank-system rolling-stock leasing companies (ROSCOs), and TOCs leased the rolling stock from these ROSCOs. If the performance of the rolling stock did not meet the agreed standard, the TOCs imposed a penalty on the ROSCOs. Since the ROSCOs did not have their

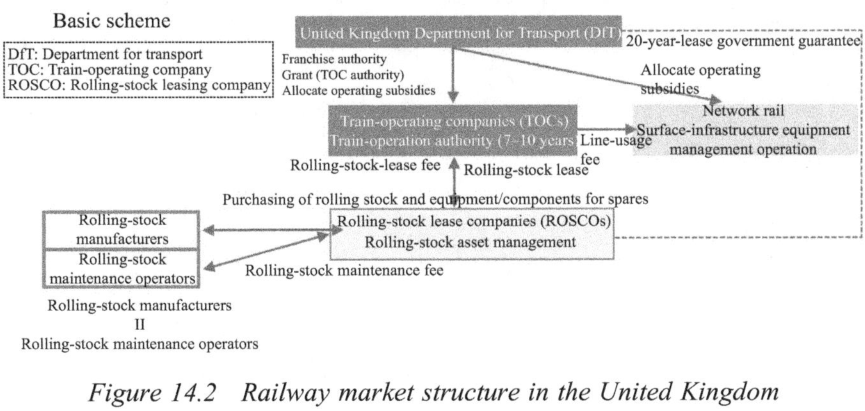

Figure 14.2 Railway market structure in the United Kingdom

own technology, they could not judge whether the unsatisfactory performance was caused by the rolling stock itself or by inadequate maintenance. Therefore, rolling-stock manufacturers were required to not only deliver the rolling stock to customers but also provide a rolling-stock maintenance package to guarantee its reliability over its entire lifetime.

14.2.2 Expansion approach for entering the UK railway market

Since the privatization of the UK railway, a number of serious accidents occurred. For example, there was a head-on train collision at Ladbroke Grove in 1999, and a train derailment at Hatfield in 2000. A chronic lack of investment in the railways was determined as the reason for these accidents. To solve this problem, the DfT led a revitalization plan for UK railways, and the high reliability and safety of Japanese railways was attracting attention.

At the same time, in an effort to avoid oversaturation of the domestic railway market, Hitachi planned to expand into foreign markets to sustain the development of its railway business [1]. To ensure their railway business expansion overseas, Hitachi hypothesized that if it could offer a rolling-stock punctuality rate of 99%, then it would be considered more than reliable enough compared to that of the UK railway.

Furthermore, Hitachi focused on the facts that, the United Kingdom is the birthplace of the train and was drawing global attention, and it still had a relatively open market because it was not fully controlled by a “BIG 3” rolling-stock manufacturing oligopoly, and there was no rolling-stock manufacturer in the country. In addition, criticism regarding low quality and late delivery from European rolling-stock manufacturers was increasing. Taking these facts into account, Hitachi decided to enter the UK market and started their business there by dispatching representatives specialized in railways. Figure 14.3 shows Hitachi’s approach for entering the UK railway market.

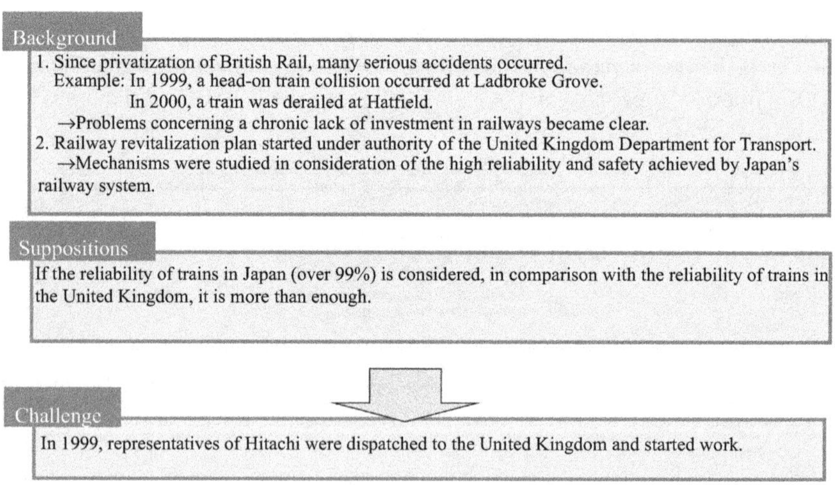

Figure 14.3 Approach for entering the UK market

14.2.3 Initial obstacles in the UK railway market

Upon deciding to enter the UK railway market, Hitachi had to overcome a number of successive failures in 2001 and 2002. The following five points were learnt from these failures.

1. Although Hitachi was recognized as a manufacturer of home appliances, almost no one recognized them as a railway manufacturer. Therefore, it was necessary to establish a "brand profile" in the railway business.
2. It was necessary to prove that participation in bidding was not superficial, and that Hitachi was prepared to take root in the UK railway market.
3. Although the excellence of Japanese railways was well known in the United Kingdom, their ability to exhibit the same performance as in Japan on the UK aging infrastructures was not believed. Therefore, it was necessary to remove the "paper train" prejudice that had formed about Hitachi's rolling stock.
4. Two questions needed to be addressed: "Does Hitachi have products adaptable to the UK railway business?" and "Does Hitachi have the know-how for railway maintenance?"
5. As Hitachi was new to the railway business in the United Kingdom, it was necessary to hire experienced local staff who could lead business development to satisfy various stakeholders appropriately in the United Kingdom.

To establish its brand, Hitachi actively held various seminars and participated in authoritative conferences in Europe. They also promoted Japanese business style by emphasizing the high quality and safety of their products. Furthermore, to demonstrate their quality, Hitachi loaded a whole set of main-circuit equipment to the existing UK rolling stock and ran a trial all over the United Kingdom on the rolling stock with a perfect driving record. Local staff, alongside the Japanese staff, was assigned to appropriate activities for the development of the business that would satisfy various stakeholders. Moreover, a framework for establishing railway maintenance know-how was developed with the aid of a railway company in Japan.

14.3 Expansion to railway maintenance business

14.3.1 Overview of the Class 395 project

In 2003, Hitachi submitted a proposal to receive orders for the Class 395 rolling stock. They got priority negotiating rights in 2004 even though Siemens and Alstom were also bidding at the same time. In 2005, a contract for the Class 395 was formally signed. Customers of this contract were rolling-stock management companies. The contract covered 29 Class 395 train sets (including 174 cars) and rolling-stock maintenance service for a maximum of 35 years.

The planned Class 395 route was a new line from Ashford to London (St. Pancras Station). Based on a new concept, the Class 395 was manufactured to run on both new high speed and existing lines while shortening the required time between London and Paris by 40 min. This improved train service for residents in commuter towns in Kent situated along the new line. Furthermore, Hitachi opened a rolling-stock maintenance depot in Ashford. Figure 14.4 shows the overview of this Class 395 project.

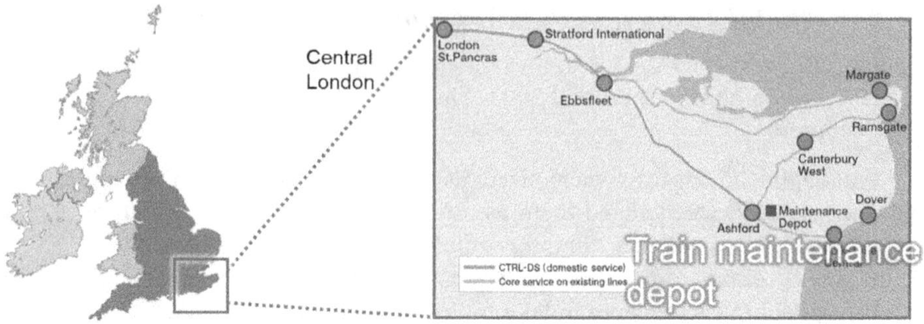

Figure 14.4 Overview of Class 395 project

14.3.2 Evaluation of rolling-stock maintenance business

To gain trust in the UK market, Hitachi had to ensure deadlines which were strictly met. In June 2009, Hitachi met their delivery deadline 6 months earlier than contracted. This achievement was significant because the 'Preview Service' was merely served, and delayed delivery was normalized in the United Kingdom. In addition, results of questionnaires showed that the majority of customers were satisfied with the Japanese quality. After the London Olympics in 2012, Hitachi received high evaluation for an accident-free operation of the rolling stock as an express shuttle, connecting central London with the Olympic venue, transporting a multitude of people including staff and visitors from all over the world. Furthermore, in December 2009 and December 2010, Hitachi also received high evaluation for successfully operating the Class 395, while most of the other railways, including the Eurostar, were suspended due to record snowfall. Moreover, in February 2011, the Class 395 achieved the highest monthly reliability of all types of trains in the United Kingdom for attaining the longest distance travelled with no accidents: 160,000 mi.

By demonstrating what Japanese quality means with the rolling stock and maintenance service, Hitachi established itself as worthy in the market and transformed to a business entity including manufacturing, delivering, and maintaining rolling stock.

14.4 Expansion to finance business

14.4.1 Overview of the IEP project

The Intercity Express Programme (IEP) was a project led by the DfT to replace the rolling stock of high-speed trains manufactured 30 and more years ago. Targeted regions were the Great Western Main Line and East Coast Main Line, both of which include electrified and non-electrified sections. In these lines, Hitachi's "Bi-Mode" technology, which combines motor and diesel-engine driving systems, was effective. After an official review, in June 2008, a proposal was submitted for the largest scale procurement project in the world (estimated at 1 trillion yen).

The crucial difference between this project and the Class 395 project was that the public–private partnership (PPP) scheme was adopted as the business model on top of its larger scale. Because of a lack of motivation in ROSCOs to replace the aged rolling stock and adopt this method, the idea was to introduce a competitive element. In this method, a special purpose company for leasing was organized mainly by rolling-stock manufacturers. The special purpose company, which granted business rights from the DfT for rolling-stock leasing services, provides maintained rolling stock to TOCs every day for a 27.5-year period. The scope of Hitachi's business covered the following four areas: leasing of rolling-stock through investment in the special purpose company, manufacturing of rolling stock, constructing of maintenance depots for the rolling stock, and maintaining of the rolling stock.

14.4.2 Application of PPP scheme

Hitachi established a joint venture leasing company, called Agility Trains, as a special purpose company with John Laing, a general UK contractor. Agility Trains gets financial collaboration from Hitachi, John Laing Ltd. and bank lending group and runs rolling-stock lease business for train operation companies (Figure 14.5). This business scheme was designed by the DfT. To run the rolling-stock leasing business, Hitachi had about 30-year operational guarantee from its investment in Agility Trains (authority granted by the DfT). It also had further financial support from the various financial institutions previously mentioned. For the rolling-stock manufacturing business, Hitachi manufactured 369 cars for the Great Western main line launched in 2017, and 497 cars will be manufactured for the East Coast main line to be launched in 2018. The Bi-Mode train type for direct services between electrified and non-electrified sections and dedicated electric-car types in

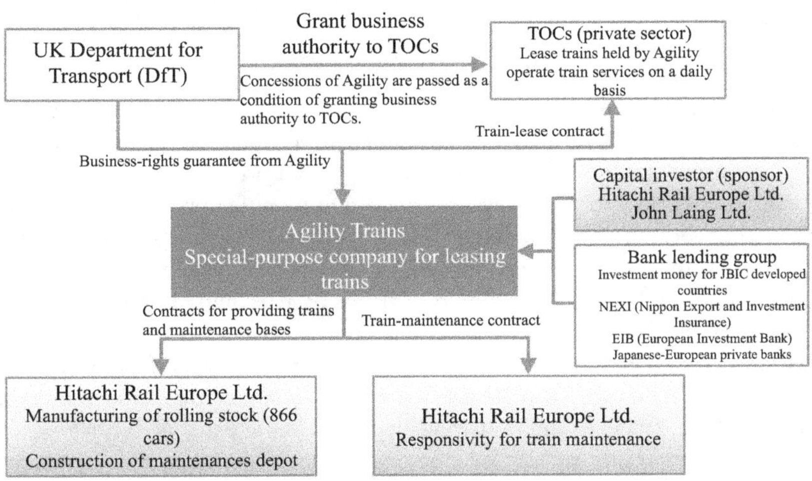

Figure 14.5 Overview of PPP business scheme

electrified sections will be introduced. Regarding rolling-stock maintenance, Hitachi constructed and will continue to maintain operations in four depots for the Great Western main line and nine depots for the East Coast main line, respectively.

14.4.3 Business expansion in the United Kingdom

Hitachi implemented a gradual business expansion using a PPP scheme as a foreign manufacturer of rolling stock. One way this was accomplished was through localization of maintenance technology. By training local maintenance staff at rolling-stock maintenance depots in the United Kingdom, manufacturing technology was localized, which also created local jobs through the construction of a rolling stock-manufacturing facility (in Newton Aycliffe). Manufacturing high-performance rolling stock locally, instead of manufacturing in Japan and exporting it, allowed for higher quality maintenance services on site. This presented a significant advancement of Japanese business infrastructure development abroad. Additionally, this scheme, where Hitachi not only sells rolling stock as the manufacturer but also is financed by Japan and the United Kingdom with leasing rights TOCs, had not been done before. Furthermore, transforming our business into the field of operations through yearly rolling-stock maintenance was one of the most innovative efforts Hitachi had made so far.

14.5 Summary and future developments

To sustain the growth of our business in a globally diverse railway infrastructure, and where social and value structures are rapidly changing, it was necessary to innovate in finance, organization, technology, and management as a single company by keeping up with the structural changes in the market. To expand its business to the global railway market, Hitachi gradually transformed their business from just rolling-stock manufacturing to maintaining and financing based on the ADS business architecture and achieved the following:

1. Acquired understanding that a more UK-based business approach was necessary to enter the UK railway market not just as a rolling-stock manufacturer but also by offering highly qualified technology as a product.
2. Secured contracts for the Class 395 in which manufacturing and maintenance services were included. Hitachi further proved its product reliability and quality. Additionally, we established a maintenance system through the depot construction in Ashford and met local needs by halving required time as a way of improving QoL.
3. Expanded business into the financial field on the basis of a PPP scheme and secured contracts with the IEP to lease, manufacture, and maintain all rolling-stock services and construct depots. This expansion further benefited the local social infrastructure by creating jobs and improving the manufacturing process of rolling stock.

In the future, to sustain railway business growth and to meet the needs of the global market, Hitachi will continue to upgrade and provide a high added value to the business model cultivated in the United Kingdom with a full-value chain.

Reference

[1] Mitsudomi, "Innovative Express – UK Intercity Express Programme," Hitachi Hyoron, 95, 1, pp. 6–15 (Jan. 2013) in Japanese.

Chapter 15

Sustainable business through alliance based concept model of management & technology of railway infrastructure

Masaki Ogata[1]

Abstract

With the rapid development of technologies such as information and communication technology (ICT) and the accompanying major social changes, it has become increasingly difficult for companies to offer highly satisfactory customer services of their own in current society where customer needs are very widely diversified. Under such circumstances, by alliances among different companies, it is important to combine individual specialties organically and create new value constantly. On the other hand, the dramatic development of ICT enables companies and customers co-create new values using social network service as common practice. In this chapter, the alliance strategy of Japan Railway East which has developed vertically integrated business model in railway business and its advantages is described. In addition, autonomous decentralized technology and assurance technology as system technologies supporting these business alliances are also described.

15.1 Introduction

Japan Railway (JR) East owns infrastructure such as land, facilities, vehicles, train control system and so on, while, at the same time, it operates and maintains its infrastructure. This type of railway can be called the vertically integrated railway system together with its management and technology. In addition, since the privatization in 1987, JR East has developed the lifestyle business as the second business which includes retail, shopping centre, restaurant, hotel, office leasing and so on. As the third business, micropayment business utilizing Suica, namely Super Urban Intelligent CArd, of the smart integrated circuit (IC) ticketing system has also been developed since 2004. This can be classified as Fintech. The fourth one is rolling stock manufacturing which can also lead to the export of it. The

[1]East Japan Railway Company, Japan

overseas business is becoming the fifth one today. Thus, JR East has been seeking for the multiple business models in order to adapt to the rapidly changing managerial environment.

On the other hand, owning infrastructure on its own is also a high risk. In recent years, the change of social environment is very fast, so effective utilization of infrastructure is an important issue for the apparatus industry which owns high cost and long service life infrastructure. Especially in the railway business, infrastructure continues to operate 24 hours a day, every day all through a year; it is very important to sustainably develop the business while brushing up the infrastructure and conforming to social change.

By the way, infrastructure is just a facility by itself, and it can continue to produce services as a sustainable infrastructure system only by appropriately operating, maintaining, managing and innovating it. We take this acronym for Management, Technology, Operation, Maintenance and Infrastructure as the MTOMI model and position this business model as the foundation of our ecosystem. In this chapter, we show the direction toward the outline of our business, business operation through MTOMI model and future public transport (PT) business.

15.2 Characteristics of JR East

15.2.1 Outline of JR East

JR East on its own rail infrastructure operates and maintains various categories of rail transport, including Shinkansen, metropolitan, suburban and regional railways as shown in Figure 15.1. More precisely, JR East has the following seven characteristics:

1. Vertical structure

   JR East owns all its railway infrastructure, operates it and maintains it as a fully integrated railway enterprise which includes all types of infrastructure, such as station, rolling stock, track, electric power supply, signalling system, information technology (IT) system and so on together with the ones of other businesses.

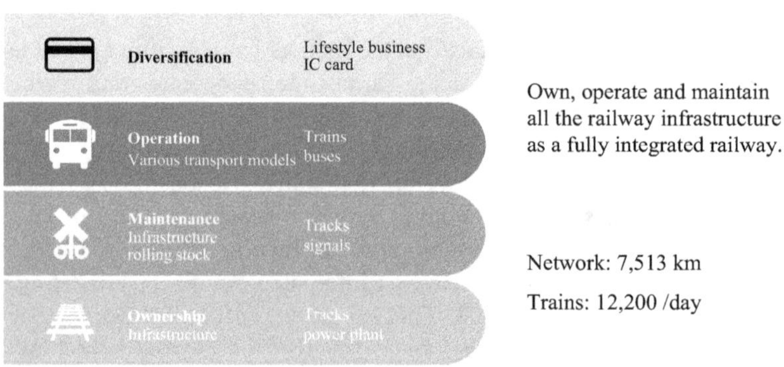

Figure 15.1 Vertical structure

Figure 15.2 Horizontal structure

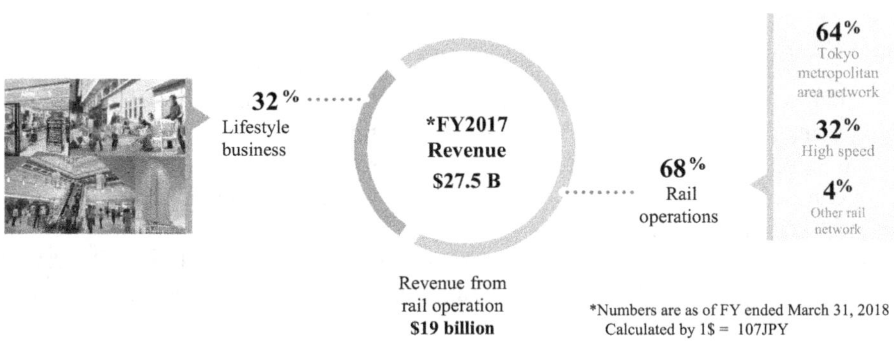

Figure 15.3 Business structure

2. Horizontal structure

   JR East owns and operates all categories of rail transport such as five-directional Shinkansen lines, metropolitan and regional railways (Figure 15.2). Thus, its horizontally integrated rail network is working organically.
3. Business structure

   JR East also manages lifestyle businesses utilizing station space, shopping centres, hotels and so on. Revenue from such non-transportation was more than 30% in total revenues of the JR East Group in FY2017 as shown in Figure 15.3.
4. Micropayment

   JR East owns a micropayment infrastructure. Its origin was the first IC card ticketing system in Japan, introduced in 2001 which 'Suica' implemented first in the Tokyo metropolitan area. In 2004, JR East launched a micropayment service (e-money) based on this IC card ticketing system and has disclosed IC card standards to other transportation companies. The total number of IC card holders is about 120 million and transactions per day surprisingly are a total of 130–150

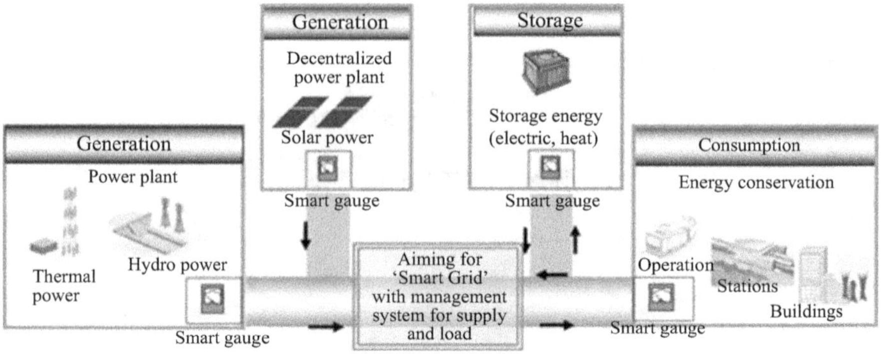

Figure 15.4 Energy chain

million, including single trip, commuter pass and e-money. The processing time is 0.2 second for passing through automated passenger gates in the station.

5. Energy chain

   JR East, in the metropolitan area, produces and consumes its own electric energy, namely owns both the 'supply' and 'demand' sides. The company owns two electric power plants, power lines substations and so on. One power plant is hydroelectric and the other thermal. Power is used by stations, trains, shopping centres, retail shops, hotels, offices and other facilities (Figure 15.4).

6. Rolling stock chain

   JR East cycles all stages in the life cycle of rolling stock, namely all through philosophy, concept making, design, manufacturing, operation, maintenance, feedback for new development, recycling and so on.

7. 'Two WITHOUTS'

   JR East has not received any subsidies from central or local governments for more than 31 years since privatization in 1987 and has not raised fares or charges except for consumption tax increases for the same period of time.

15.2.2 The quantity changing the quality and the essence

In 2012, total operating distance of rails worldwide was almost 2.9 trillion passenger-km, with Japan's rails accounting for 9% [1]. In terms of the number of passengers, Japan accounts for 29% share in the world's 30.7 billion. Japan, as one single country, shares almost the same portion as the entire European or Asia Pacific region including China and India.

Quantitative features in JR East are shown in Figure 15.5. The number of passengers (17.5 million per day) and high expectations from users and society in general have improved the quality of railways in Japan, such as higher frequency, larger capacity as well as safer, more punctual and quicker service. More precisely, in terms of frequency, JR East operates Shinkansen trains at 4-min headway maximum and conventional trains at 2-min headway. In terms of capacity, a single

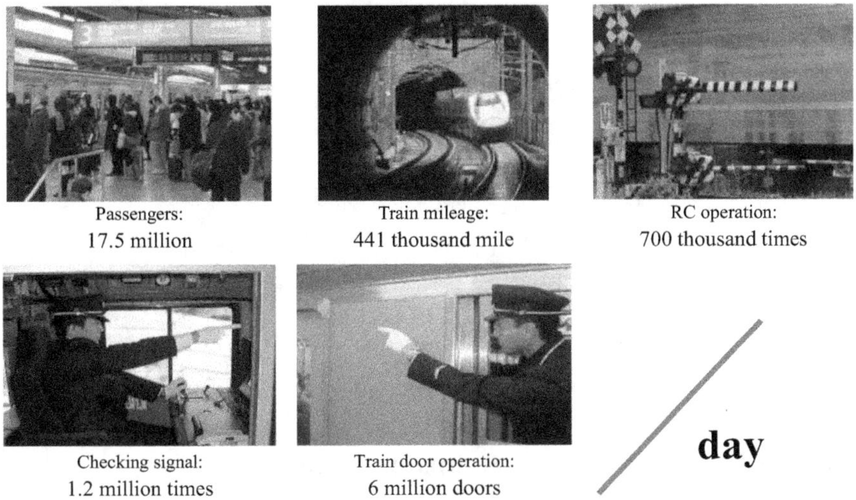

Figure 15.5 Quantitative features in JR East

commuter train in the metropolitan area of JR East can carry 3,000–4,000 passengers. In terms of safety, the most important in our management, Shinkansen fatalities on board by the internal cause have remained at zero since 1964 when Shinkansen operation started. In terms of punctuality, average time delay per Shinkansen train from 2009 to 2013 was only 46 seconds including case of natural disasters. In terms of quickness, an automated ticket gate reads and writes fare data on 'Suica' in 0.2 second. JR East must sustain and/or improve these five qualities simultaneously. Thus, as stated above, and also as my belief and theory, the quantity changes the quality and the essence.

Suica, or IC ticket cards, can be mentioned as another example of the fact that 'quantity changes quality'. Suica users were few in 2001 when Suica was introduced. Paper-based tickets were chiefly being used at that time. The number of Suica users has been increasing gradually since then, and today more than 90% of passengers in the Tokyo metropolitan area use Suica, including 'Mobile Suica' by which passengers can use a smartphone or cell phone as a ticket card instead of an plastic IC card.

Supposing 100% of passengers have and use IC cards, ticket-vending machines or money would not be needed. Therefore, new space can be generated in place of the area of current vending machines and the burden and cost of circulating money could be reduced. Various possibilities in diversified use of space can be created. For example, as in Ueno station, the necessity of ticket-vending machines has decreased thanks to the wide use of Suica. Then the space of some ticket vending machines has been already converted to commercial space. Use of space in stations will be completely restructured in the near future (Figure 15.6). Thus, again the quantity changes the quality and the essence.

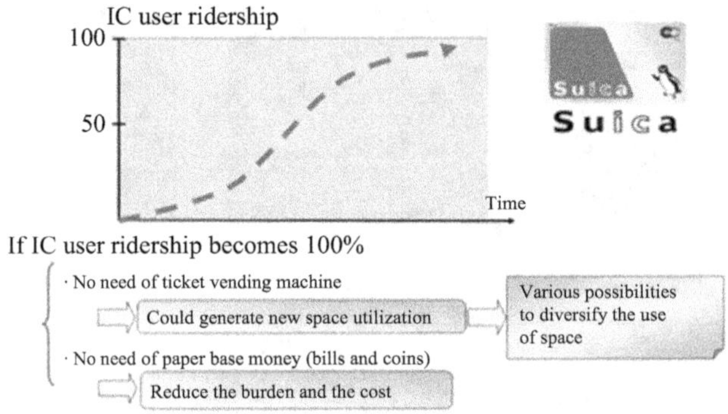

Figure 15.6 Potential of Suica

15.3 Railway infrastructure business and technology concept model

15.3.1 Essence of JR East

JR East has more than 31 years of history since its privatization. It is neither merely a railway company nor a transportation company. It is an integrated technology and service industry (Figure 15.7). In this context, 'integrated' means JR East owns all the infrastructure of railway, operation and maintenance as full integrated rail model. 'Technology' means that JR East owns all the fields of technology such as operations, rolling stocks, signalling systems, communication, tracks, electric house supply, architecture and construction and so on. 'Service' means that JR East provides society and customers with values and services based upon infrastructure with operation and maintenance.

15.3.2 What is 'infrastructure'?

In railway industry, in general, there are some definitions of infrastructure. For example, one is an infrastructure that does not include train operations, or 'two-tiered system', and the other is an infrastructure that includes train operations, or 'vertical integration' of infrastructure, operation and maintenance. Many European railway operators are the former (Figure 15.8), and many Japanese railway companies are the latter (Figure 15.9). JR East not only owns land, structure and rolling stock, but also operates and maintains them. The important thing here is that this vertical integration of infrastructure must provide customers and society with the selectable resources and the integrated technology and service.

15.3.3 What is 'service'?

Service is provided to customers from the integrated technology and service industry based upon customers' demand. Therefore, two-way communication exists between customers and the industry as shown in Figure 15.10. 'Service' is totally different from 'product goods' as it is produced and consumed simultaneously, cannot be stocked, has no distributable sample, is not replaceable or is not uniform.

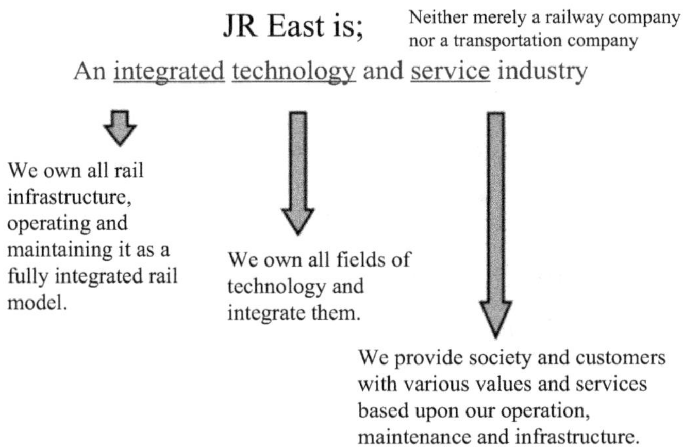

Figure 15.7 Essence of JR East

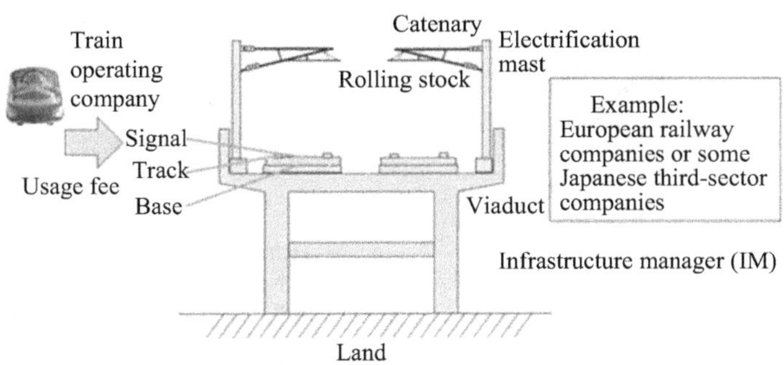

Figure 15.8 Two-tiered system

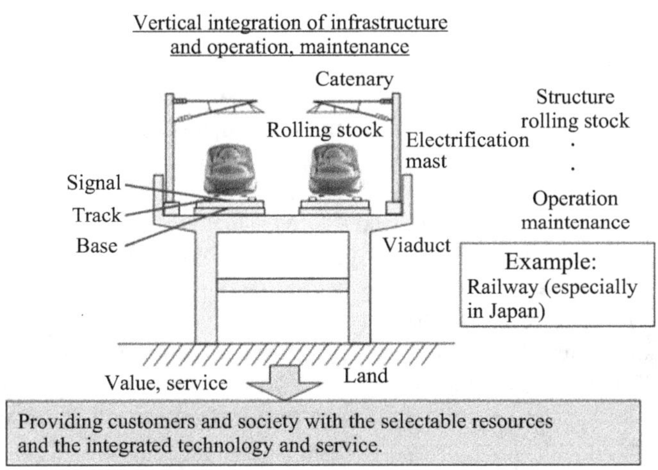

Figure 15.9 Vertical integration

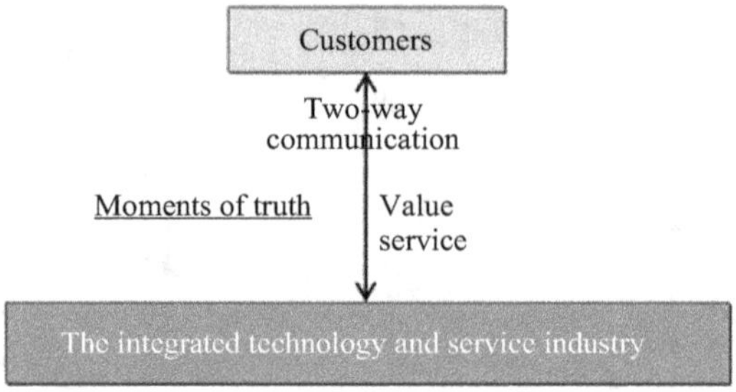

Figure 15.10 Relationship between customers and the integrated technology and service industry

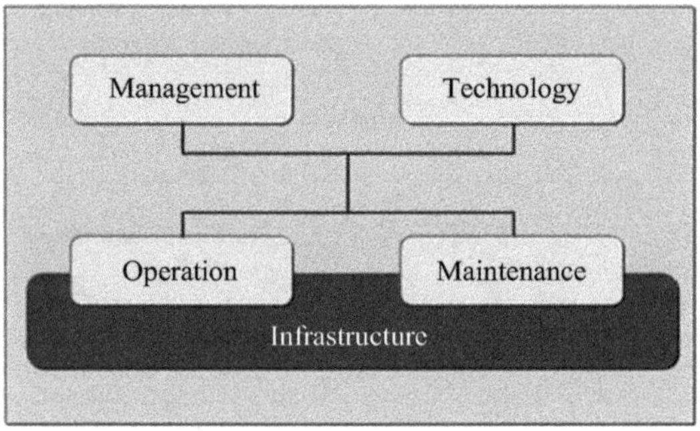

Figure 15.11 MTOMI model

Moments at which customers and employees encounter each other can be called 'moments of truth' [2]. Quality of service and the company are evaluated at this moment. Hospitality is the important mutual relationship between customers and well-disciplined employees. Hospitality is only realized at the moment of truth. If it fails, it cannot be corrected or changed.

15.3.4 MTOMI model

What is the integrated technology and service industry that provides excellent value and service to society and customers in these 'moments of truth'?

JR East is a vertically integrated company owning infrastructure together with operation, maintenance, management and technology. Figure 15.11 shows the relationship among these elements, and the 'MTOMI model' can be abbreviated as based upon initials of each element (Management, Technology, Operation, Maintenance, Infrastructure).

As shown in Figure 15.12, the MTOMI model provides values and services to customers, society and nations through the moments of truth with well-disciplined employees. The integrated technology and service industry also owns its clear management and concrete policy for service; 'Integration of Service and Infrastructure' is achieved based upon these three elements described above such as MTOMI model which can be considered as effective function of services, concrete management policy for service and moments of truth.

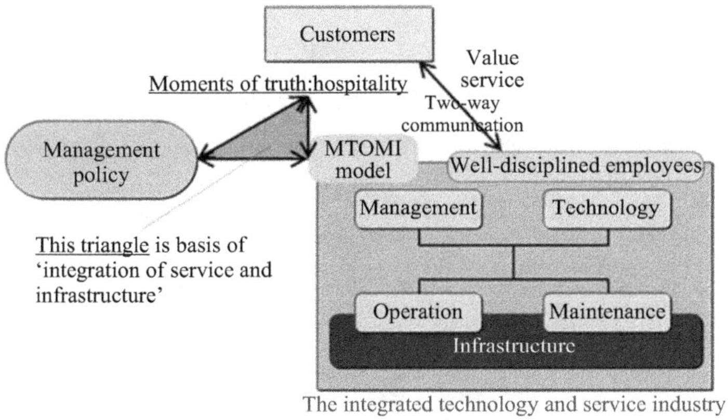

Figure 15.12 Integration of service and infrastructure

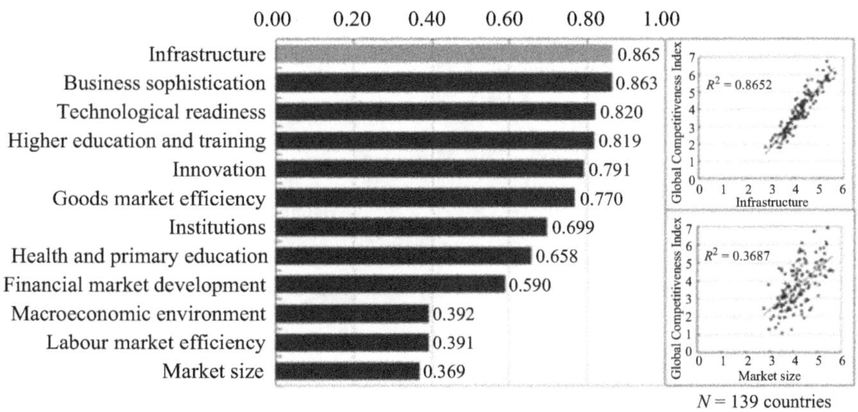

Calculated by JR East based on 'The Global Competitiveness Report 2010-2011', World Economic Forum

Figure 15.13 Determination coefficients in the global competitiveness index

15.3.5 Advantage of MTOMI model

A social infrastructure company with an organically integrated MTOMI model can provide excellent value and service to society, community and customers. According to 'The Global Competitiveness Report 2010–11' (Figure 15.13), infrastructure ranked at the top in determination coefficients in the global competitiveness index out of 12 factors, such as business sophistication, technological readiness, higher education and training, innovation and others. Thus, infrastructure is most important to a nation's global competitiveness.

As mentioned in Section 15.3.2, vertically integrated infrastructure, including operation and maintenance, provides society and customers with values and services such as safety, comfort, large capacity and so on, as shown in Figure 15.14.

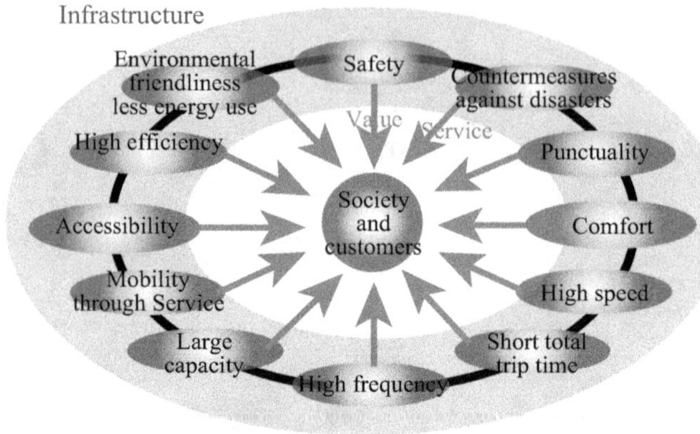

Figure 15.14 Infrastructure providing value and services to society and customers

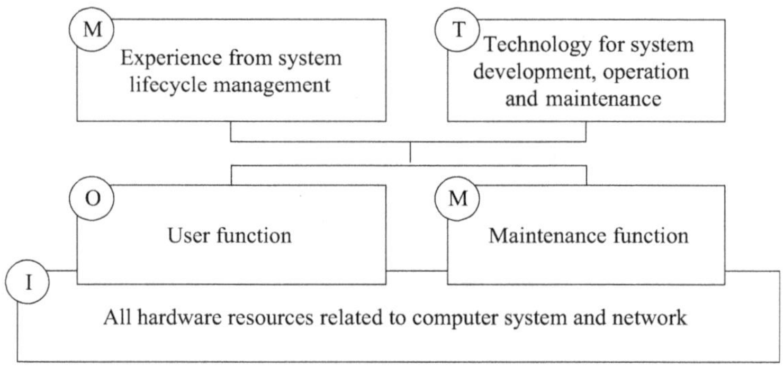

Figure 15.15 The MTOMI model of business infrastructure

15.3.6 The MTOMI model from the viewpoint of computer and communication system

Today, in almost all businesses, the computer systems are indispensably supporting them. Whereas the MTOMI model is a vertically integrated business model, the corresponding computer and communication systems also configure an MTOMI model. Figure 15.15 shows the MTOMI model of the computer and communication systems, which is the base of business infrastructure. In this figure, the infrastructure 'I' corresponds to all the necessary hardware resources to carry out the business. Operation 'O' and maintenance 'M' are the resources for user functions in the system, namely software and middleware. Then management 'M' and technology 'T' are actualized and grown based upon the computer and communication systems.

In the business alliances, the good coordination of multiple computer and communication systems as business infrastructure is the key of success. The coordination and cooperation among multiple large-scale computer and communications systems make business alliances effective, functional and efficient. In these cases, the introduction of the autonomous decentralized system (ADS) is the bases for the multiple systems configuration [3]. Now the actual business alliances utilizing ADS are explained in the following. In the Section 4, three types of business alliances are shown and concrete examples of system alliances along with the examples of actual business alliance are explained.

15.4 Alliances based on MTOMI model

In this section, an alliance strategy of JR East based on the MTOMI model is described.

15.4.1 Requirements for good alliance (What is alliance?)

The alliance is aimed at the cooperation by a group that consists of various organizations and various persons. Therefore, trust brought by mutual communication, commitment with each other and control of conflict is necessary to establish win–win relationship. JR East has potential to make the alliance with any business fields beyond the railway industry because the business model, the MTOMI model mentioned in the former section, in which JR East owns infrastructures, conducts operation and maintenance by itself, is unique and powerful while storing experiences and knowledge.

15.4.2 Classification of alliance (three models of alliance)

Three models of alliance and the examples at JR East are described.

15.4.2.1 Inside a company (Inside B model)

There are both a facility department, which constructs and maintenances JR East's infrastructures, and an operation department, which plans and conducts the train operation exploiting the infrastructures of JR East. They are basically a vertical management division. A control centre supervising the daily operation maintenance and many operation and maintenance depots in the actual field are in the relation of vertical specialization from the viewpoint of an organization structure. In JR East, the common system for this organization was developed and the smooth cooperation among different vertical management divisions by using the system ensures safety, stable operation, maintenance and improvement of customer service.

15.4.2.2 Company and company (B2B model)

After privatization, JR East has diversified and strengthened multiple businesses related to the railway business for synergetic effect. Experiences and knowledges for the businesses also have been brought to JR East by owning new business infrastructures and conducting operation and maintenance. In this way, the extension of the MTOMI to the out of traditional field has increased the possibility of new alliances. Specific example is described later, but drastic extension of the market can be made by establishing the alliance with companies in the way that the one with MTOMI model transferring its business model to the other without such MTOMI model.

15.4.2.3 B2C model

A company provides information to customers, and customers act autonomously on the basis of the information. The company creates and provides a new service with the feedback on the action. This sequence yields a positive synergetic effect. The relation between the company and the customers can be regarded as a kind of alliances, in which the company creates new and innovative services with the customers together. Development of information and communication technology (ICT) – artificial intelligence (AI), Internet of Things (IoT), big data and so on – can enhance heightened interactivity and real-time communication, and consequently, synergetic effect can be obtained more easily.

15.5 Three types of business alliance based upon MTOMI model

In this section, three types of alliance cases on business based on MTOMI model are described.

15.5.1 ATOS (Inside Business; InB model)

The departments which are in charge of operation, rolling stock and ground facility, respectively, are individually independent business units of JR East. The roles of the operation department are train operation such as operation planning, signal and route control, traffic fleet control, passenger information and so on. On the other hand, the roles of the rolling stock and facility departments are the design, manufacture or construction, maintenance and so on. Since the operations are carried out by the designed timetable, it is very essential and necessary for safe and reliable operation that all the related information must be shared in any departments.

In the Greater Tokyo metropolitan area, within a radius of 100 km from Tokyo central business district, the traffic volume is extremely dense, maybe the densest in the world; consequently, the frequency of trains are extraordinarily high. In the past, the management of train operation was executed by human capability. Their change of timetables, turnarounds of trains, modifications of train slots and so on due to the operational disturbance were executed by human brain and hands. The modified information had to be transmitted through phonic information via telephone. These modifications of train information were done by the station staff members.

To solve these issues, JR East introduced ATOS, an abbreviation of 'autonomous decentralized transport operation control system', that is a large-scale train operation and maintenance management system for the Greater Tokyo metropolitan area in JR East. This system was first introduced in Chuo Line between Tokyo and Kofu in 1996, the city is approximately 100-km west to Tokyo. Later this system was expanded to 24 lines in Tokyo metropolitan area [4].

Figure 15.16 shows the concept and configuration of Autonomous Decentralized System (ADS) of ATOS. In ATOS, operation every day is controlled at the central train control centre. Rescheduling of train operation due to the disruption is only controlled by dispatcher in the centre. The information changed rescheduled time table is instantly transmitted to the programmed route control computer in each

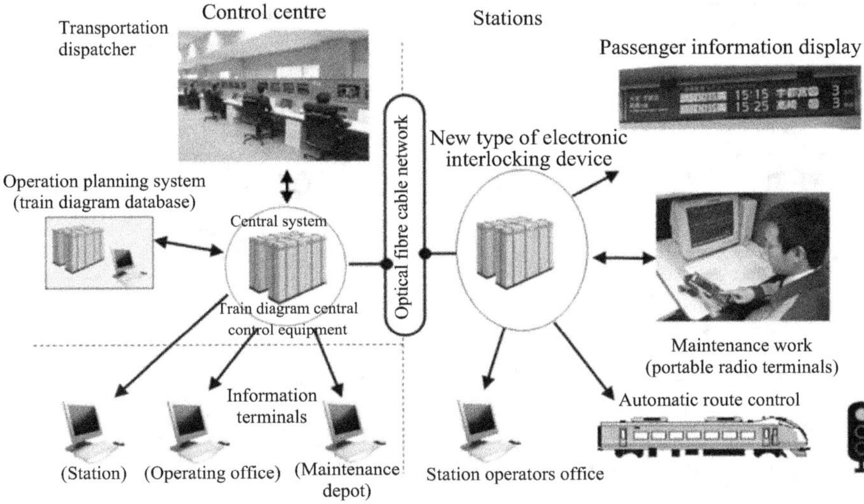

Figure 15.16 Autonomous decentralized system for ATOS

station. Furthermore, signals and routes are controlled automatically, and latest information for ground facility maintenance and for passenger is updated automatically on ATOS. By the ATOS, the time required back to normal operation has been shortened, and safety of the ground facility maintenance has been improved. At the same time, the passenger information is instantly and accurately transmitted to the display on board and in the stations. Furthermore, to the application of smartphone today which have raised customer satisfaction significantly. As mentioned in the earlier section, an individual department in the company is a business unit based upon each respective MTOMI model, but the comprehensive MTOMI model of the company can be formed by the agile and effective alliance of multiple MTOMI models of each department.

Figure 15.17 shows the comprehensive MTOMI model formed by the alliances of multiple MTOMI model. In ATOS, signal and route control which has been conventionally built and maintained by the signalling group of ground facilities department. And, operation done by operation department were integrated with ADS as the key technology. The core concept of the ADS is the common data field for multiple departments which is the innovative architecture of the data. Consequently, thus, the train operation and the maintenance work can be controlled in the same system with the common data field [3].

15.5.2 Suica ('Super Urban Intelligent CArd')

This example is Business-to-Business type alliance. Suica is an IC card with which passengers can pass quickly and smoothly ticket gates. This means passengers do not have to insert their cards in a ticket slot of a passenger gate, and all they need to do is pass their cards over a card reader/writer at a gate. The system was first introduced in the Tokyo metropolitan area as the first transport IC card system in Japan, in 2001.

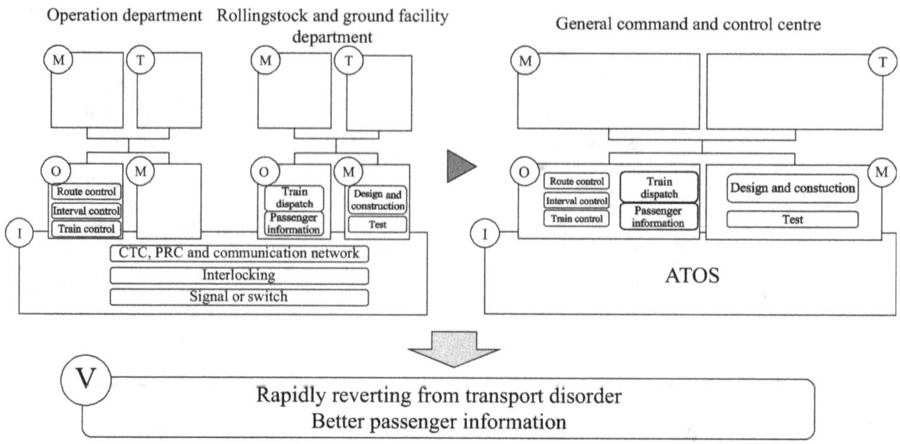

Figure 15.17 The comprehensive MTOMI model formed by multiple MTOMI models of each department

Effective coverage was expanded to Tohoku and Niigata areas later. Because the Suica IC card standard has been disclosed to other transportation companies, many railway companies have adopted the same IC card system as Suica. Consequently, mutual service has expanded with nine other participating transportation IC cards, including JR West's ICOCA and PASMO (major private railway companies aggregated in the Tokyo metropolitan area) and others. The total number of the ten transportation IC cards holder reached about 100 million in 2015.

Suica is an integrated system of automatic-gate control, payment service called micropayment, limited express tickets, first-class car tickets, many necessary information related and so on. Since this single card enables passengers to receive various services provided on the railway and retail platform, customers have drastically increased. Mobile Suica service, with which the function of Suica is available on a mobile phone has been launched since 2006. More convenient service, like the payment through the Internet, has been also developed.

The first Suica application for smartphone was only for Android operating system (OS) only, but in 2016, mobile phone Suica also has got available on iPhone and later Apple Watch. The web payment of Suica has been also developed and expanded as more convenient service.

Transaction time is only 0.2 second, and, today, the system processes 130–150 million transactions per day (Figure 15.18).

Suica system is such a large-scale system that we adopted an ADS concept for online expansion, fault tolerance and online maintenance [5,6], as shown in Figure 15.19. As mentioned above, Suica is not only the one of the most important technological innovations including many new ideas, concepts and technologies but also a social innovation creating new values to make big changes in people's lifestyles.

Figure 15.20 shows comprehensive MTOMI model formed by multiple MTOMI models of multiple operators. As mentioned earlier in this section, Suica was first

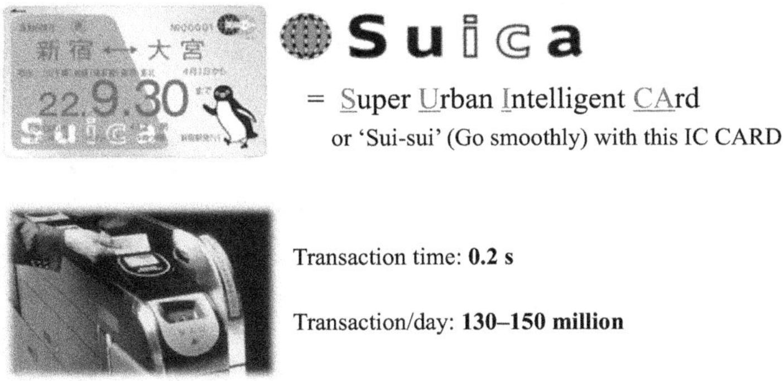

Figure 15.18 Suica

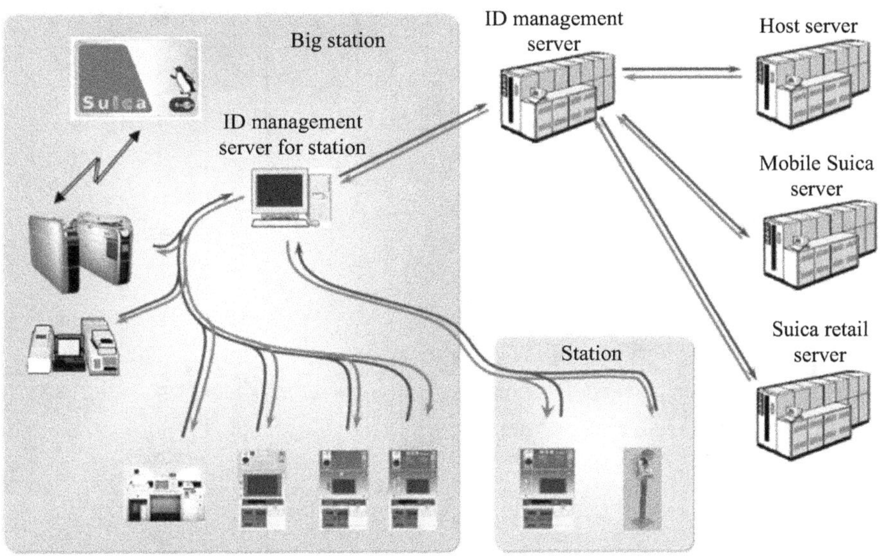

Figure 15.19 Autonomous decentralized system for Suica

introduced in the Greater Tokyo metropolitan area of JR East on 18 November 2001. At that time, Suica was nothing more than a single MTOMI model, namely business platform only for JR East. Completion of mutual usage all over Japan was in 2013. This mutual usage can be considered as the common business platform for 200 different operators. Namely, the comprehensive MTOMI model of this common business platform has been formed by the agile and effective business alliance of multiple MTOMI models of multiple operators.

15.5.3 JR East Train Info App

JR East Train Info' App (JR East App) is JR East's free smartphone app for both iOS and Android users. The concept of this app is 'Everything about JR East railway in

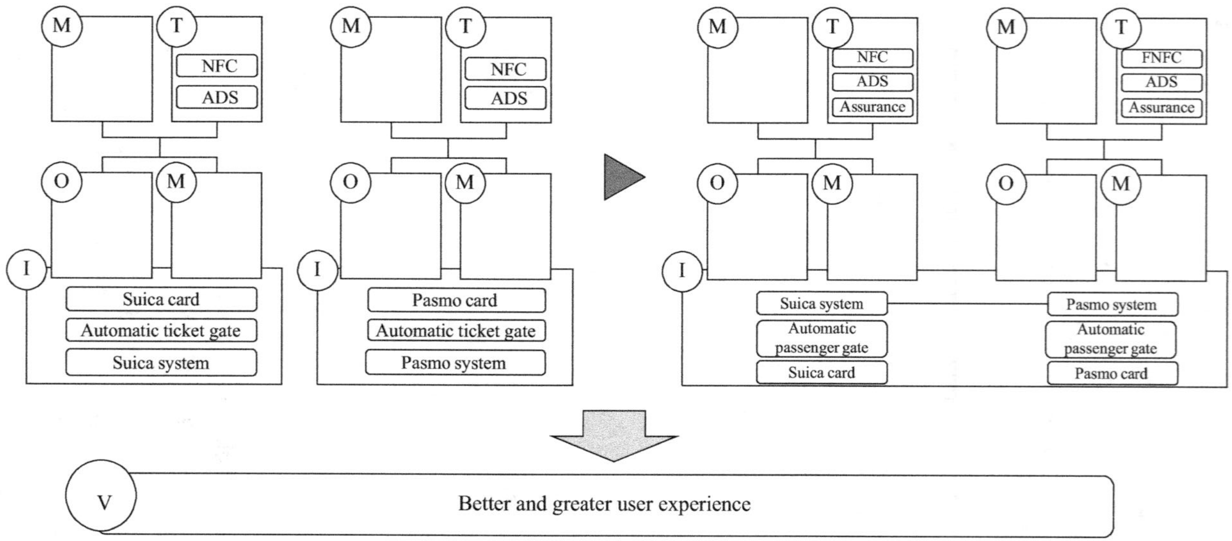

Figure 15.20 Comprehensive MTOMI model formed by multiple MTOMI models of multiple operators

one app'. Passengers can easily view JR East train status and station facilities on their smartphones. JR East App is available in both Japanese and English. The Japanese version was launched on 10 March 2014, and more than 3.6 million have been downloaded as of October 2018. English version was launched on 20 March 2015, and more than 0.35 million have been downloaded as of October 2018.

User interface design is most important for smartphone apps. We took great care in designing the top page. Station information is displayed at the top and the nearest station automatically appears by GPS. Passenger can choose from around 150 major JR East stations. Weather and train operation information for 17 major JR East railway lines can be seen at all times on the left side. On the right side, they can view the information they need: train information, Yamanote Line Train Net, departure information, stops information, station maps and others (Figure 15.21).

Route search provides information on train connecting from departure station to destination station (Figure 15.22).

JR East App is for viewing information on train service. Details are available by selecting a specific area. Yamanote Line Train Net is a featured function of the

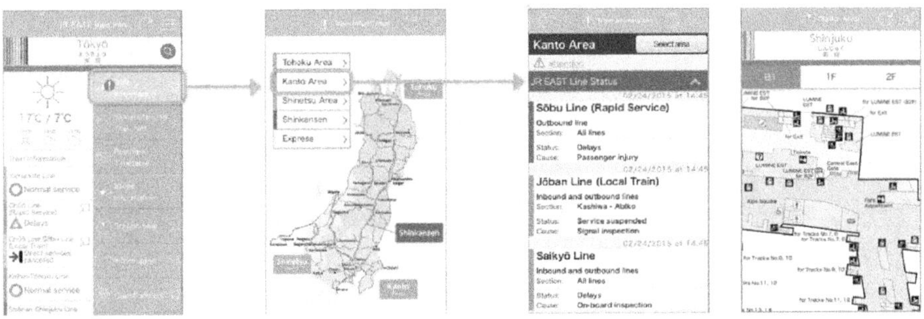

Figure 15.21 JR East Train Info

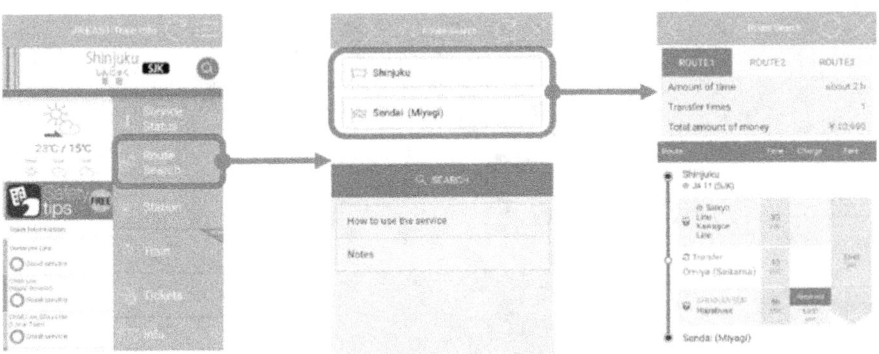

Figure 15.22 Route search function

JR East App. They can view the real-time positions of Yamanote Line trains and carriage conditions (congestion and temperatures for each car) in real time as shown in Figure 15.23.

Departure information provides information on trains departing from Tokyo, Shinjuku, Ueno and Shinagawa in real time. This information is the same as the electric signboard displaying train departures in stations. Stop information shows possible routes departing from the stations displayed on the top page and stops along the routes. Station maps has details station information on maps.

These information contents have been continuously updating. As latest update, train location providing service was expanded. Train location provides information on trains, such as timetable of selected station, train location, destination and delay of selected line in real time. This function was originally targeted only for the line of the Tokyo metropolitan area, but now it is provided for the line of the whole JR East as shown in Figure 15.24.

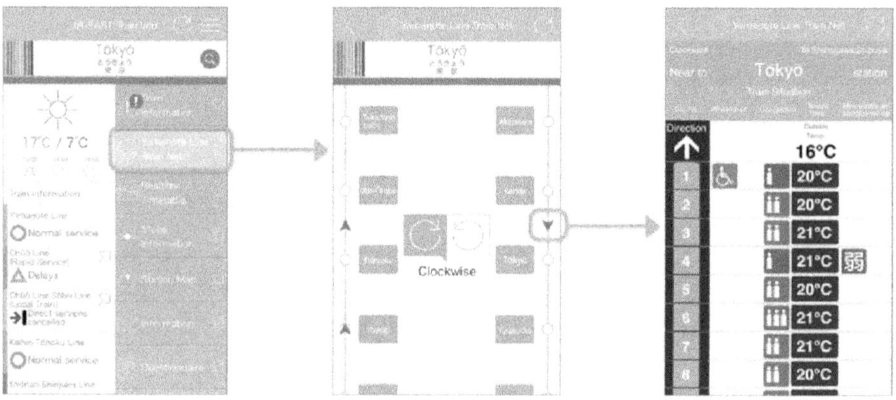

Figure 15.23 Yamanote Line Train Net

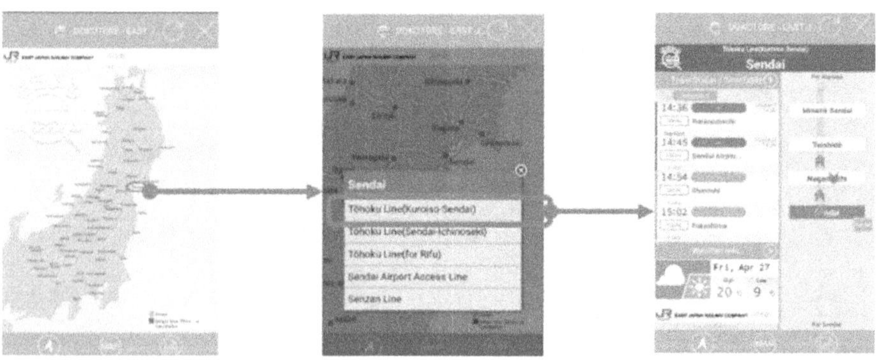

Figure 15.24 Train location

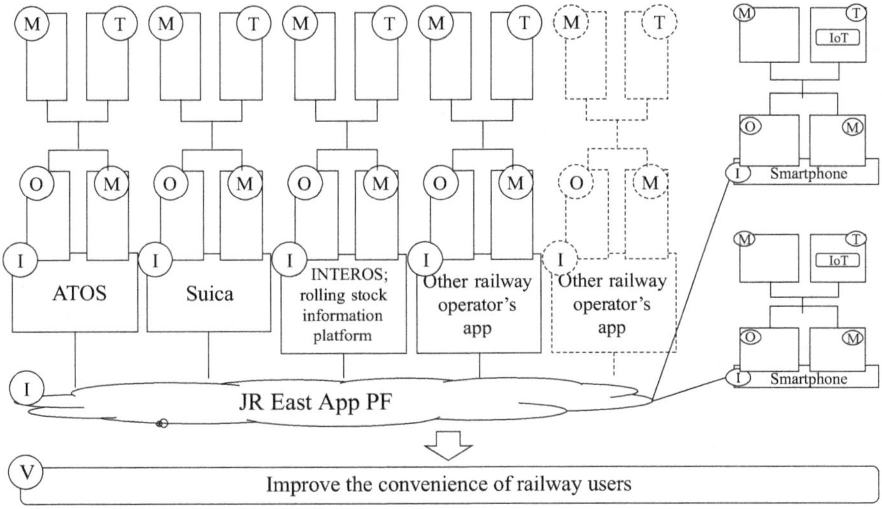

Figure 15.25 The comprehensive MTOMI model formed by multiple information to the customer plus multiple MTOMI model of business partner

On the other hand, collaboration among railway operators is proceeding. Applications like JR East App have been providing by major private railway operator or subway operator. Collaboration among JR East App and other railway company's apps has started in October 2016. These collaborations are gradually expanding. Recently, collaboration with Tokyo Metro has started on November 2017. Currently, in this collaboration, cross-reference of train location and timetable with major railway operators in Tokyo Metropolitan area is realized.

JR East App is a cloud-type service. Each service can be called from the platform of JR East App installed on a personal smartphone. Each service is an autonomous unit based on independent MTOMI. Also, JR East App on personal smartphone is also an autonomous unit based on MTOMI. It also carries out information collaboration with passenger smartphone applications provided by other railway operators in the metropolitan area and is working to improve the convenience of railway users in the metropolitan area (Figure 15.25).

15.6 Future business mode

Although the MTOMI model itself is business model originated in a self-contained way, a new MTOMI model organically combined together with a heterogeneous business can create new values and services. In other words, we have shown in the previous section that the alliance is an important key in bringing about change in future business. In this section, we will look at the future image of the public transportation business by the alliance [7].

15.6.1 Innovation concepts of public transportation

15.6.1.1 Concept

Nowadays, the management environment is exponentially, rapidly changing. The following six critical issues can be considered to be exponentially, rapidly changing in a large scale in the world.

- Social economic trends
  - Instantly propagating economy
  - Business market (population, polarization, ageing society, urbanization and so on)
- Management and technology environment
  - Globalization
  - ICT utilization
- Natural environment
  - Climate changes
  - Increase of natural disasters

The following three concepts of the mission of PT are very important to adapt exponentially, rapidly changing critical issues surrounding the management environment.

1. PT as the social infrastructure can sustainably innovate by itself, while knowing the genuine needs of the customers, communities and societies in the vis-a-vis communication.
2. Sustainable innovation of PT can raise the mobility of people and the quality of life.
3. Social innovation can be achieved by PT innovation.

As far as PT will follow these concepts, we, PT, will be surely able to adapt to critical issues, which I have previously mentioned, and grow sustainably, I strongly believe. I would like to repeat these concepts at every moment everywhere.

15.6.1.2 Proposal

Being based upon the concepts, I have mentioned that I would like to make the following three proposals:

1. In order to adapt to the rapidly changing environment, we, PT, will need to enhance further innovations in PT management and PT technology.
2. In terms of driving innovations, it is very crucial to enhance smooth coordination and collaboration among all the stake holders of PT, which I like to phrase 'open innovation of public transport in the globalization'.
3. I would like to propose that we should advocate the value of PT and the value and benefits that PT can provide to the society, community and customers.

15.6.1.3 Future visions about PT

Shorten total trip time by alliance

Traditionally, each PT has owned infrastructure for itself and has been responsible for large-volume high-speed transport between origin and destination station. Then, personal mobility such as private cars, taxis, bicycles and the like has been responsible for travel from the departure place of the traveller to an origin station and from destination station to the final arrival place. However, whenever people meet a transit, a waiting time and some other redundant arises. Although it is possible to shorten the time of each mode, it is always required to shorten the total trip time.

From the past, in railway, shortening of a trip time is achieved by reducing the time onboard.

However, shortening of the waiting time between different transport modes has not been taken into account so much, so the competitiveness of PT has gradually declined over long periods of time against vehicles capable of transport without transit.

However, automobiles have a problem of safety, high environmental burden, truncation of weak traffic villagers. Therefore, for travellers, it is necessary to shorten the total trip time by personal mobility and the alliance of PT from the viewpoint that 1 min on board is equivalent to 1 min of the transit time and one for the others (Figure 15.26).

In recent years, as a new personal mobility, car sharing, Uber and so on are gaining attention. These provide transport of first/last 1 mi at a cheaper price than private cars or taxis. For these first and last 1 minute, it is also important to combine existing traditional PT operators and new mobility service provider. However, as shown in Figure 15.27, the main part of the larger demand, namely trunk line, will continue to depend upon PT and mass transit.

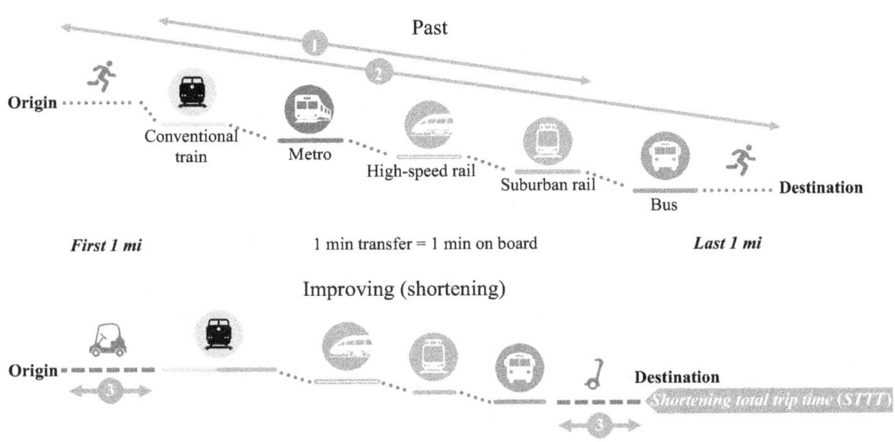

Figure 15.26 Shorter total trip time model

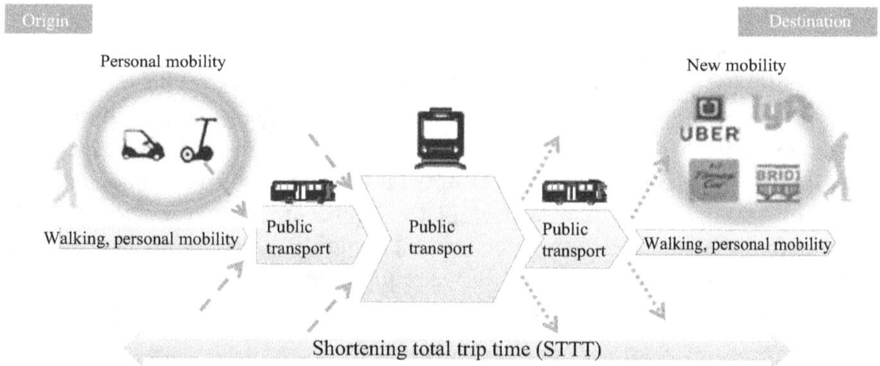

Figure 15.27 Alliance with new mobility

PT at the heart of the new mobility world

A role that PT should play in the new transport model, where a traveller can move from origin to destination seamlessly with continuous mobility by PT and personal mobility mentioned above. With the increasing urbanization of the planet, along with the limitation of space in today's cities, PT combined with a wide mixture of new mobility services is the viable solution for further mobility. Whereas PT is trying to cut the urban car dependency, new mobility services, such as on-demand, car/bike/ride sharing, autonomous driving and so on, are emerging with challenging innovation. In these trends, four following concepts are very critical and important:

1. PT must innovate itself with customer-oriented methods and technology to which the new mobility services are moving forward.
2. PT has to actualize 'door-to-door service' or 'shorter total trip time' with further better and stronger intermodal cooperation.
3. PT had better collaborate with the new mobility services as supports, especially in the first and last 1 mi. In Paris, 65% of Uber trips start or end within 200 m of a metro station. These function well together with PT.
4. In terms of capacity of the trunk segment with the heaviest traffic volume in cities, only PT can provide it. In this context, PT will continuously innovate, grow and play important roles in the rapidly changing environments.

The above concept is exact embodiment of 'mobility as a service' namely MaaS. It is also an idea shown in Figure 15.28 when taking future automatic driving and sharing of automobiles into consideration. In short, the field of automobiles progresses towards automatic driving and sharing and that of PT looks towards the same direction. Then both fields will be in harmony.

15.6.2 Information business

JR East already has a huge amount of data in its daily railway operation. We have train operation data, Suica data, ticket data and more. The data volume was too

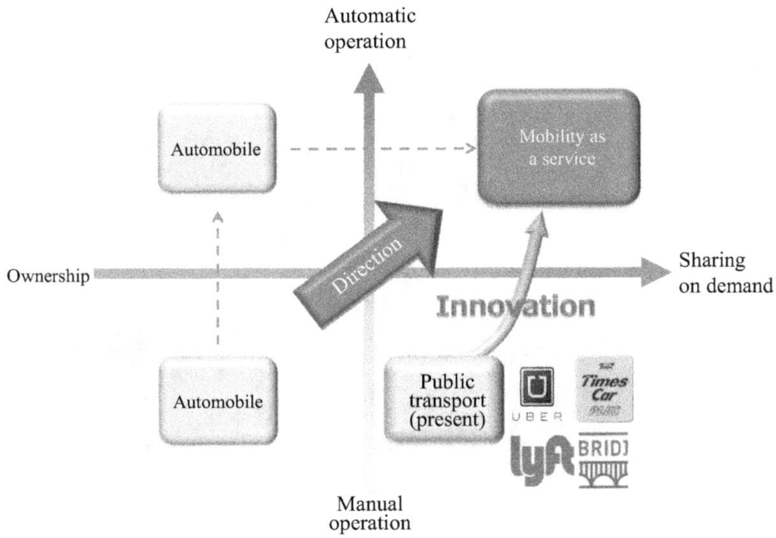

Figure 15.28 Mobility as a service

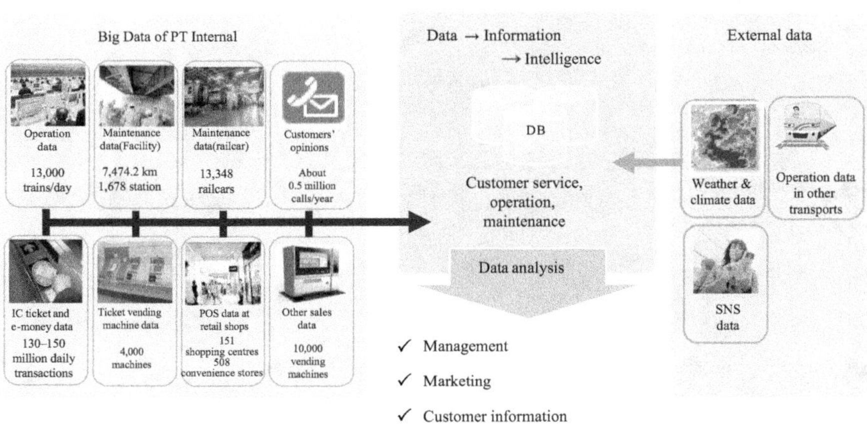

Figure 15.29 Model of information business

huge to analyse several years ago, but now we can very efficiently analyse them thanks to technological progress in the fields of computation, database, storage, AI, IOT and so on. Associating and analysing these data and extrinsic data like information of weather and climate in an organized way can yield positive effects to management and marketing (Figure 15.29).

Figure 15.30 shows one example of big data analysis, visualization of train location, congestion and delay information using data of automated gate, passenger load and train location.

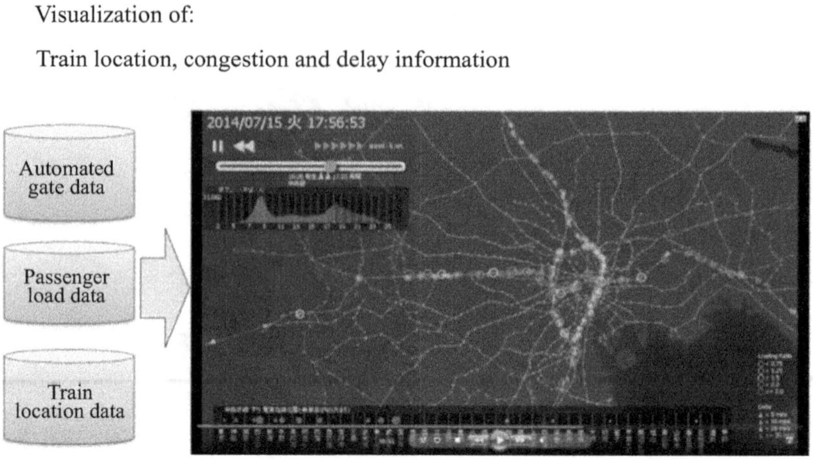

Figure 15.30 Big data analysis

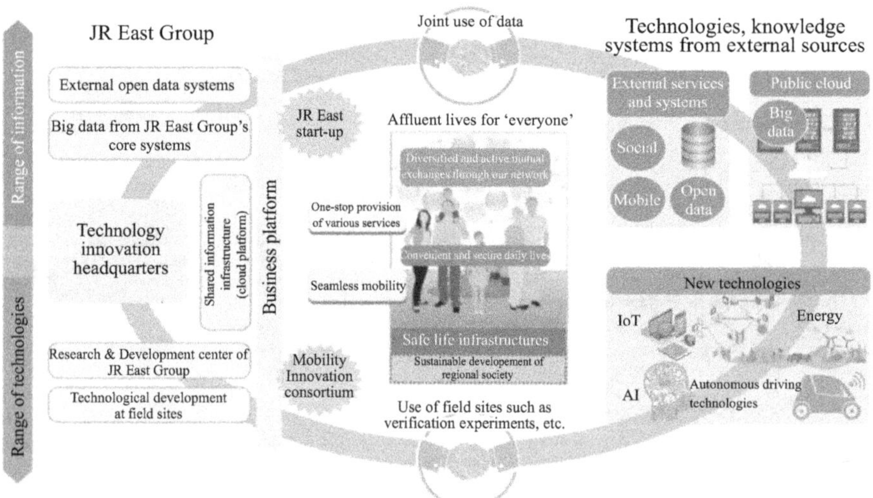

Figure 15.31 Technology innovation head quarters. Copyright © 2018 East Japan Railway Company [8]

Information business has a little effect when only JR East conducts it. Reform of overall value chain by making alliances with various enterprises or customers maximizes their utilities.

JR East established technology innovation headquarters on June 2018 to create new value by integrating technology and information. In this organization, common information, infrastructure for utilizing various big data and clouds has been created. And by further expanding the range of coalitions with external networks by joint use of data and use of field sites such as verification experiments, JR East will realize affluent lives for 'everyone' (Figure 15.31) [8].

Table 15.1 Advanced ICT fits PT well and vice versa

PT issues	Solution: advanced ICT
Safety	IoT, sensor
Security	Graphic recognition
Various issues to optimize	AI
Numerous equipment and devices management	IoT, sensor, big data
Huge user datasets	Big data
Real time information for customers	Mobile device
Face-to-face guidance	Robot
Danger, Dirty, Difficult (3Ds)	Robot

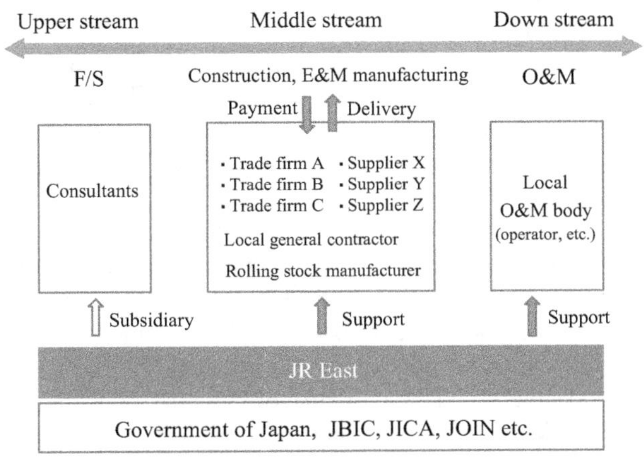

Figure 15.32 Global business

PT is the business that handles many kinds of information. Connecting and analysing these information in an organized way can produce various values. State-of-the-art ICT technologies are indispensable for it (Table 15.1).

15.6.3 Global business

The framework on participation of JR East to foreign railway projects is described as follows. In general, a railway project consists of multiple phases: feasibility study, construction of an infrastructure, design of systems and rolling stocks, manufacture, installation, operation and maintenance. JR East has experiences and knowledges of all the phase with its MTOMI model and, therefore, can support the project totally as well as partly. Especially, Japan International Consultants for Transportation (JIC) was established by ten Japanese major railway operators including JR East for foreign consultation businesses. JIC also succeeded consultations, functions and works from Japan Railway Technical Service (JARTS) (Figure 15.32).

15.7 Conclusion

In this paper, the business model of JR East named MTOMI model and its business development based on the MTOMI model are described. The railway is a huge size of aparatus industry, with the huge investment and the long construction period of new facilities. It can be a risk for pure private enterprises in modern society where the speed of change is fast. In fact, in many countries other than Japan, particularly in Europe, it is common to adopt a vertical separation model where a company that owns and maintains infrastructure and a company that operates trains are split into separate companies. On the other hand, at the time of privatization, JR East sets a concept that railway is integrate technology and service industry which owns the infrastructure, carries out the operation and maintenance of infrastructure by itself. This business model encourages JR East business improvement and innovation to provide good customer services. Thus, the MTOMI model brought a positive spiral to the business development of JR East.

The MTOMI model is an autonomous business unit. By organically alliancing them, it is further possible to provide services more effectively than they exist alone. In Section 15.5, three examples of alliance model are shown as InB, B2B and B2C; its effects and future strategy of alliance with new mobility services and oversea businesses are also explained in section 15.6.

Finally, to realize the alliance of MTOMI models, it is necessary to be able to freely combine the social infrastructure system which is the execution basis of each MTOMI in order to make the alliance of MTOMI models cooperate organically. In order to realize such a system, utilization of autonomous decentralized technology is effective. JR East introduced an ADS for core business such as transport operation control system for Tokyo metropolitan area (ATOS), Suica. JR East internalizes the experiences and knowledge of the ADS itself, thereby making it possible to develop new alliances.

JR East always seeks for open strategy with the homogeneous industry and the alliances with heterogeneous industries.

References

[1] UIC, "International Railway Statistics 2012", 2012.

[2] J. Carlzon, "Moments of Truth", HarperBusiness, New York, 1989.

[3] K. Mori, "Introduction of Autonomous Decentralized System", Morikita Pub. Co., Tokyo, 2006.

[4] F. Kitahara, K. Kamijiou, Y. Kakurai, K. Bekki, K. Kare, and K. Kawano, "Phased-in Construction Method of ATOS", IEEE Proc. of ISADS1999, pp. 415–424, 1999.

[5] A. Shiibashi, "Autonomous Decentralized High-Speed Processing Technology and the Application in an Integrated IC Card Fixed-Line and Wireless System", IEICE Trans., vol. E88-D, no. 12, pp. 2699–2707, 2005.

[6] A. Shiibashi, Y. Maruyama, M. Yamana, and K. Mori, "Multi-Layered Data Consistency Technology in IC Card Ticket System", ADSN, Toronto, Canada, pp. 58, 2007.

[7] K. Mori, "Concept-Oriented Research and Development in Information Technology", Wiley Series in Systems Engineering and Management, John Wiley & Sons, Inc., Hoboken, NJ, 2013.

[8] East Japan Railway Company, "JR East Group Management Vison 'Move UP' 2027". Available at: https://www.jreast.co.jp/e/investor/moveup/pdf/all.pdf, 2018.

Chapter 16

Smart cities, IOT, Industrie 4.0/Industrial Internet, cyber-physical systems: concepts, burdens and business models

Radu Popescu-Zeletin[1]

Abstract

In this chapter, we analyze the main frameworks for developing cyber-physical systems, their interrelationships and impact. The technological history and their main concepts are described here. The main burdens for a practical use are identified. Based on an existing development on a smart city project, a reference model for smart city is identified. The business model characteristics and their implications for the governance of such operational system are addressed. The chapter concludes with recommendations on future-required developments.

16.1 Introduction

Around 10–15 years ago in Berlin, the IBM CEO Samuel J. Palmisano announced a new IBM R&D program on "smart planet" based on the idea that every object on the earth valued more than 5$ will have an Internet address. The IBM observation on the technological development at that time can be seen in Figure 16.1 in which technological support for a smarter planet is summarized.

The main observation here is that only interconnectivity and instrumentation cannot create the "smart" technology of the future. In parallel with the Internet, penetration for interconnection technological developments is required for distributed processing ("smartness"). Additionally, governance, legal frameworks, business models and life cycle have to be developed for operational systems in a real world.

In the field of distributed processing ("smartness"), technological progress can be witnessed in the same time period of the Internet development. Concepts, standards and products have to be mentioned in the last decades, and a lot of results

[1]ICAM GmbH, Germany

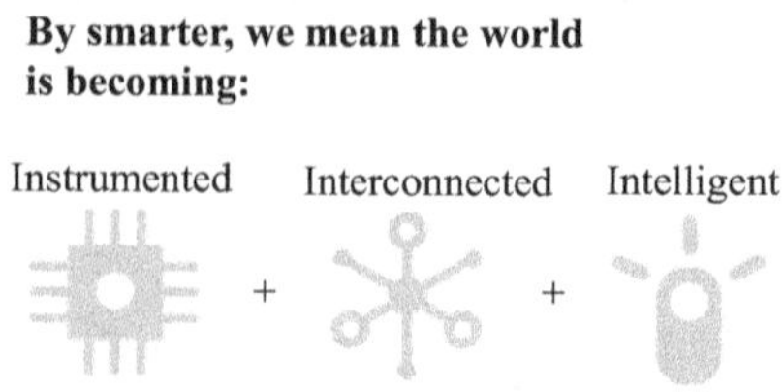

Figure 16.1 The "smart planet" formula

can be identified in the efforts to develop "smartness." Some of them should be mentioned and remembered also due to the lesson learned in the past.

- Autonomous decentralized systems concepts [1],
- Open distributed processing (ODP) framework [2],
- Super distributed objects (by OMG) [3] and
- Computer-integrated manufacturing (CIM and derivate) [4].

The technological developments are based on the Internet penetration on one hand, the advances in microelectronics, the storage capacity for huge data volumes (big data) and the advances in analytical data processing on the other. These developments characterize the past decade, and their in time synchronized technological advances allow to solve and propose solutions for complex problems and domains.

The present directions toward "smart" solutions for complex systems are as follows:

- Cyber-physical systems (CPSs),
- Internet of Things (IOT),
- Industrie 4.0 and Industrial Internet and
- Smart cities.

One may observe that these domains are attractive also because they have a "hype" character understandable in general by the majority of the population and on the other hand due to the fact that the main technologies are already developed and have no long-term research and development agenda.

These popular domains have also created a huge number of ideas and business models in the field, providing an additional "hype." National, industrial and European initiatives on society digitization have been started and become a part of the political agendas of the governments in different countries.

Worldwide-development programs have been started, but after these years, the results are in general not satisfactory. We will try to analyze the different concepts and identify the problems for which solutions have to be developed for operational systems.

From the beginning, one may observe that the problems to be solved are not mainly technological but in the governance, operation, legal frameworks and business cases of these complex systems.

16.2 Cyber-physical systems

CPSs are networks of ITC subsystems with mechanical and electronical components over a data communication infrastructure like Internet (Wikipedia). The

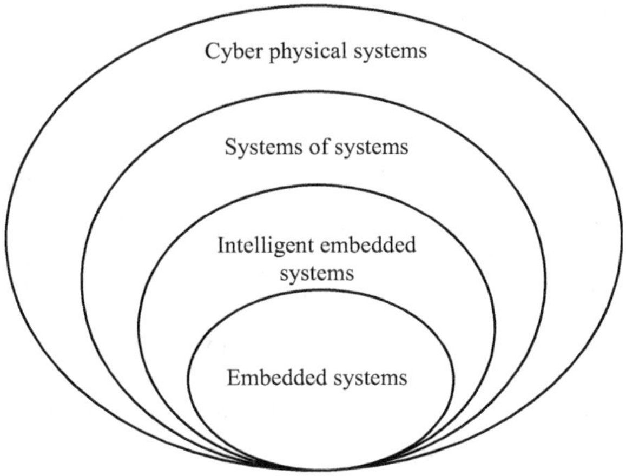

Figure 16.2 Composability of cyber-physical system

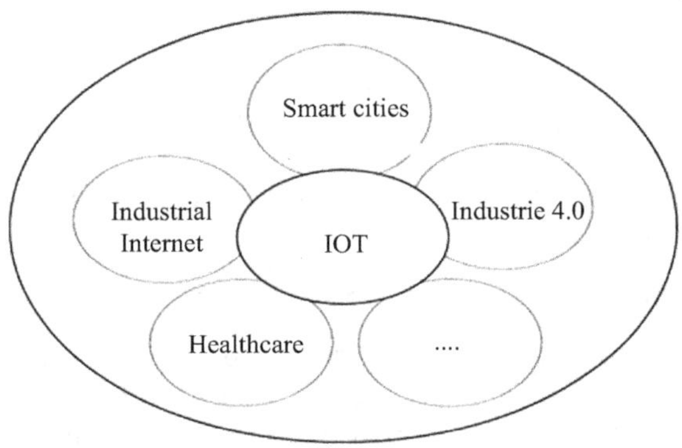

Figure 16.3 Cyber-physical systems

approach of addressing complex domains (smart planet) was to start in defining a general architecture for these complex systems (CPSs). The architecture provides the composition rules from embedded systems to CPSs (see Figure 16.2).

CPSs do not describe specific technologies but rather provide a framework to compose/decompose a complex problem in elements which identify the composability of the CPSs (Figure 16.3).

Here one may also observe that such systems are complex and in governance of different authorities. Interoperability of different levels and security solutions is main technological problems to be solved by adopting the framework. Major burden in introducing such systems are bound to the complexity, legislation and their controllability. For all, security and survivability of the operational systems are the main technological problems, and different solutions for different domain requirements are underway.

In general, the openness degree and the complexity of a system define the dimension of the security problem to be solved.

Due to the heterogeneity and openness of the CPSs, security becomes a non-deterministic polynomial (NP) complex problem with aspects like the following:

- Security against attackers from everywhere due to the open character of CPS,
- Safety of the controlled systems (like critical infrastructures) and
- Security of the Intellectual Property Rights.

There is no CPS secure system. Security can be improved, but we never achieve a complete secure CPS system (fata morgana effect). The lessons learned today in the first developed projects in CPS are as follows:

- It is better to learn how to live in an insecure cyber system rather than hope that technology will provide a secure system.
- We need security "instruction leaflets" for products and services we are using and integrate in the CPS solution. They will allow to evaluate the risk and derive attacker models for the envisaged solution.
- One of the solutions promoted here is "security by design," but this has drawbacks.
    - In CPSs "secured by design" Internet, and the systems to integrate have to be redefined.
    - CPSs require an NP complex security framework which is impossible to manage and implement in an operational systems (integrating legacy systems).
    - Securities (technological, governance, laws, etc.) which are independent of countries borders.

What should be done in order to improve security in CPS?

- Systems to provide the identity of persons, objects, services, and everything.
- Identity can be seen as the bridge between real world and cyberspace.
- Certification of everything (components) in the communication space of the CPSs.
- End-to-end authentication in order to provide trust and responsibility.
- Different levels of security for different applications.
- Provide "security instruction leaflets" and learn to evaluate the **RISC** of your application in a certain system (up to physical separation).

16.3 Internet of Things

From an infrastructural viewpoint, the INTERNET developments in the last decades have had an exponential development in throughput and number on end points in the world. This penetration required more and more end-point devices and services to be offered to the consumer instrumentation.

The world IPfication is a reality, and per day new devices are attached to Internet. TELCos have had a major role in this deployment, and new technologies are under way to support this trend (fifth generation (5G)). We will not present here the expected deployments numbers, one may find in a lot of forecasting or fact studies, but rather concentrate on the weak points in this development. The rapid development was mainly driven by providing to the customer incomplete solutions from an application/service point of view. The major accent here is to provide communication infrastructure due to the fact that telecommunication provider has control on the governance of the offer and supports and monetizes on mass communication.

The IOT developments in the past years can be seen as a bottom-up development starting with "instrumentation of devices" and their "interconnection" in the "smart planet" formula.

The development and operation of 5G mobile networks will boost the IOT technologies, and the IOT global cellular market is expected to reach 15 billion USD by 2030.

Paramount here is the provision of solutions in a very complex environment to ensure secure and interoperable end-to-end developments. Data security and privacy must be a major design parameter for IOT solutions considering the integration of information technology (IT) and operational technology infrastructure [5].

A comprehensive methodology for developing secure IOT solutions was developed by the Cloud Security Alliance (CSA). Thirteen steps in the design and development of secure solutions have been identified by analyzing software/hardware products available today [6].

The aims of the framework developed by CSA are as follows:

- Protect consumer privacy,
- Protect business data and limit the exposure of sensitive information,
- Safeguard against IOT products being used in distributed denial of service (DDOS) attacks and
- Safeguard against damage or harm resulting from compromised IOT solutions [6].

One may observe that IOT products have been mainly introduced by the TELCOs under the pressure of low costs. Providing secure IOT systems is expensive and should consider the solution domain and the RISC evaluation.

Again, ID systems, certification, security, and governance are the main burdens to be addressed in an operational environment.

16.4 Industrie 4.0 and Industrial Internet

In 1973, Josef Harrington introduced CIM for industrial manufacturing. A lot of efforts and systems can be witnessed in the past in developing standards and products in this space.

CIM is the integration of total manufacturing enterprise by using integrated systems and data communication coupled with new managerial philosophies that improve organizational and personnel efficiency (Wikipedia).

CIM and derivatives address island systems with physical separation, closed systems (even in the enterprise, no communication with the outside world).

Germany has initiated a governmental program for the digitization of the industrial manufacturing domain under the name of INDUSTRIE 4.0. IT considers

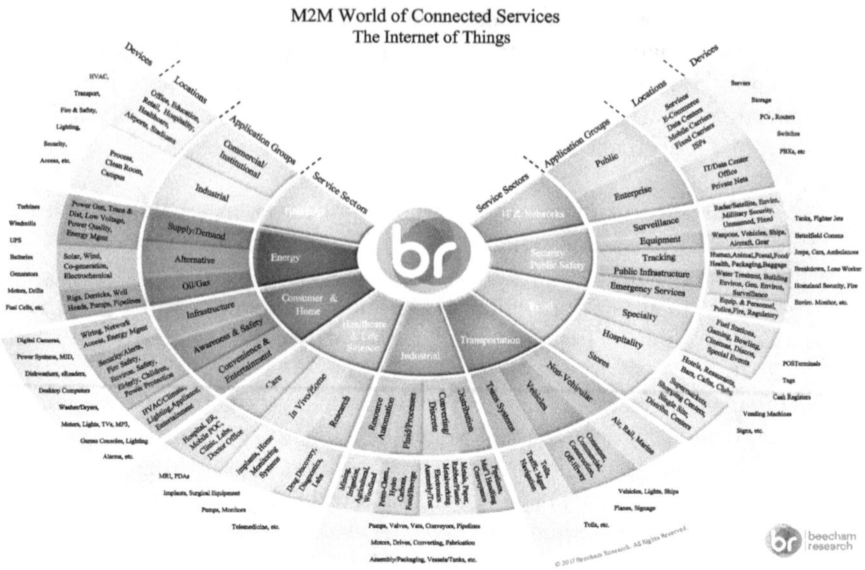

Figure 16.4   The Internet of Things world

it as the basis for the fourth industrial revolution in which "smartness" for industrial environments and products is promoted. The distribution of intelligence in such CPSs is captured in a reference model RAMI 4.0 (see Figure 16.4). The model also captures governance, standards and life cycle in a common framework which can be instantiated by different industrial domains.

The reference model proposed by the INDUSTRIE 4.0 consortium encompass three planes and is a good starting point to address related standards in designing, developing and operating industrial solutions.

In one plane, the composability of the industrial manufacturing system from product, field device, control device, station, station, work center, enterprise and the connected world are captured and their functionality is defined. It is similar to the CPS framework a composability framework for industrial environments.

The second plane describes the hierarchical views (asset, integration, communication, functionality and business), and similarity with the ODP views in the past is evident.

The third plane deals with the operational aspects of the solution to be designed like maintenance and production in an industrial manufacturing system.

The integration of the different elements is provided by encapsulation techniques and standards for data structures and communication.

The reference model is used for different CPS domains like energy, automotive, logistics, etc. and provides a framework for existing domain specific standards.

Main standards considered are IEC industrial standards; Internet is playing a minor role in this environment but pose the main problems in providing real-time and security-required solutions.

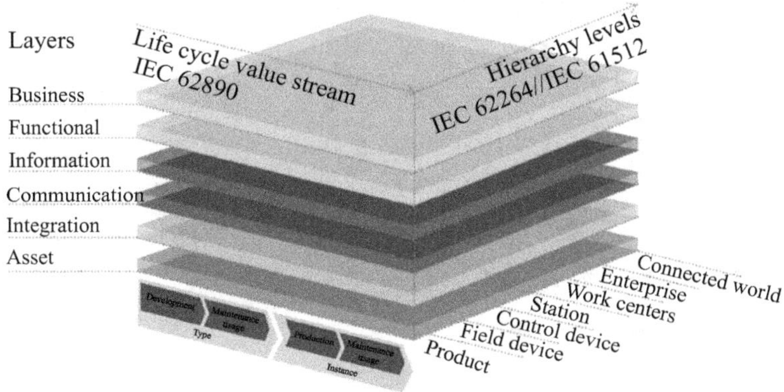

Figure 16.5 The reference architectural model Industrie 4.0 (RAMI 4.0)

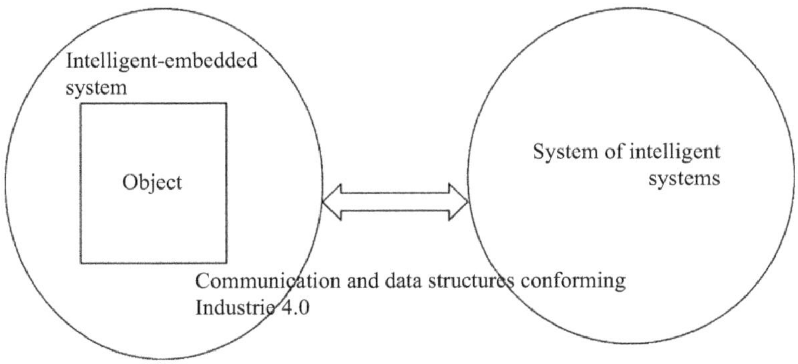

Figure 16.6 Encapsulation of elements in Industrie 4.0

Similar technological developments can be witnessed in the Open Fog Consortium (www.openfogconsortium.org) providing encapsulation in IOT and introducing edge computing for CPSs (Figures 16.5 and 16.6).

Industrial Internet (IIOT) program initiated by OMG is promoting test beds for specific Industrial Internet domains.

The motto of IIC is "Things are coming together"; let us try them [7].

Several projects have been started worldwide.

One major direction is the architecture of systems including the definition of several classes of gateways as encapsulation technique for the different components in the solution designed. Main industrial test beds at present are as follows:

- Track and trace,
- Communication and control test bed for microgrid application,
- Edge intelligence,
- Factory operation visibility and intelligence,
- High-speed network,

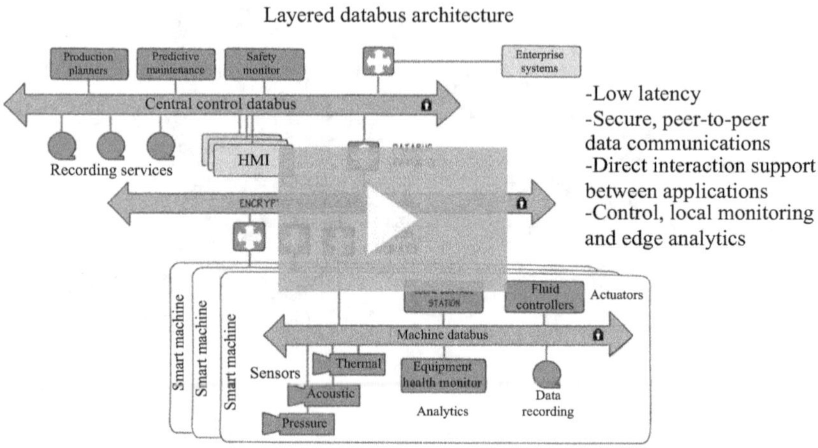

Figure 16.7 Blue print for the Industrial Internet [7]

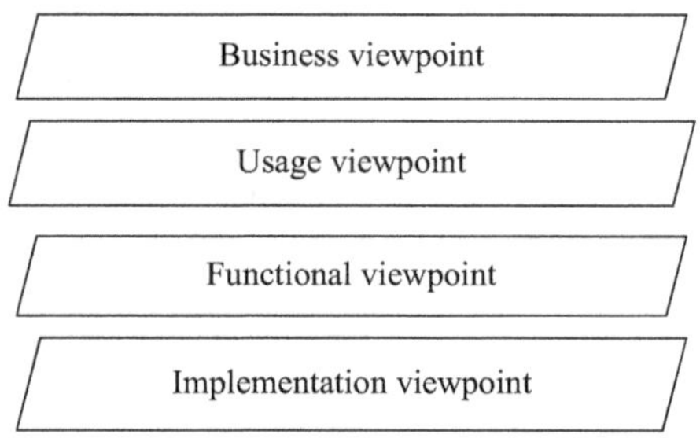

Figure 16.8 The IIOT planes [7]

- Future Industrial Internet (INFINITE) and
- Condition monitoring and predictive maintenance.

Similar to Industrie 4.0 reference model, IIOT address different views of the systems conforming to their specification encompassing the whole life cycle of the solution envisaged and conforming with the IIOT specifications (Figure 16.7). Models and standards are developed in the different hierarchical views of the IIOT system (Figure 16.8).

The business viewpoint identifies the stakeholders, their business concerns, values and objectives of the IIOT system in its business and regulatory concerns.

The usage viewpoint focuses on the expected system usage representing the sequence of activities for achieving the fundamental system capabilities.

The functional viewpoint focuses on the functional components of an IIOT system, their functionality and interrelation.

The implementation viewpoint addresses the technologies required to implement the functional viewpoint, communication and lifecycle procedures [7].

16.5 Smart cities

A good example of CPSs development which one may observe is the history of smart cities development. Probably all mayors in the world and citizen in the major cities had the impression they understand what a smart city is, and every city started to plan and discuss it, hopping on the holy grail in the cities management and development. Yes, the cities' problems are major due to the rapid increase of the populations towards the limited resource a city has to share. Cities of over 10 million inhabitants become normal all over the world and mainly in countries and continents with raising economies like China, India, Africa and South America. The resources to be shared like water, energy, air, roads and transportation in a city are limited and do not grow with the speed of the population growth.

Controlling and managing the resources in a smart city is the major benefit in the smart city vision. The expectation that a city becomes smart in short time, by investing money in some projects, is unrealistic. This is also due to the election periods in the cities compared with the required long-term plan in integrating heterogeneous systems and organizational frameworks in a smart city.

Cities like Barcelona, Amsterdam, Berlin, etc. have embraced a project-oriented view in the development of their smart cities programs, and very few results have more than a show ware character. Politically, this is a good approach in order to make visible what can be done, but sustainability of the solutions developed are always bound to economical (business models) and organizational aspects which are little or not covered by the technological solutions proposed.

Smart cities is a global, long-term development plan of a city including legal, organizational and economical aspects based on well-defined goals. Smart city is a city development master plan which allows integration of different projects at different point in time. It is a CPS in which legacy systems under control of different private and public organizations share communication and component infrastructure, data, services and allow the development of new applications.

In aiming to a master plan for a smart city, different steps (nontechnical and technical) have to be considered:

- Define a strategy and targets for the city and provide the required legislation to support the master plan,
- Set a common integration IT platform (reference model) and define the standards to be used and
- Define in this framework pilot projects and stakeholders with well-defined targets and time frame.

Legacy systems in a city have been developed by utilities organizations (water, energy, traffic, fire brigade, police, etc.) which are in a rapid development of their digitization process. It is unrealistic to think that new city infrastructure could be the solution for a master plan. The integration of the legacy systems is the only way to achieve value added and a master plan for a smart city. Integration implies sharing of infrastructure component and communication, sharing data for new applications and services but also common governance, political will and

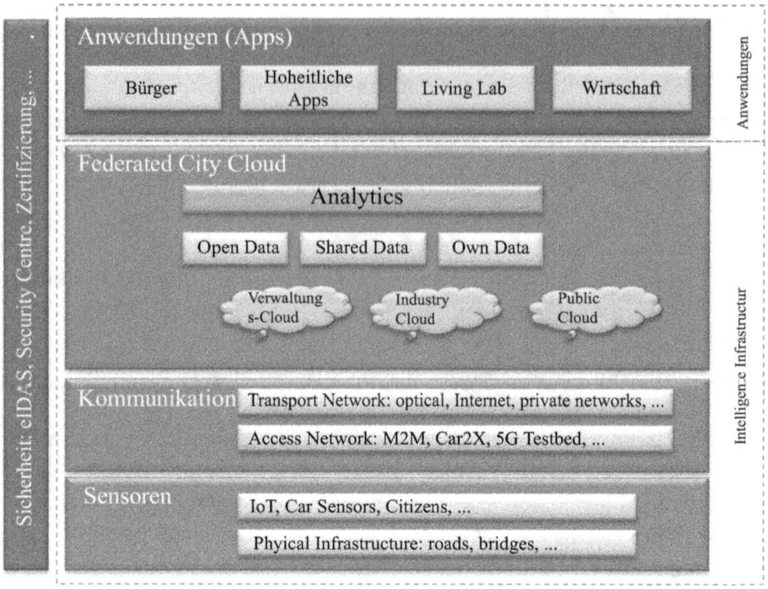

Figure 16.9 Reference Model for Smart Cities CPS Integration [8]

legislation. A possible reference model was developed by CISCO/ICAM white paper [8] for Berlin. It identifies the main classes of customers for city digitization: citizens, industry, living labs for spinoff companies in application area and the utilities of the city. Integration is done at different levels: communication infrastructure, data, sensors and information sources. It is an integration model of existing and new CPSs under the governance of different stakeholders.

It identifies different levels of integration, security and certification procedures based on the different requirements of the different application domains. It provides the architecture of a master plan in the city development and the framework for long-term planning and governance between city administration, utilities, citizens, etc. (Figure 16.9).

There are four integration levels of CPSs identified in a general smart city blue print:

1. Sensors: identifying the possible information sources including IOT, cars as a sensor, and citizens (utilities may share components like smart meters for energy, water, heating, etc.).
2. Communication infrastructure: including public and private-owned networks.
3. Data and information level in a federated city cloud infrastructure: Here based on the criticality of data, one may observe that different levels are required: open data, shared data between city agencies based on governance rules (e.g., police, fire brigades, hospitals). Analytics have an important role for data integration and common applications [9].

4. The application plane: offered to citizens, city planners industries, utilities involved in the city management and operation. Berlin, having a political aim in supporting new spinoff companies, requires infrastructure for living labs on which new ideas and application can be developed and integrated.

16.6 Conclusions

This chapter identifies the relevant and major concepts underway in the field of today's efforts for developing CPSs. We have presented here the main trends in the development of complex solutions for a "smart planet." A lot of parallel developments have been worked out, but there is little coordination between these overlapping technologies.

One may observe that technological solutions are there or underway, but solution for governance, security of complex systems, business models for their operation have a major role in this field and will play a major role in the future.

The major developments required for the future in all these frameworks are as follows:

1. Security, safety and privacy are of major concern due to the fact that CPSs provide an open environment for unknown attacks worldwide. New techniques like blockchain may be a solution for the future moving from secure CPS to secure transactions. How and where this can be applied is an interesting research and development issue.
2. The provision of identity systems in order to integrate certified elements is a very important element in designing CPS.
3. A RISC evaluation system is a part of designing a secure CPS.
4. The elements of the CPS have to be certified, and certification procedures have to accompany the development of CPS.
5. Governance, regulation and lows have a major impact in the operation of CPS and in general are country specific. CPSs do not end at the border of a specific country and different solutions may be required.
6. Special attention should be paid to data analytics and its distribution in the solution architecture [10].
7. Business models are critical in the CPS solution. It is at the present time not evident if costs for the provision of secure CPS are accepted by the costumers.

References

[1] Kinji Mori, Special issue on distributed computing systems, Computer, Vol. 24, Nov 1991. Autonomous Distributed Systems.

[2] ISO/IEC CD 10746, Reference model for open distributed processing, July 1994.

[3] OMG, Platform Independent Model for super distributed objects (SDO), OMG, 2008, www.omg.org/SDO/.

[4] Joseph Harrington, Computer Integrated Manufacturing, Krieger Pub Co, 1979, ISBN 978-0882758565.
[5] The INTEL IOT platform – architecture specification white paper on Internet of Things, www.intel,com/white papers.
[6] CSA Cloud Security Alliance: Future proofing the connected world: 13 steps to developing secure IOT products, 2016, www.cloudsecurityalliance.org.
[7] OMG, The Industrial Internet of Things, volume G1: reference architecture, IIC:PUB:G1:V1.80:20170131, 2016.
[8] ICAM/CISCO, "White paper on smart cities" 2017 on request, www.icam-gmbh.com.
[9] OMG, Industrial analytics: the engine driving the IIOT revolution, IIC: WHT:V1.1:PB:20170329, 2016.
[10] Industrial Internet Consortium, Industrial analytics: the engine the IIOT revolution, IIC:WHT:IN2:V1.1.

Conclusion

ADS (Autonomous Decentralized System) proposed in 1977 has been consistently advanced in technology and its application for 40 years in the rapidly changing technology field. During these years, innovation of technologies and business are achieved under unpredictably and rapidly changing socioeconomic situations in these days. The major trend is generated by four factors in structural changes in the world.

1. Structural change
   - (i) In society, birthrate is declining and aging people increases.
   - (ii) In value, unspecified majority value is gradually taken by specified personal value.
   - (iii) In business, commoditization of product has been progressed and competitiveness in business is more dependent to service.
   - (iv) In technology, target of research and development is focused from component and system to system of systems.

   These changes are strongly and mutually related, and they successively moved from (i) to (iv).
2. Value

   This book shows the change of the values which are as follows:
   - (i) Cost–performance ⇒ life-cycle cost

     This change of values from cost–performance to life-cycle cost is urged by the social structural change of 1(i). At the era of population explosion, culture has been quickly progressed, and economic growth can be attained by abundant workforce. But in low population, consumption of products decreases, and then sustainable society is required, in which life-cycle cost of product and system is highly evaluated to eliminate waste and utilize clean energy 1(iv).
   - (ii) Optimality ⇒ fairness

     This change from optimality to fairness comes from structural change 1(ii) from majority requirements to specific individual requirements. As variety in requirements of users increases, one measure of optimality for the many users based on majority rule cannot be accepted, but fairness for the heterogeneous and multiple types of the requirements has to be evaluated.
   - (iii) Reliability ⇒ assurance

     Reliability is important for any system. Even if society unpredictably changes 1(i) and value structural changes 1(ii), system such as infrastructure is expected to adapt its functionality to the changing situations. This property of adaptation to continue the operation under the dynamic and heterogeneous situations is assurance 1(iv).

(iv) Correctness ⇒ resilience

The planed and scheduled activity clarifies a specification of technology and application, and then they are correctly developed according to the specification. But the situation in market and business changes unpredictably in 1(ii) from the predetermined specification, the system is required to flexibly modify the original structure and function and to keep the provision of the service to the users 1(iii). Therefore, the system has to have the property of resilience.

(v) Majority ⇒ diversified

As business structure change 1(i) and 1(iii) from product to service, the systems with the various missions have to be connected for providing the services for various users. In this system of systems, each system is developed for the multiple values 1(iv). But the service is generated from the heterogeneous systems such as manufacturing, logistics, market analysis, sales, operation and maintenance systems in the supply chain and, in some case, competitor and allies systems. In the service system being integrated from heterogeneous systems, the diversified values coexist and the service is produced by their integration.

3. Viewpoint

The system and technology are innovated by changing the viewpoint.

(i) Normality ⇒ abnormality

Conventionally, socioeconomic situation is changed slowly within local area. But recently, technology and business are changing rapidly and globally. The system has been connected to the others and become larger and more complex by the socioeconomic dynamic situation. In this large and complex system, it has to have the viewpoint that the system includes anytime abnormal parts. As living thing, it is almost impossible to assume that the system is complete and normal. The technology under the viewpoint of abnormality in system is different from under the viewpoint of normality.

4. Design

Plan based top-down approach has to be changed to situation based bottom-up approach in the unpredictable socioeconomic situation.

More structural changes will be emerged in the world. But it is important to have the firm concept of technology and to achieve the chain of technology and market under the unpredictable environment. The ADS trend of technologies and applications shows that the concept-oriented approach is valuable to achieve the successive innovation in the chain of technology and market.

These trends 1, 2, 3 and 4 of the ADS have been shown in this book. Furthermore, the evolutional topics of technologies and applications will be accelerated in:

(i) Healthcare system and its network for preventive medicine and nursing service,

(ii) Finance system for global value exchange,

(iii) Management organization by strategic unit for global and changing market and human investment and

(iv) Natural disaster control system for its prevention and recovery.

This new trend is based on the structural changes and then the evolution in the society and technology unceasingly occurs. In conclusion, the following two points are principles.

(a) Technology without concept is reckless; and concept without technology is arid.

(b) Technology without operation is illusion; operation without technology is time killing.

Index

www.ingramcontent.com/pod-product-compliance
Lightning Source LLC
LaVergne TN
LVHW011910200926
848484LV00002B/579